Electronics Pocket Reference

Electronics Pocket Reference
Third Edition

Edward Pasahow

McGraw-Hill

New York San Francisco Washington, D.C. Auckland Bogotá
Caracas Lisbon London Madrid Mexico City Milan
Montreal New Delhi San Juan Singapore
Sydney Tokyo Toronto

McGraw-Hill

A Division of The ***McGraw-Hill*** *Companies*

3 4 5 6 7 8 9 0 DOC/DOC 0 4 3 2 1 0

ISBN 0-07-134700-3

The sponsoring editor for this book was Steve Chapman, the editing supervisor was Andrew Yoder, and the production supervisor was Sherri Souffrence. It was set in Times Roman by Lisa M. Mellott through the services of Barry E. Brown (Broker—Editing, Design and Production).

Printed and bound by R.R. Donnelley & Sons Company.

This one's for Charlie, Feather, Ginger,
Bander S, Varmint, Quasimodo, and S. Fratz.
Each helped in his or her own way.

Contents

Preface

The third edition of the *Electronics Pocket Reference* continues in the tradition of its predecessors by providing a unique compendium of the latest information on all aspects of electronics in a handy format. The compact size of this book facilitates carrying it in a pocket, briefcase, or workbox.

Among the new technologies represented are the Internet, World Wide Web, trusted computer systems, digital versatile disk (DVD), Ethernet, Digital Subscriber Lines (DSL), videoconferencing, Universal Serial Bus (USB), Fibre Channel, IEEE-1394 (Firewire) bus, Transmission Control Protocol (TCP)/Internet Protocol (IP), Windows NT, Novell NetWare, Synchronous Optical Network (SONET), Extended Industry Standard Architecture (EISA), Peripheral Component Interconnect (PCI) bus, Small Computer System Interface (SCSI), Video Electronic Standards Association (VESA) Local Bus (VLB), UNIX, and a variety of computer display, keyboard, and graphics standards.

The purpose of this manual remains to provide a single source of up-to-date electronics information for engineers, technicians, students, and hobbyists. Perhaps the best way to characterize its content is to describe what is not found between these covers. Space is not wasted on page after page of trigonometric and logarithmic tables (though both trigonometry and logarithms are covered). Anyone working in electronics today will have a pocket calculator or desktop or laptop computer that readily computes such functions. Instead of those traditional tables, this manual provides the formulas, data lists, and diagrams that supply straightforward solutions to electronics circuits and problems. Both calculators and computers can apply these solution methods. Related material is gathered into a single topic, so all the information you need for understanding a particular circuit is usually found within a few adjacent pages.

The range of topic coverage spans the spectrum of modern electronic applications and technologies. Coverage begins with the general laws of electronics. Included are Ohm's law, voltage and current, impedance, resonance, time constants, and power. Circuit analysis techniques are thoroughly described. Robust network methodologies such as Kirchhoff's law, Thevenin's theorem, Norton's theorem, Millman's theorem, superposition, and substitution are fully explained.

The following two sections cover passive and active devices. Parameters and data for resistors, capacitors, inductors, transformers, diodes, transistors, integrated circuits, and optoelectronic devices are detailed. The liner circuit section fully covers operational amplifier (op amp), timer, and trigger characteristics.

Filter designs suitable for active and passive elements show how to construct a variety of practical circuits. Among the many filters to choice from are 5- and 7-element Chebyshev filters. Power supply configuration, voltage regulation, and rectifier circuit descriptions demonstrate how a number of applications can be handled. The power (voltage and frequency) used around the world is tabulated, so next time you travel outside the US you can be prepared with the correct converter. Precise measurements of electronic values using bridges and meters comprise another series of circuit designs.

The ongoing fusion of computers and communications has created a need for people working in these fields to have ready access to a full complement of data on analog and digital equipment-including software. The various types electromagnetic emissions used in radio, television, communications, and time standards for worldwide coverage are tabulated. Television and radio standards are described. Topics range from frequency and wavelength to Doppler effects; amplitude, frequency, and pulse modulation; the electromagnetic spectrum; and error detection and correction. Coverage extends to antenna characteristics and designs and transmission lines. Important communications codes are found here also.

Another key grouping of topics is formed by the digital circuits and computer sections. Number systems, Boolean algebra, gate theory, and flip-flops are incorporated. Specifics on computer keyboards, displays, peripherals, buses, and connectors can be readily located. Information on data files and UNIX is provided as well.

New to this edition is the section on networking. The discussion of this topic includes everything from simple modem and serial data communications to the latest in networking technologies. The latter coverage includes such vital topics as the Internet, Universal Serial bus, Fibre Channel, NetWare, Windows NT, and Firewire.

Remaining sections provide mathematical tables, complete schematic symbology definitions, and conversion formulas. Frequently used terminology, relationships, and algorithms are found in these sections. Properties of materials and physical constants, with an emphasis on insulators, conductors, and cables, occupy a succeeding section. A summary of electrical safety and first aid concludes the manual.

I wish to acknowledge the generous support that the following manufacturers and trade associations have provided in allowing their materials to be used in this book.

Brand-Rex Electronics and Industrial Cable Division

Electronics Industries Association

General Instruments

Motorola, Inc.

National Semiconductor

Texas Instruments

Without their help, it would have been impossible to compile such a complete set of electronic data.

EDWARD PASAHOW

List of Figures

List of Tables

1

Definitions and Equations

1-1 ELECTRICAL UNITS

Ampere (A) One ampere is the constant current flowing in two parallel conductors one meter apart in free space that would produce a force of 2×10^{-7} newtons per meter of length.

Coulomb (C) One coulomb is the charge that is moved in one second by a current of one ampere.

Farad (F) One farad is the capacitance of a capacitor that produces a potential of one volt between the plates when charged to one coulomb.

Henry (H) One henry is the inductance of a coil that has one volt induced in it when the current varies uniformly at one ampere per second.

Joule (J) One joule is the work done by a force of one newton acting over a distance of one meter.

Ohm (Ω) One ohm is the resistance between two points of a conductor that produces a current of one ampere when one volt is applied between these points.

Siemens (S) One siemens is the conductance between two points of a conductor that produces a current of one ampere when one volt is applied between these points. Reciprocal of the ohm. Formerly mho (℧).

Volt (V) One volt is the potential difference between two points in a wire carrying one ampere when the power dissipated between these points is one watt.

Watt (W) One watt is the power which produces energy at one joule per second.

Weber (Wb) One weber is the magnetic flux that produces an electromotive force of one volt in a circuit of one turn as the flux changes uniformly from maximum to zero in one second.

1-2 OHM'S LAW IN DIRECT CURRENT CIRCUITS

$$V = IR \quad I = \frac{V}{R} \quad R = \frac{V}{I}$$

where V = voltage, V
I = current, A
R = resistance, Ω

Figure 1-1 shows the application of Ohm's law in dc circuits.

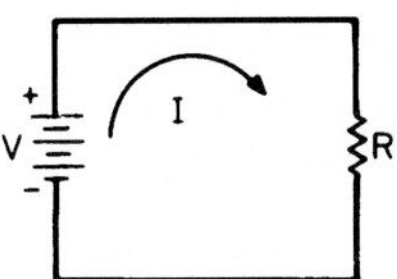

FIG. 1-1 Ohm's law in a dc circuit.

1-3 OHM'S LAW IN ALTERNATING CURRENT CIRCUITS

$$V = IZ \quad I = \frac{V}{Z} \quad Z = \frac{V}{I}$$

where V = voltage, V
I = current, A
Z = impedance, Ω

Figure 1-2 illustrates the involvement of Ohm's law in ac circuits.

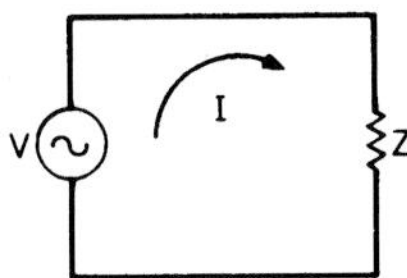

FIG. 1-2 Ohm's law in an ac circuit.

1-4 AVERAGE, RMS, AND PEAK VOLTAGE AND CURRENT

Table 1-1 can be used to find the equivalent forms of *sinusoidal* voltages or currents only. For other waveforms, refer to Fig. 1-3.

TABLE 1-1 Average, RMS, Peak and Peak-to-Peak Value Conversion for Sinusoidal Waves

Known value	Multiplication factor to find			
	Average	RMS	Peak	Peak-to-peak
Average	1.0	1.11	1.57	3.14
RMS	0.9	1.0	1.414	2.828
Peak	0.637	0.707	1.0	2.0
Peak-to-peak	0.32	0.3535	0.5	1.0

1-5 REACTANCE

The formulas in this section refer only to sinusoidal waves, except as noted. Other waveforms must be decomposed into fundamental and harmonic sine waves.

Angular Frequency, ω, in radians/s:

$$\omega = 2\pi f \quad f = \frac{\omega}{2\pi}$$

where f = frequency, Hz

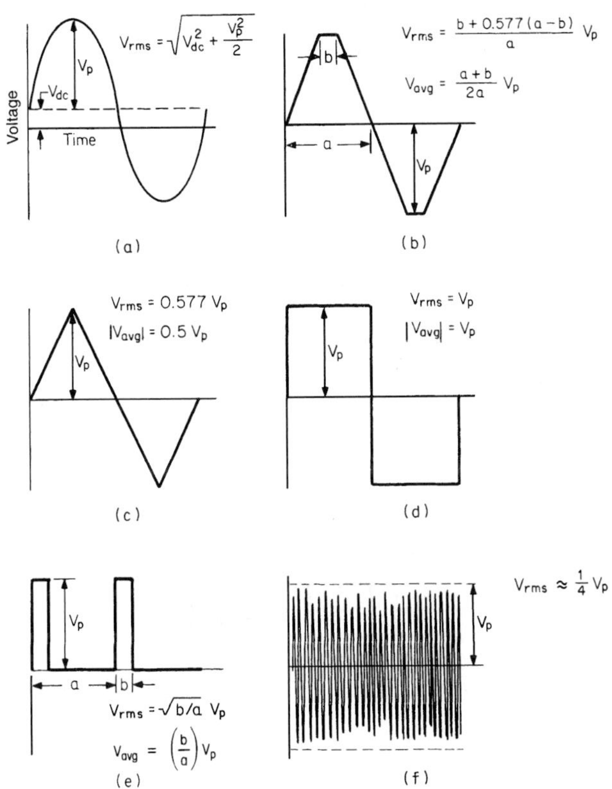

FIG. 1-3 Average, rms, and peak values of nonsinusoidal signals: *(a)* sine wave with dc bias, *(b)* trapezoid, *(c)* triangular wave, *(d)* square wave, *(e)* pulse train, and *(f)* white noise.

Period and Frequency
(applies to any periodic wave)

$$T = \frac{1}{f} \quad f = \frac{1}{T}$$

where T = period, s
f = frequency, Hz

Reactance Formulas

$$X_C = \frac{1}{2\pi fC} = \frac{1}{\omega C} \quad X_L = 2\pi fL = \omega L$$

where X_C = capacitive reactance, Ω
X_L = inductive reactance, Ω
f = frequency, Hz
C = capacitance, F
L = inductance, H

Other Forms of Reactance Formulas

$$V = IX \quad I = \frac{V}{X} \quad X = \frac{V}{I}$$

1-6 CONDUCTANCE

Conductance is the reciprocal of resistance.

$$G = \frac{1}{R}$$

where G = conductance, S
R = resistance, Ω

1-7 SUSCEPTANCE

Susceptance is the reciprocal of reactance.

$$B = \frac{1}{X}$$

where B = susceptance, S
X = reactance, Ω

1-8 IMPEDANCE

Impedance is the result of resistance and reactance in an ac circuit. The phase angle θ between the voltage and current ranges between $-90°$ and $+90°$.

Impedance in a Series Circuit

$$Z = \sqrt{R^2 + X^2} \quad \theta = \tan^{-1}\frac{X}{R} \quad F = \cos\theta = \frac{R}{Z}$$

where Z = impedance, Ω
R = resistance, Ω
X = reactance, Ω
θ = phase angle, degrees
F = power factor (ratio of true power to apparent power)

Impedance in a Parallel Circuit

$$Z = \frac{RX}{\sqrt{R^2 + X^2}} \quad \theta = \tan^{-1}\frac{R}{X} \quad F = \cos\theta = \frac{Z}{R}$$

where Z = impedance, Ω
R = resistance, Ω
X = reactance, Ω
θ = phase angle, degrees
F = power factor (ratio of true power to apparent power)

Impedance also can be expressed in terms of conductance and susceptance in a parallel circuit.

$$Z = \frac{1}{\sqrt{G^2 + B^2}}$$

where G = conductance, S
B = susceptance, S

Impedance in Circuit Configurations.
Impedance values in various circuit configurations are given in Table 1-2.

1-9 ADMITTANCE

Admittance is the reciprocal of impedance.

$$Y = \frac{1}{\sqrt{R^2 + X^2}} = \frac{1}{Z} = \sqrt{G^2 + B^2}$$

where Y = admittance, S
Z = impedance, Ω
R = resistance, Ω
X = reactance, Ω
G = conductance, S
B = susceptance, S

TABLE 1-2 Impedance in Various Circuit Configurations

Z	θ	Circuit
R	0°	 R
$R_1 + R_2 + R_3 \cdots$	0°	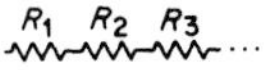 R_1 R_2 R_3 ···
X_L	90°	 X_L
$X_{L1} + X_{L2} + X_{L3} \cdots$	90°	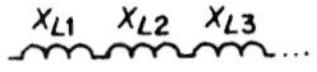 X_{L1} X_{L2} X_{L3} ···
X_C	−90°	 X_C
$X_{C1} + X_{C2} + X_{C3} \cdots$	−90°	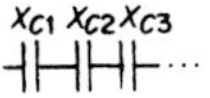 X_{C1} X_{C2} X_{C3} ···
$\sqrt{R^2 + X_L{}^2}$	$\tan^{-1} \frac{X_L}{R}$	 R X_L
$\sqrt{R^2 + X_C{}^2}$	$\tan^{-1} \frac{X_C}{R}$	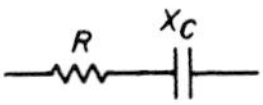 R X_C
If $X_L > X_C$ $Z = X_L - X_C$	90°	
If $X_C > X_L$ $Z = X_C - X_L$	−90°	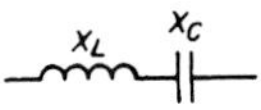 X_L X_C
If $X_L = X_C$ $Z = 0$	0°	

TABLE 1-2 Impedance in Various Circuit Configurations (Continued)

Z	θ	Circuit
$\sqrt{R^2+(X_L-X_C)^2}$	$\tan^{-1}\frac{X_L-X_C}{R}$	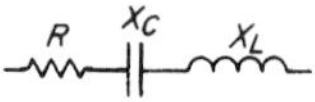 R, X_C, X_L
$\frac{1}{\frac{1}{R_1}+\frac{1}{R_2}+\frac{1}{R_3}+\cdots}$	0°	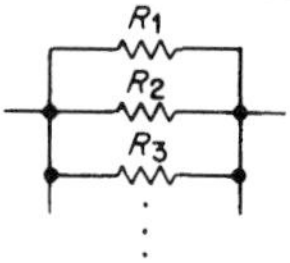 R_1, R_2, R_3
$\frac{1}{\frac{1}{X_{L1}}+\frac{1}{X_{L2}}+\frac{1}{X_{L3}}+\cdots}$	90°	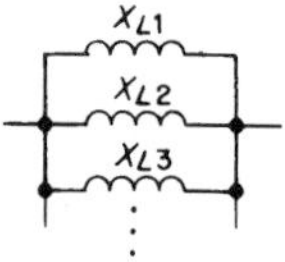 X_{L1}, X_{L2}, X_{L3}
$\frac{1}{\frac{1}{X_{C1}}+\frac{1}{X_{C2}}+\frac{1}{X_{C3}}+\cdots}$	−90°	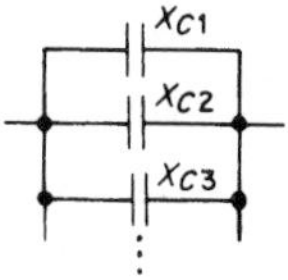 X_{C1}, X_{C2}, X_{C3}
$\frac{RX_L}{\sqrt{R^2+X_L^2}}$	$\tan^{-1}\frac{R}{X_L}$	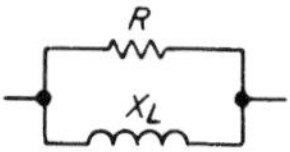 R, X_L
$\frac{RX_C}{\sqrt{R^2+X_C^2}}$	$\tan^{-1}\frac{R}{X_C}$	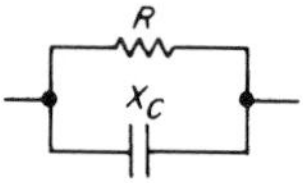 R, X_C

TABLE 1-2 Impedance in Various Circuit Configurations (Continued)

Z	θ	Circuit
$\dfrac{RX_L X_C}{\sqrt{X_L^2 X_C^2 + R^2 (X_L - X_C)^2}}$	$\tan^{-1} \dfrac{R(X_L - X_C)}{X_L X_C}$	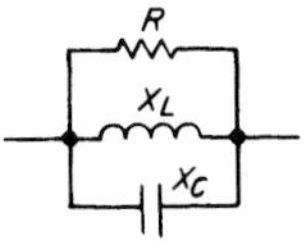 R, X_L, X_C
$R_2 \sqrt{\dfrac{R_1^2 + X_L^2}{(R_1 + R_2)^2 + X_L^2}}$	$\tan^{-1} \dfrac{X_L R_2}{R_1^2 + X_L^2 + R_1 R_2}$	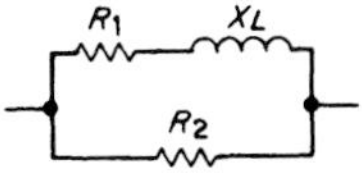 R_1, X_L, R_2
$X_C \sqrt{\dfrac{R^2 + X_L^2}{R^2 + (X_L - X_C)^2}}$	$\dfrac{X_L (X_C - X_L) - R^2}{R X_C}$	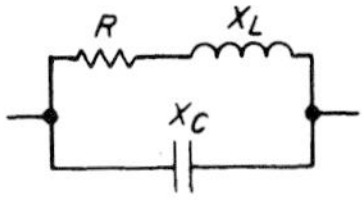 R, X_L, X_C

1-10 RESONANCE

A resonant frequency is reached when the capacitive and inductive reactances in a tuned circuit are equal. A series *LC* circuit has zero reactance at the resonant frequency. A parallel *LC* circuit has infinite reactance at the resonant frequency.

The resonant frequency formula is:

$$f_r = \frac{1}{2\pi \sqrt{LC}}$$

where f_r = resonant frequency, Hz
L = inductance, H
C = capacitance, F

By transposing, you can find the inductance or capacitance to achieve resonance at a particular frequency:

$$L = \frac{1}{4\pi^2 f_r^2 C} \quad C = \frac{1}{4\pi^2 f_r^2 L}$$

The ratio of reactance to resistance is called the *Q factor; Q* is, therefore, a ratio of the energy stored to energy dissipated per cycle in the circuit.

Series *RL* Circuit

$$Q = \frac{2\pi f_r L}{R}$$

Series *RC* Circuit

$$Q = \frac{1}{2\pi f_r RC}$$

The Q value of a tuned circuit is related to bandwidth

$$Q = \frac{f_r}{B}$$

where B = bandwidth, Hz

The bandwidth is the frequency interval between the half-power points in Fig. 1-4.

$$B = f_2 - f_1 \quad B = \frac{f_r}{Q} \quad Q = \frac{f_r}{f_2 - f_1}$$

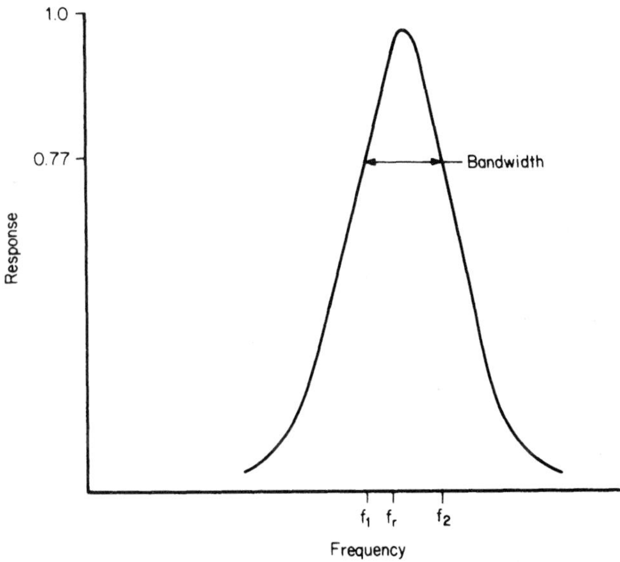

FIG. 1-4 Bandwidth and half-power points in a resonant circuit.

Universal Resonance Curve for Constant Q

Figure 1-5 provides plots of the ratio of current and phase angle at frequency f to current at resonant frequency f_r as a function of

$$x = \frac{Q(f - f_r)}{}$$

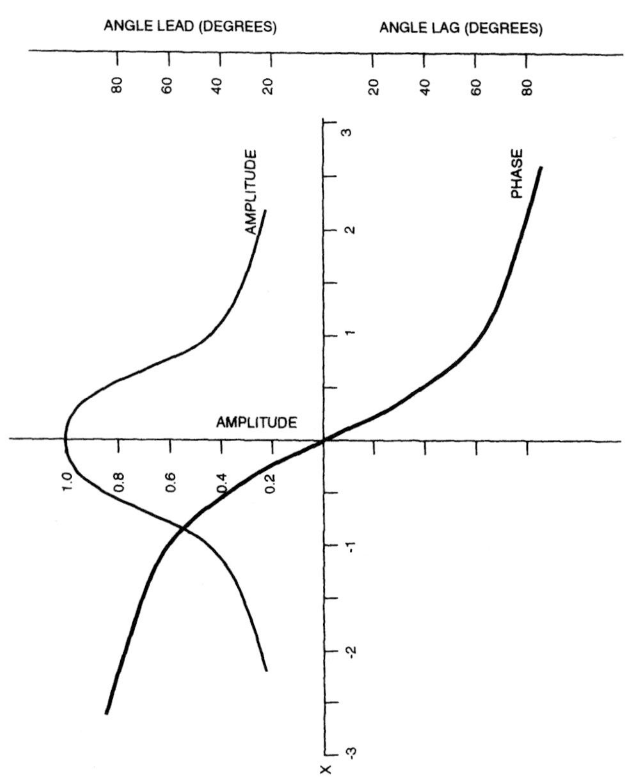

FIG. 1-5 Universal resonance curve.

where $f_r = \dfrac{1}{2\pi\sqrt{LC}}$

1-11 TIME CONSTANTS

The time required for voltage or current to reach 63.2 percent of the final value is as follows:

In *RC* Circuits, $\tau = RC$

In *RL* Circuits, $\tau = \dfrac{L}{R}$

In Charging Circuits,

$$v(t) = V_f - (V_f - V_i)e^{-t/\tau} \quad i(t) = I_f - (I_f - I_i)e^{-t/\tau}$$

If Circuit Is Initially discharged,

$$v(t) = V_f(1 - e^{-t/\tau}) \quad i(t) = I_f(1 - e^{-t/\tau})$$

In Discharging Circuits:

$$v(t) = V_i e^{-t/\tau} \quad i(t) = I_i e^{-t/\tau}$$

where R = resistance, Ω
C = capacitance, F
L = inductance, H
t = time, s
V_f = final voltage, V
V_i = initial voltage, V
$i(t)$ = time varying current, A

e = base of natural logarithms

$v(t)$ = time varying voltage, V

I_f = final current, A

I_i = initial current, A

TABLE 1-3 Percentage of Final Voltage or Current versus Time Constants

τ	Charging circuit (initially discharged), %	Discharging circuit, %
0.1	9.5	90.5
0.2	18.1	81.9
0.3	25.9	74.1
0.4	33.0	67.0
0.5	39.3	60.7
1.0	63.2	36.8
2.0	86.5	13.5
3.0	95.0	5.0
4.0	98.2	1.8
5.0	99.3	0.7

EXAMPLE: Find the voltage in a charging circuit after two time constants if the initial voltage is 0.7 V and the final voltage is 5 V.

$$
\begin{aligned}
v(t) &= 5 - (5 - 0.7)e^{-2\tau/\tau} \\
&= 5 - 4.3e^{-2} \\
&= 4.42 \text{ V}
\end{aligned}
$$

The transient voltages and currents in typical circuits are shown in Figs. 1-6 through 1-9.

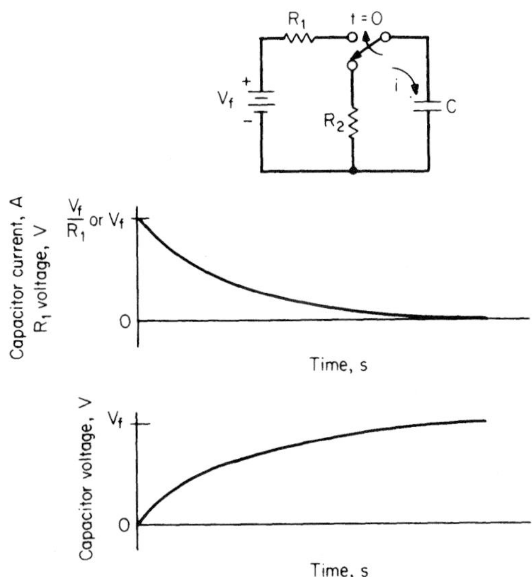

FIG. 1-6 Charging capacitor.

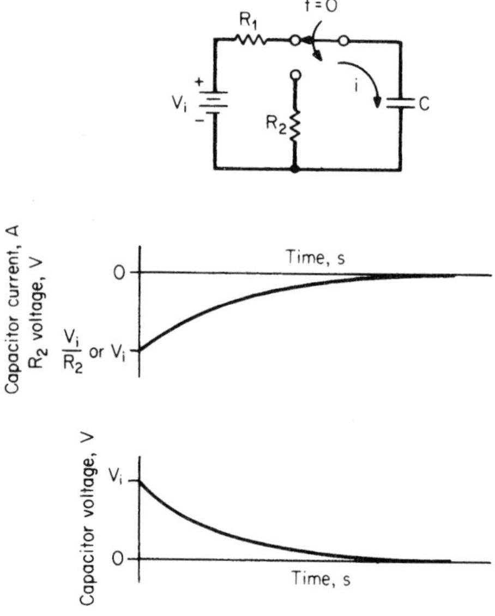

FIG. 1-7 Discharging capacitor.

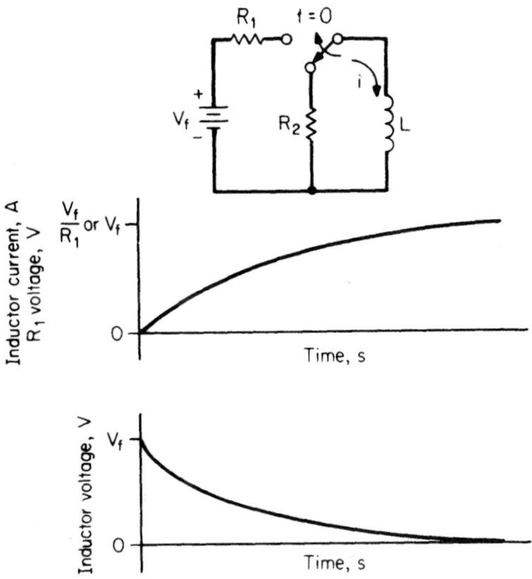

FIG. 1-8 Charging inductor.

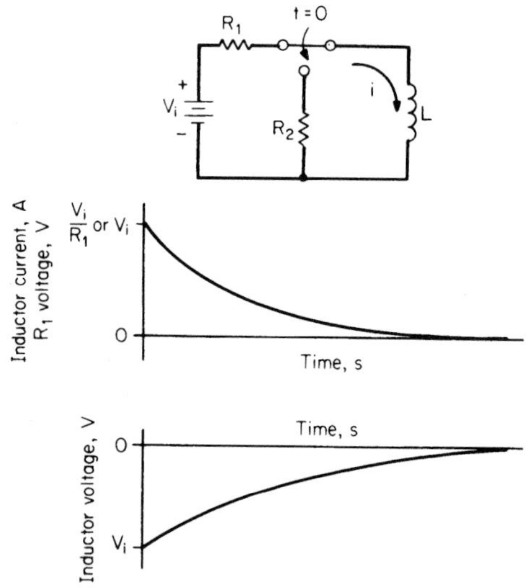

FIG. 1-9 Discharging inductor.

1-12 POWER IN DIRECT CURRENT CIRCUITS

$$P = IV \quad I = \frac{P}{V} \quad V = \frac{P}{I} \quad R = \frac{P}{I^2}$$

$$P = I^2R \quad I = \sqrt{\frac{P}{R}} \quad V = \sqrt{PR} \quad R = \frac{V^2}{P}$$

$$P = \frac{V^2}{R}$$

where P = power, W
V = voltage, V
I = current, A
R = resistance, Ω

1-13 POWER IN ALTERNATING CURRENT CIRCUITS

$$P = IV \cos\theta \quad I = \frac{P}{V\cos\theta} \qquad V = \frac{P}{I\cos\theta} \quad Z = \frac{P}{I^2\cos\theta}$$

$$P = I^2 Z \cos\theta \quad I = \sqrt{\frac{P}{Z\cos\theta}} \quad V = \sqrt{PZ\cos\theta}$$

$$P = \frac{V^2}{Z}\cos\theta \qquad Z = \frac{V^2\cos\theta}{P}$$

where θ = phase angle, degrees

1-14 KIRCHHOFF'S LAWS

Current. The sum of all electric currents flowing toward a circuit junction is zero.

$$i_1 + i_2 + i_3 + \ldots + i_N = 0$$

$$I_1 + I_2 + I_3 + \ldots + I_N = 0$$

Voltage. The sum of all voltages around a closed circuit is zero.

$$v_1 + v_2 + v_3 + \ldots + v_N = 0$$

$$V_1 + V_2 + V_3 + \ldots + V_N = 0$$

1-15 THEVENIN'S THEOREM

Thevenin's theorem states that any linear, two-terminal network will behave in a manner equivalent to a single-voltage source in series with a single impedance. To find the equivalent circuit:

1. Determine the voltage, V_T, between the two terminals.
2. Short-circuit all voltage sources and calculate the impedance Z_T between the two terminals.
3. Replace the original network with Z_T in series with V_T.

EXAMPLE: Consider the circuit in Fig. 1-10*a*.

$$V_T = \frac{R_2}{R_1 + R_2} V$$

$$= \left(\frac{300}{400}\right)(10)$$

$$= 7.5 \text{ V}$$

$$Z_T = \frac{R_1 R_2}{R_1 + R_2}$$

$$= \frac{(100)(300)}{100 + 300}$$

$$= 75\ \Omega$$

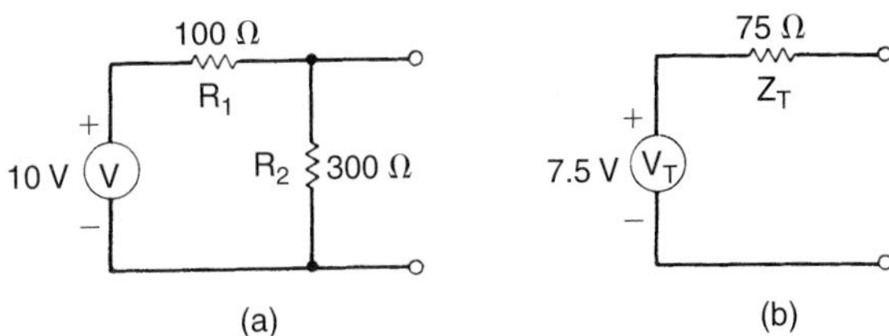

FIG. 1-10 Finding the Thevenin equivalent circuit. *(a)* original circuit and *(b)* equivalent circuit.

1-16 NORTON'S THEOREM

Norton's theorem states that any linear, two-terminal network will behave in a manner equivalent to a single-current source in parallel with a single impedance. To find the equivalent circuit:

1. Determine the current I_N flowing through a short circuit across the two terminals.
2. Open circuit all current sources and calculate the impedance Z_N between the two terminals.
3. Replace the original network with Z_N in parallel with I_N.

EXAMPLE: Find the Norton theorem equivalent for the circuit in Fig. 1-10*a*.

$$I_N = \frac{V}{R_1}$$

$$= \frac{7.5}{75}$$

$$= 100 \text{ mA}$$

$$Z_N = \frac{R_1 R_2}{R_1 + R_2}$$

$$= \frac{(100)(300)}{100 + 300}$$

$$= 75\ \Omega$$

The equivalent circuit appears in Fig. 1-11.

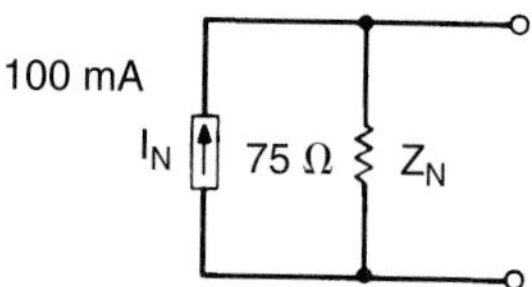

FIG. 1-11 Finding the Norton equivalent circuit.

The conversion between Thevenin and Norton equivalent circuits is easily accomplished.

$$I_N = \frac{V_T}{Z_T}$$

$$V_T = I_N Z_N$$

$$Z_T = Z_N$$

1-17 SUPERPOSITION THEOREM

The superposition theorem makes it possible to find the solution of networks with multiple voltage or current sources. The voltage or current response in a linear, bilateral network can be found by summing the response produced by each source acting by itself.

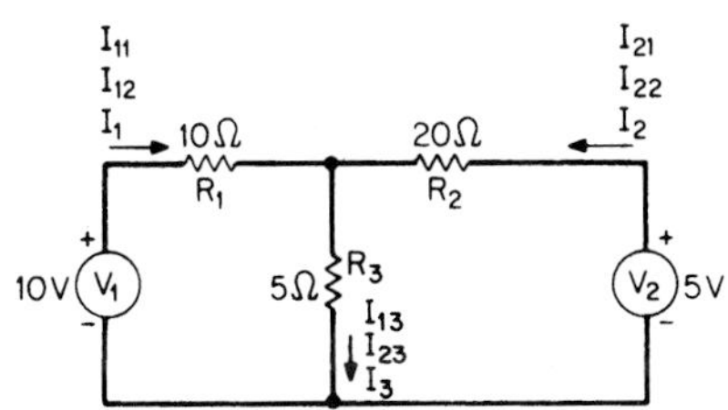

FIG. 1-12 Demonstration of the superposition theorem.

EXAMPLE: Find the currents in each branch of Fig. 1-12. The input impedance for V_1 is

$$Z_{1\text{in}} = 10 + \frac{(5)(20)}{5 + 20}$$

$$= 14\ \Omega$$

With only source V_1 operating,

$$I_{11} = \frac{V_1}{Z_{1\text{in}}} = \frac{10}{14} = 714\ \text{mA}$$

$$I_{12} = \frac{-R_3}{} I_{11}$$

$$= \frac{-5}{} (714 \times 10^{-3})$$

$$= -\ 143 \text{ mA}$$

$$I_{13} = \frac{R_2}{R_2 + R_3} I_1$$

$$= \frac{20}{20 + 5} (714 \times 10^{-3})$$

$$= 571 \text{ mA}$$

Following this same procedure for source V_2,

$$Z_{2\text{in}} = R_2 + \frac{R_1 R_3}{R_1 + R_3}$$

$$= 20 + \frac{(10)(5)}{10 + 5}$$

$$= 23.3\ \Omega$$

The currents are

$$I_{22} = \frac{V_2}{Z_{2\text{in}}} = \frac{5}{23.3} = 214 \text{ mA}$$

$$I_{21} = -\frac{R_3}{R_1 + R_3} I_{22}$$

$$= -\frac{5}{10+5}(214 \times 10^{-3})$$

$$= -71 \text{ mA}$$

$$I_{23} = \frac{R_1}{R_1 + R_3} I_{22}$$

$$= \frac{10}{10+5}(214 \times 10^{-3})$$

$$= 143 \text{ mA}$$

The total currents are then found by adding individual currents.

$$I_1 = I_{11} + I_{21} = 714 - 71 = 643 \text{ mA}$$

$$I_2 = I_{12} + I_{22} = -143 + 214 = 71 \text{ mA}$$

$$I_3 = I_{13} + I_{23} = 571 + 143 = 714 \text{ mA}$$

1-18 SUBSTITUTION THEOREM

The substitution theorem states that a branch in a network carrying a current of I_B, with a voltage across the branch of V_B, can be substituted for another branch which produces the same voltage when carrying the same current. Figure 1-13 shows several branches that are interchangeable by this theorem.

1-19 MILLMAN'S THEOREM

Millman's theorem states that parallel voltage or current sources in series with an impedance can be replaced by an equivalent

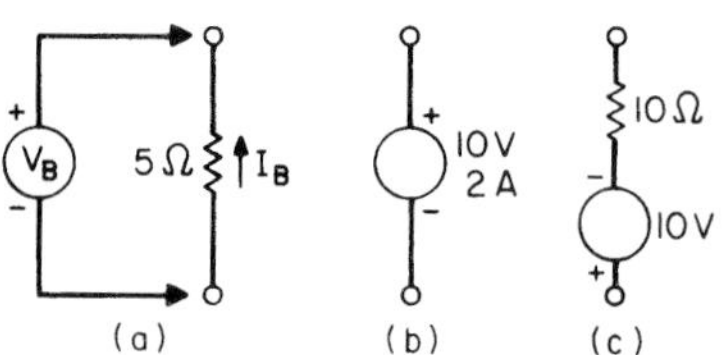

FIG. 1-13 Examples of the substitution theorem in a branch where $V_B = 10$ V, $I_B = 2$ A.

voltage or current source and series impedance (see Fig. 1-14). The equations for finding the equivalent circuit elements are

$$Z_M = \frac{1}{Y_1 + Y_2 + Y_3 + \ldots}$$

$$V_M = \frac{V_1Y_1 + V_2Y_2 + V_3Y_3 + \ldots}{Y_1 + Y_2 + Y_3 + \ldots}$$

$$I_M = V_1Y_1 + V_2Y_2 + V_3Y_3 + \ldots$$

where $Y_1 = \dfrac{1}{Z_1}$

$$Y_2 = \frac{1}{Z_2}$$

$$Y_3 = \frac{1}{Z_3}$$

$$\vdots$$

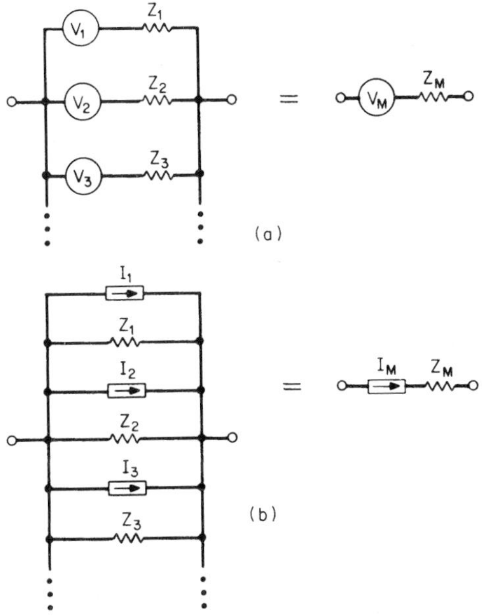

FIG. 1-14 Examples of Millman's theorem: *(a)* voltage sources and *(b)* current sources.

1-20 THREE-TERMINAL CONFIGURATIONS

In three-terminal circuits (often used in three-phase power networks), the components are frequently connected in a wye or delta configuration as shown in Fig. 1-15. The equivalent impedance in the corresponding circuit can be found readily from the following equations:

$$Z_A = \frac{Z_1Z_2 + Z_1Z_3 + Z_2Z_3}{Z_2}$$

$$Z_B = \frac{Z_1Z_2 + Z_1Z_3 + Z_2Z_3}{Z_1}$$

$$Z_C = \frac{Z_1Z_2 + Z_1Z_3 + Z_2Z_3}{Z_3}$$

$$Z_1 = \frac{Z_AZ_C}{Z_A + Z_B + Z_C}$$

$$Z_2 = \frac{Z_BZ_C}{Z_A + Z_B + Z_C}$$

$$Z_3 = \frac{Z_AZ_B}{Z_A + Z_B + Z_C}$$

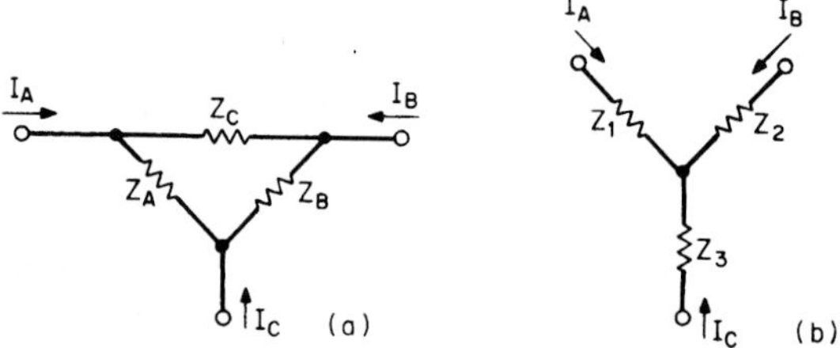

FIG. 1-15 Three-terminal circuits: *(a)* delta connection and *(b)* wye connection.

1-21 CIRCUIT ANALYSIS BY LOOP EQUATIONS

If a circuit is too complex to analyze with one of the simpler techniques, loop (also called *mesh*) equations can be applied. (Another form of analysis, using nodal equations, is described in Sec. 1-22.) This analysis can be applied only to planar networks (that is, where no wires are crossing). The steps are:

1. Draw loop currents on the circuit (see Fig. 1-16). Conventional current (flow from positive to negative) is used in this example.
2. Write the loop equations. For each loop, the product of the currents and impedance are summed and set equal to the algebraic sum of voltage sources in the loop. Opposing currents (arrows which point in opposite directions, as in Z_2 in Fig. 1-16) have a negative sign.
3. Solve the set of simultaneous equations that result.

(See determinants in Chapter 11.)

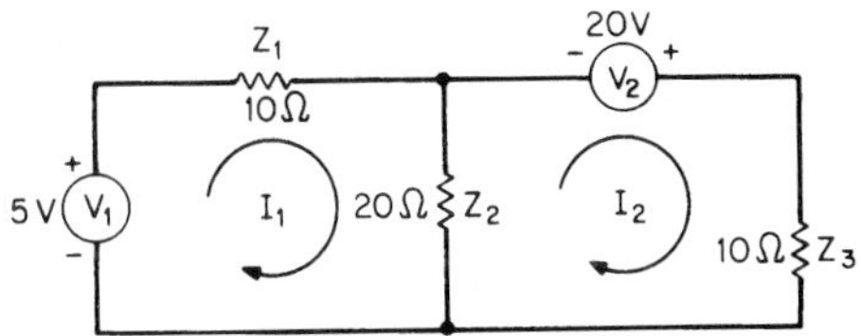

FIG. 1-16 Loop current analysis.

	Algebraic sum of voltage sources		Main current		Opposing current
Loop 1	V_1	=	$I_1Z_1 + I_1Z_2$	−	I_2Z_2
Loop 2	V_2	=	$I_2Z_2 + I_2Z_3$	−	I_1Z_2

Substituting values:

$$5 = 30I_1 - 20I_2$$

$$20 = -20I_1 + 30I_2$$

The solutions of these equations are the currents:

$$I_1 = 1.1 \text{ A}$$

$$I_2 = 1.4 \text{ A}$$

1-22 CIRCUIT ANALYSIS BY NODE EQUATIONS

Node equations permit us to solve the voltages between nodes in a circuit. The steps are:

1. Identify the node voltages (see Fig. 1-17).

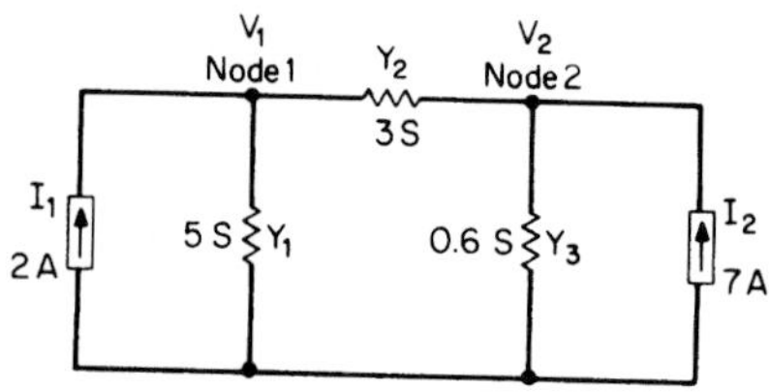

FIG. 1-17 Node voltage analysis.

2. Write the node equations. For each node, the products of the voltages and admittances of all branches joined at the node are summed and set equal to the algebraic sum of current sources attached to the node. (Subtract the product of voltages and admittances in branches not connected to the reference node.)
3. Solve the set of simultaneous equations that result.

(See determinants in Chapter 11.)

Current Sources

	Algebraic sum of current sources		Branches connected to reference node		Other branches
Node 1	I_1	=	$V_1Y_1 + V_1Y_2$	−	V_2Y_2
Node 2	I_2	=	$V_2Y_2 + V_2Y_3$	−	V_1Y_2

Substituting values, we have

$$2 = 8V_1 - 3V_2$$

$$7 = -3V_1 + 3.6V_2$$

Solving the equations, we obtain the voltages

$$V_1 = 1.4\ \text{V}$$

$$V_2 = 3.1\ \text{V}$$

2

Passive Components

2-1 RESISTANCE EQUATIONS

Series Resistors

$$R_T = R_1 + R_2 + R_3 + \dots$$

where R_T = equivalent total resistance, Ω
R_1, R_2, R_3 = component resistors, Ω

Parallel Resistors

$$R_T = \frac{1}{1/R_1 + 1/R_2 + 1/R_3 + \dots}$$

$$G_T = G_1 + G_2 + G_3 + \dots$$

where G_T = equivalent total conductance, S
G_1, G_2, G_3 = component conductance, S

Two Parallel Resistors

$$R_T = \frac{R_1 R_2}{R_1 + R_2}$$

2-2 RESISTOR COLOR CODES

The resistor color code is shown in Fig. 2-1.

Military Standard Codes

EXAMPLE: Variable, wirewound, precision resistor, 3 W, ±25 ppm/°C, 30.5 Ω, ±2% tolerance, 0.1%/1000 h.

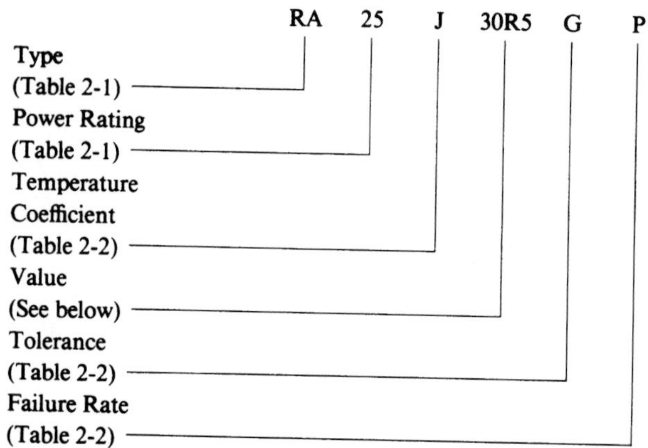

Value: Three or four digits (first two or three significant). The last digit gives the number of trailing zeros; alternatively, the letter "R" can indicate a decimal point, and all numbers are significant (no multiplier).

COLOR	1ST DIGIT	2ND DIGIT	MULTIPLIER	TOLERANCE (percent)
Black	0	0	1	
Brown	1	1	10	
Red	2	2	100	
Orange	3	3	1,000	
Yellow	4	4	10,000	
Green	5	5	100,000	
Blue	6	6	1,000,000	
Violet	7	7	10,000,000	
Gray	8	8	100,000,000	
White	9	9	1,000,000,000	
Gold			.1	5
Silver			.01	10
No color				20

(a)

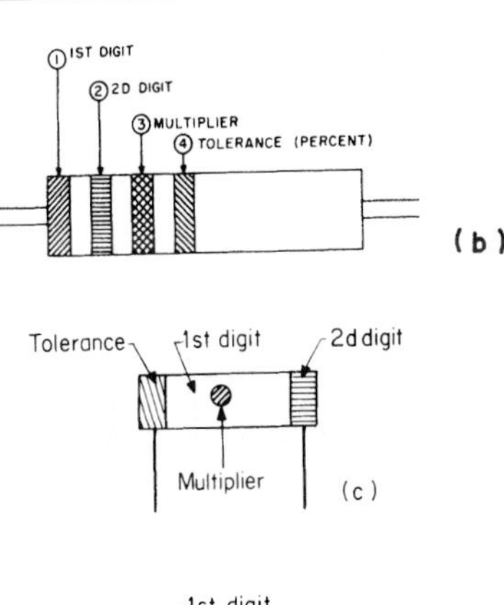

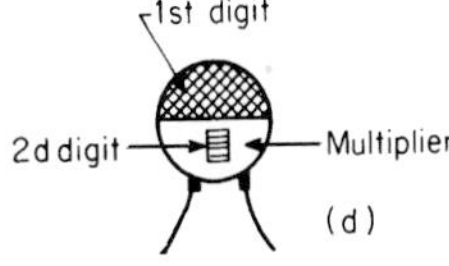

FIG. 2-1 Resistor color code: *(a)* color code; *(b)* composition, wirewound, or film resistor; *(c)* body-end dot; and *(d)* miniature resistor. (From *Basic Electronics, Bureau of Naval Personnel*)

TABLE 2-1 Military Standard Resistor Code, Part 1

Type		Power	
Code*	Meaning	Code	Value, W
RA	Variable, wirewound precision	20	2
		25	3
		30	4
RB	Fixed, wirewound precision	08	0.5
		16	0.666
		17	1
		18	1.5
		19	2
		52	1
		53	0.5
		55	0.25
		56	0.125
		57	1
		58	2
		70	0.5
		71	0.25
RC	Fixed, composition	05	0.125
		07	0.25
		09	0.5
		20	0.5
		30	1
		32	1
		41	2
		42	2
RD	Power, film, noninductive	31	7
		33	13
		35	25
		37	55
		39	115
		60	1
		65	2
		70	4

TABLE 2-1 Military Standard Resistor Code, Part 1 (Continued)

Type		Power	
Code*	Meaning	Code	Value, W
RE	Power, wirewound, with heatsink	60	7.5
		65	20
		70	25
		75	50
		77	100
		80	250
RL	Fixed, film	07	0.25
		20	0.5
		32	1
		42	2
RN	Fixed, film, high stability	05	0.125
		50	0.05
		55	0.1
		60	0.125
		65	0.25
		70	0.5
		75	1
RP	Variable, power, wirewound	10	25
		11	12
		15	50
		16	25
		20	75
		25	100
		30	150
		35	225
		40	300
		45	500
		50	750
		55	1000
RV	Variable, composition	01	0.25
		04	2
		05	0.5
		06	0.333

TABLE 2-1 Military Standard Resistor Code, Part 1 (Continued)

Type		Power	
Code*	Meaning	Code	Value, W
RW	Fixed, power, wirewound	55	5
		56	10
		67	5
		68	10
		69	2.5
		70	1
		74	5
		78	10
		79	3
		80	2.25
		81	1

* The letter "R" following the type code indicates that the component meets established military standard reliability levels.

TABLE 2-2 Military Standard Resistor Code, Part 2

Temperature coefficient		Tolerance		Failure rate	
Code	Coefficient, ppm/°C	Code	Tolerance, %	Code	Rate, %/1000 h
J	±25	F	±1	M	1
E	±25	G	±2	P	0.1
H	±50	J	±5	R	0.01
C	±50	K	±10	S	0.001
K	±100				
O	±100				

2-3 CAPACITOR EQUATIONS

Series Capacitors

$$C_T = \frac{1}{1/C_1 + 1/C_2 + 1/C_3 + \ldots}$$

where C_T = equivalent total capacitance, F
C_1, C_2, C_3 = component capacitors, F

Two Series Capacitors

$$C_T = \frac{C_1 C_2}{C_1 + C_2}$$

Parallel Capacitors

$$C_T = C_1 + C_2 + C_3 + \ldots$$

Charge Storage

$$Q = CV$$

$$C = \frac{Q}{V}$$

$$V = \frac{Q}{C}$$

where Q = charge, C
C = capacitance, F
V = voltage, V

Energy Storage

$$W = \frac{CV^2}{2}$$

$$C = \frac{2W}{V^2}$$

$$V = \sqrt{\frac{2W}{C}}$$

where W = energy, J
C = capacitance, F
V = voltage, V

Parallel-Plate Capacitance

$$C = \frac{8.855(N-1)KA}{d}$$

$$K = \frac{Cd}{8.855(N-1)A}$$

$$A = \frac{Cd}{8.855(N-1)K}$$

$$d = \frac{8.855(N-1)KA}{C}$$

$$N = \frac{Cd}{8.855KA} + 1$$

where C = capacitance, pF
N = number of plates
K = relative dielectric constant (see Table 2-3)
A = plate area, m^2
d = thickness of dielectric, m

Instantaneous Charging or Discharging

$$C = I\frac{\Delta t}{\quad} \qquad I = C\frac{\Delta V}{\Delta t}$$

$$\Delta V = I\frac{\Delta t}{\quad} \qquad \Delta t = C\frac{\Delta V}{I}$$

where C = capacitance, F
I = current, A
Δt = time interval, s
ΔV = change in voltage, V

Constant Charging or Discharging

$$C = \frac{It}{V}$$

$$V = \frac{It}{C}$$

$$I = \frac{CV}{t}$$

$$t = \frac{CV}{I}$$

where C = capacitance, F
I = current, A
t = time, s
V = voltage, V

TABLE 2-3 Relative Dielectric Constants*

Substance	Dielectric Constant
Air	1.0006
Crown glass	6†
Ethyl alcohol	25.7
Hydrogen	1.0003
Mica	7†
Paper (dry)	3.5†
Paper (oiled)	3.5†
Paraffin	2.25†
Petroleum oil	2.1†
Polyethylene	2.26‡
Polystyrene	2.55‡
Porcelain	6.5†
Rubber	3†
Teflon	2.1‡
Titanium oxides	$10-10^4$†
Vacuum	1.0000
Water	78

* Values given for 20°C, atmospheric pressure, and frequency of less than 1 MHz.
† Extremely variable.
‡ Values apply for all frequencies.

2-4 CAPACITOR COLOR CODES

Color codes for various types of capacitors are shown in Figs. 2-2 through 2-5.

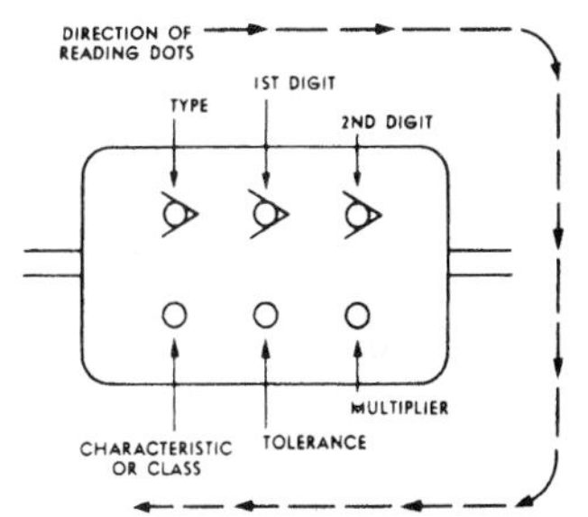

TYPE	COLOR	1ST DIGIT	2ND DIGIT	MULTIPLIER	TOLERANCE (PERCENT)	CHARACTERISTIC OR CLASS
JAN, MICA	BLACK	0	0	1.0		APPLIES TO
	BROWN	1	1	10	±1	TEMPERATURE
	RED	2	2	100	±2	COEFFICIENT
	ORANGE	3	3	1,000	±3	OR METHODS
	YELLOW	4	4	10,000	±4	OF TESTING
	GREEN	5	5	100,000	±5	
	BLUE	6	6	1,000,000	±6	
	VIOLET	7	7	10,000,000	±7	
	GRAY	8	8	100,000,000	±8	
EIA, MICA	WHITE	9	9	1,000,000,000	±9	
	GOLD			.1		
MOLDED PAPER	SILVER			.01	±10	
	BODY				±20	

FIG. 2-2 Dot color code for mica and molded paper capacitors. *(Bureau of Naval Personnel)*

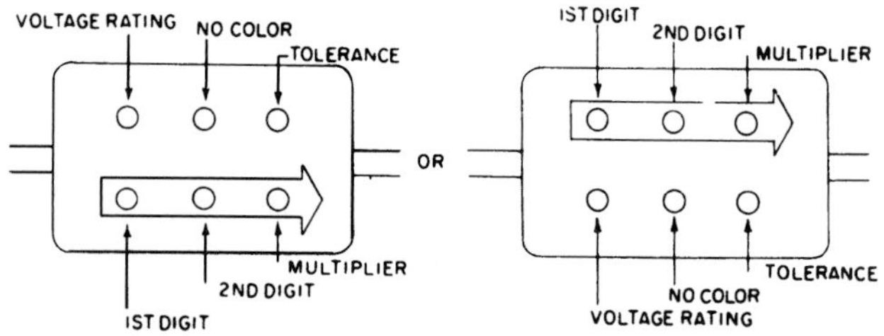

COLOR	1ST DIGIT	2ND DIGIT	MULTIPLIER	TOLERANCE (PERCENT)	VOLTAGE RATING
BLACK	0	0	1.0		
BROWN	1	1	10	±1	100
RED	2	2	100	±2	200
ORANGE	3	3	1,000	±3	300
YELLOW	4	4	10,000	±4	400
GREEN	5	5	100,000	±5	500
BLUE	6	6	1,000,000	±6	600
VIOLET	7	7	10,000,000	±7	700
GRAY	8	8	100,000,000	±8	800
WHITE	9	9	1,000,000,000	±9	900
GOLD			0.1		1000
SILVER			0.01	±10	2000
BODY				±20	*

* WHERE NO COLOR IS INDICATED, THE VOLTAGE RATING MAY BE AS LOW AS 300 VOLTS.

FIG. 2-3 Five-dot color code for capacitors (dielectric not specified). *(Bureau of Naval Personnel)*

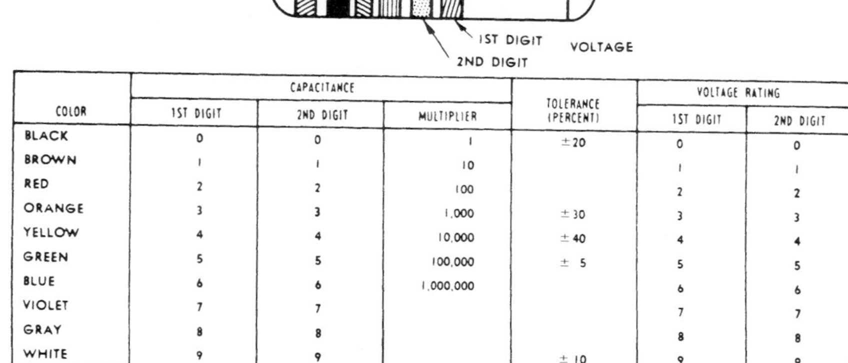

COLOR	CAPACITANCE			TOLERANCE (PERCENT)	VOLTAGE RATING	
	1ST DIGIT	2ND DIGIT	MULTIPLIER		1ST DIGIT	2ND DIGIT
BLACK	0	0	1	±20	0	0
BROWN	1	1	10		1	1
RED	2	2	100		2	2
ORANGE	3	3	1,000	±30	3	3
YELLOW	4	4	10,000	±40	4	4
GREEN	5	5	100,000	± 5	5	5
BLUE	6	6	1,000,000		6	6
VIOLET	7	7			7	7
GRAY	8	8			8	8
WHITE	9	9		± 10	9	9

FIG. 2-4 Six-band color code for tubular paper dielectric capacitors. *(Bureau of Naval Personnel)*

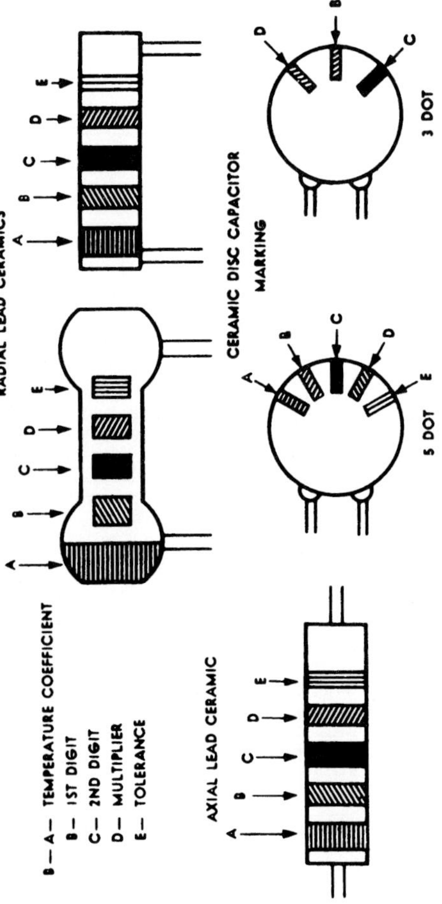

FIG. 2-5 Color codes for ceramic capacitors. *(Bureau of Naval Personnel)*

COLOR	1ST DIGIT	2ND DIGIT	MULTIPLIER	TOLERANCE		TEMPERATURE COEFFICIENT *
				MORE THAN 10 pf (IN PERCENT)	LESS THAN 10 pf (IN pf)	
BLACK	0	0	1.0	±20	±2.0	0
BROWN	1	1	10	± 1		—30
RED	2	2	100	± 3		—80
ORANGE	3	3	1,000			—150
YELLOW	4	4	10,000			—220
GREEN	5	5		± 5	±0.5	—330
BLUE	6	6				—470
VIOLET	7	7				—750
GRAY	8	8	0.01		±0.25	+30
WHITE	9	9	0.1	±10	±1.0	+120 TO —750 (EIA)
						+500 TO —330 (JAN)
SILVER						+100 (JAN)
GOLD						BYPASS OR COUPLING (EIA)

* PARTS PER MILLION PER DEGREE CENTIGRADE.

FIG. 2-5 Color codes for ceramic capacitors. *(Bureau of Naval Personnel)* (Continued)

2-5 INDUCTOR EQUATIONS

Series Inductors

$$L_T = L_1 + L_2 + L_3 + \dots$$

where L_T = equivalent total inductance, H

L_1, L_2, L_3 = component inductors, H

Parallel Inductors

$$L_T = \frac{1}{1/L_1 + 1/L_2 + 1/L_3 + \dots}$$

Two Parallel Inductors

$$L_T = \frac{L_1 L_2}{L_1 + L_2}$$

Energy Storage

$$W = \frac{LI^2}{2}$$

$$L = \frac{2W}{I^2}$$

$$I = \sqrt{\frac{2W}{L}}$$

where W = energy, J

L = inductance, H

I = current, A

Instantaneous Charging or Discharging

$$L = V\frac{\Delta t}{\ }$$

$$V = L\frac{\Delta I}{\Delta t}$$

$$\Delta I = V\frac{\Delta t}{L}$$

$$\Delta t = L\frac{\Delta I}{V}$$

where L = inductance, H
V = voltage, V
Δt = time interval, s
ΔI = change in current, A

Constant Charging or Discharging

$$L = \frac{Vt}{I}$$

$$I = \frac{Vt}{L}$$

$$V = \frac{LI}{t}$$

$$t = \frac{LI}{V}$$

where L = inductance, H
V = voltage, V
t = time, s
I = current, A

Mutual Inductance

$$L_T = L_1 + L_2 + 2M \text{ (aiding fields)}$$

$$L_T = L_1 + L_2 - 2M \text{ (opposing fields)}$$

where L_T = total inductance, H
L_1, L_2 = component inductance, H
M = mutual inductance, H

Coupling Coefficient

$$k = \frac{M}{\sqrt{L_1 L_2}}$$

where k = coupling coefficient
M = mutual inductance, H
L_1, L_2 = component inductance, H

Flux Linkage

$$L = \frac{N\Phi}{}$$

where L = inductance, H
N = number of turns of the coil
Φ = magnetic flux, Wb
i = instantaneous current, A

Coil Formulas

Long Coil (Fig. 2-6*a*):

$$L = \frac{\mu N^2 A}{\quad}$$

Short coil (Fig. 2-6*a*):

$$L = \frac{\mu N^2 A}{l + 0.45d}$$

where L = inductance, H
μ = permeability ($4\pi \times 10^{-7}$ for air)
N = number of turns
A = cross-sectional area of coil, m^2
l = length of coil, m
d = diameter of coil, m

Toroidal Coil with Rectangular Cross Section (Fig. 2-6*b*):

$$L = \frac{\mu N^2 h}{2\pi} \ln \frac{d_2}{d_1}$$

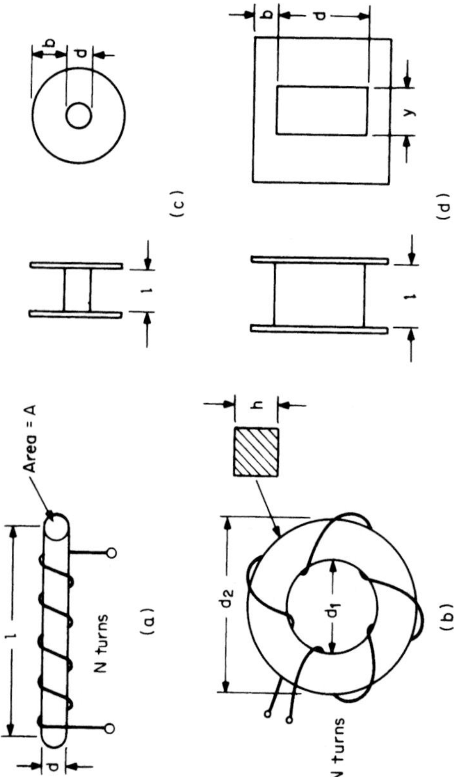

FIG. 2-6 Coil geometries: (*a*) long coil or short coil, (*b*) toroidal coil with rectangular cross section, (*c*) circular air-core coil, and (*d*) rectangular air-core coil.

where L = inductance, H
μ = permeability ($4\pi \times 10^{-7}$ for air)
N = number of turns
h = thickness, m
d_1, d_2 = inner and outer diameter, m

Circular Air-Core Coil (Fig. 2-6*c*):

$$L = \frac{0.07(RN)^2}{6R + 9l + 10b}$$

where L = inductance, μH
$R = \frac{d}{2} + \frac{b}{2}$
N = number of turns
d = core diameter, inches
b = coil buildup, inches
l = length, inches

Rectangular Air-Core coil (Fig. 2-6*d*):

$$L = \frac{0.07(CN)^2}{1.908C + 9l + 10b}$$

where L = inductance, μH
$C = d + y + 2b$
d = core height, inches
y = core width, inches
b = coil buildup, inches
l = length, inches

Magnetic Core Coil (no air gap)

$$L = \frac{0.012N^2\mu A}{}$$

where L = inductance, μH
N = number of turns
A = effective cross-sectional core area, cm^2
l_c = magnetic path length, cm
l_g = gap length, cm
μ = magnetic permeability (see Table 2-4)

TABLE 2-4 Relative Magnetic Permeability

Material	Permeability (μ)	
	Low signal level	High signal level
Air	1.000	1.000
Ferrites		
3B7	2,300	1,900
3C8	2,300	1,900
TI	2,300	1,900
W-03	2,300	1,900
3E2A	5,000	1,800
3E3	12,000	1,900
3D3	750	1,500
4C4	125	600
1Z2	15	—
Iron-nickel alloy	3,000	20,000
Powdered iron	125	127
Silicon iron	400	40,000

Magnetic Core Coil (air gap)

$$L = \frac{0.012N^2A}{l_g + l_c/\mu}$$

2-6 TRANSFORMER EQUATIONS

Turns Ratio

$$T_R = \frac{N_s}{N_p}$$

where N_s = secondary number of turns
N_p = primary number of turns

Voltage/Current Ratios

$$T_R = \frac{V_s}{V_p} = \frac{I_p}{I_s}$$

where V_p, V_s = primary and secondary voltage, V
I_p, I_s = primary and secondary current, A

Impedance Ratio (Ideal Transformer with Unity Coupling)

$$T_{\mathrm{R}} = \sqrt{\frac{Z_s}{Z_p}}$$

where Z_p = impedance reflected into the primary side with a secondary load of Z_s

2-7 HAND RULES

A variety of rules to determine the direction of parameters in electrical circuits is given in Table 2-5.

TABLE 2-5 Hand Rules

Rule	Hand	Thumb	Fingers
Field around a conductor	Right	Direction of conventional current	Curl around conductor in direction of magnetic lines of force (north to south)
Solenoid magnetic polarity	Right	Points to north pole of solenoid	Curl around coil in direction of conventional current
Induced current	Right	Direction of motion of the conductor	Index finger—direction of magnetic lines of force (north to south); middle finger—direction of induced conventional current
Force on a moving charge	Left	Direction of force	Index finger—direction of magnetic lines of force (north to south); middle finger—direction of conventional current

TABLE 2-5 Hand Rules (Continued)

Rule	Hand	Thumb	Fingers
Electromagnetic wave	Right	Direction of E field	Index finger—direction of propagation; middle finger—direction of H field

3

Active Components

3-1 STANDARD SEMICONDUCTORS

Standard semiconductors are listed in Table 3-1.

TABLE 3-1 Standard Semiconductors

Device	Type	Characteristics	Designator
Diode	Signal	75 V, 20 mA	1N914A, 1N914B
	Rectifier	400 V, 1 A	1N4004
Rectifier	High-voltage	1000 V, 1 A	1N4007
	High-current	100 V, 35 A	1N1184
	Fast-switching	200 V, 1 A	1N4935
Triac		400 V, 4 A	2N6073A, 2N6073B
SCR	High-power	600 V, 16 A	2N6404
	Low-power	200 V, 0.8 A	2N5064
Transistor	NPN	40 V, 350 mW	2N4400
	PNP	40 V, 350 mW	2N4402
	Low-power NPN	60 V, 5 W	2N3053
	Low-power PNP	60 V, 5 W	2N4036
	High-power NPN	60 V, 115 W	2N3055
	High-power PNP	60 V, 115 W	2N5875
	High-voltage NPN	250 V, 5 W	2N3440
	High-voltage PNP	250 V, 5 W	2N5416
	FET, N-channel	25 V, 350 mW	2N3819

3-2 LOW-FREQUENCY TRANSISTOR COMMON-EMITTER (CE) HYBRID PARAMETERS

Various types of transistor amplifiers are shown in Fig. 3-1.

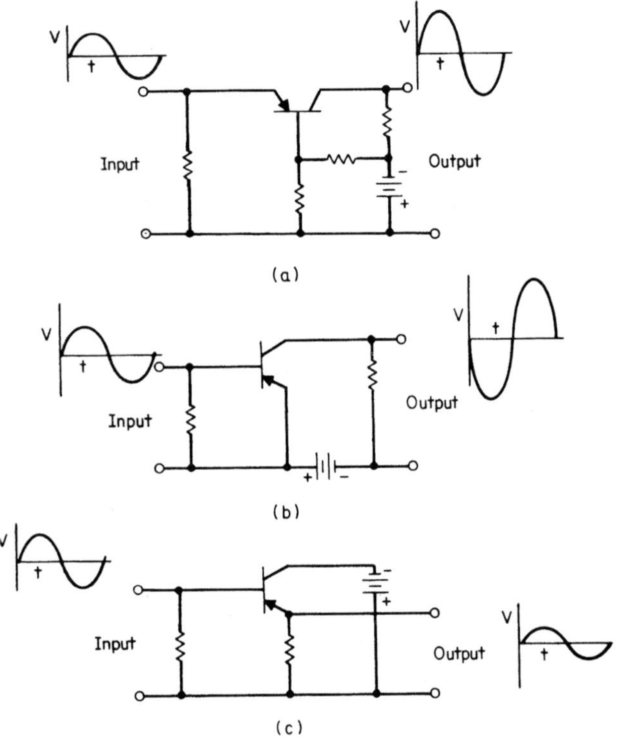

FIG. 3-1 Transistor amplifiers: *(a)* common-base amplifier; *(b)* common-emitter amplifier; and *(c)* common-collector amplifier.

Input Resistance, Output Short-Circuited

$$h_{ie} = \left(\frac{\Delta V_{BE}}{\Delta I_B} \right)_{V_{CE} = K}$$

Reverse Open-Circuit Voltage Amplification Factor

$$\mu_{re} = \left(\frac{\Delta V_{BE}}{\Delta V_{CE}} \right)_{I_B = K}$$

Forward Short-Circuit Current Amplification Factor

$$\alpha_{fe} = \left(\frac{\Delta I_C}{\Delta I_B} \right)_{V_{CE} = K}$$

Output Conductance, Input Open

$$h_{oe} = \left(\frac{\Delta I_C}{\Delta V_{CE}} \right)_{I_B = K}$$

where V_{BE} = base-emitter voltage, V
V_{CE} = collector-emitter voltage, V
I_B = base current, A
I_C = collector current, A
K = constant value

3-3 LOW-FREQUENCY TRANSISTOR COMMON-BASE (CB) HYBRID PARAMETERS

See Fig. 3-1.

Output Resistance, Output Short-Circuited

$$h_{ib} = \left(\frac{\Delta V_{EB}}{\Delta I_E} \right)_{V_{BC} = K}$$

Reverse Open-Circuit Voltage Amplification Factor

$$\mu_{rb} = \left(\frac{\Delta V_{EB}}{\Delta V_{CB}} \right)_{I_E = K}$$

Forward Short-Circuit Current Amplification Factor

$$\alpha_{fb} = \left(\frac{\Delta I_C}{\Delta I_E} \right)_{V_{CB} = K}$$

Output Conductance, Input Open

$$h_{ob} = \left(\frac{\Delta I_C}{\Delta V_{CB}} \right)_{I_E = K}$$

where V_{EB} = emitter-base voltage, V
V_{CB} = collector-base voltage, V
I_E = emitter current, A
I_C = collector current, A
K = constant value

3-4 LOW-FREQUENCY TRANSISTOR COMMON-COLLECTOR (CC) HYBRID PARAMETERS

See Fig. 3-1.

Input Resistance, Output Short-Circuited

$$h_{ic} = \left(\frac{\Delta V_{BC}}{\Delta I_B} \right)_{V_{EC} = K}$$

Reverse Open-Circuit Voltage Amplification Factor

$$\mu_{rc} = \left(\frac{\Delta V_{BC}}{\Delta V_{EC}} \right)_{I_B = K}$$

Forward Short-Circuit Current Amplification Factor

$$\alpha_{fc} = \left(\frac{\Delta I_E}{\Delta I_B} \right)_{V_{EC} = K}$$

Output Conductance, Input Open

$$h_{oc} = \left(\frac{\Delta I_E}{\Delta V_{EC}} \right)_{I_B = K}$$

where V_{BC} = base-collector voltage, V
V_{EC} = emitter-collector voltage, V
I_E = emitter current, A
I_B = base current, A
K = constant value

3-5 TRANSISTOR AMPLIFIER EQUATIONS

Current Gain

$$A_i = \frac{-\alpha_{fx}}{h_{ox}R_L + 1}$$

Voltage Gain

$$A_v = \frac{-\alpha_{fx}R_L}{\qquad}$$

Power Gain

$$G_P = \frac{\alpha_{fx}^{\,2}R_L}{\qquad}$$

Input Resistance

$$r_i = \frac{h_{ix} + (h_{ox}h_{ix} - \alpha_{fx}\mu_{rx})R_L}{1 + h_{ox}R_L}$$

Output Resistance

$$r_o = \frac{h_{ix} + R_g}{h_{ox}h_{ix} - \mu_{rx}\alpha_{fx} + h_{ox}R_g}$$

where R_L = load resistance, Ω

R_g = source resistance, Ω

x (in subscripts) = e, b, or c as appropriate for common-emitter, common-base, or common-collector amplifier, respectively

See Table 3-2 for conversion of hybrid parameters.

TABLE 3-2 Conversion of Hybrid Parameters

CE to CB	CE to CC	CB to CE	CB to CC
$h_{ib} = \frac{h_{ie}}{1 + \alpha_{fe}}$	$h_{ic} = h_{ie}$	$h_{ie} = \frac{h_{ib}}{1 + \alpha_{fb}}$	$h_{ic} = \frac{h_{ib}}{1 + \alpha_{fb}}$
$\mu_{rb} = \frac{h_{ie}h_{oe}}{1 + \alpha_{fe}} - \mu_{re}$	$\mu_{rc} = 1 - \mu_{re} \approx 1$	$\mu_{re} = \frac{h_{ib}h_{ob}}{1 + \alpha_{fe}} - \mu_{rb}$	$\mu_{rc} = \frac{1 - h_{ib}h_{ob}}{1 + \alpha_{fb}} + \mu_{rb}$
$\alpha_{fb} = \frac{-\alpha_{fe}}{1 + \alpha_{fe}}$	$\alpha_{fc} = -1 - \alpha_{fe}$	$\alpha_{fe} = \frac{-\alpha_{fb}}{1 + \alpha_{fb}}$	$\alpha_{fc} = \frac{-1}{1 + \alpha_{fb}}$
$h_{ob} = \frac{h_{oe}}{1 + \alpha_{fe}}$	$h_{oc} = h_{oe}$	$h_{oe} = \frac{h_{ob}}{1 + \alpha_{fb}}$	$h_{oc} = \frac{h_{ob}}{1 + \alpha_{fb}}$

3-6 TYPICAL TRANSISTOR PARAMETER VALUES

These values are estimates to be expected for typical transistors.

Common emitter	Common base	Common collector
$h_{ie} = 1950\ \Omega$ $\mu_{re} = 575 \times 10^{-6}$ $\alpha_{fe} = 49$ $h_{oe} = 24.5\ \mu S$	$h_{ib} = 39\ \Omega$ $\mu_{rb} = 380 \times 10^{-6}$ $\alpha_{fb} = -0.98$ $h_{ob} = 0.49\ \mu S$	$h_{ie} = 1950\ \Omega$ $\mu_{rc} = 1$ $\alpha_{fc} = -50$ $h_{oc} = 24.5\ \mu S$

3-7 OTHER TRANSISTOR EQUATIONS

$$\alpha \approx \frac{I_C}{I_E} \qquad \beta = \frac{\alpha}{1 - \alpha}$$

where α = fraction of emitter current that reaches the collector
β = current amplification factor
I_C = collector current, A
I_E = emitter current, A

3-8 TRANSISTOR PACKAGING

Figure 3-2 shows metal and plastic cases for lead-mounted bipolar transistors and power transistors.

3-9 LIGHT-EMITTING DIODES (LED)

Packaging for typical LEDs is shown in Fig. 3-3. Decoding of multisegment LED display is shown in Figs. 3-4 and 3-5.

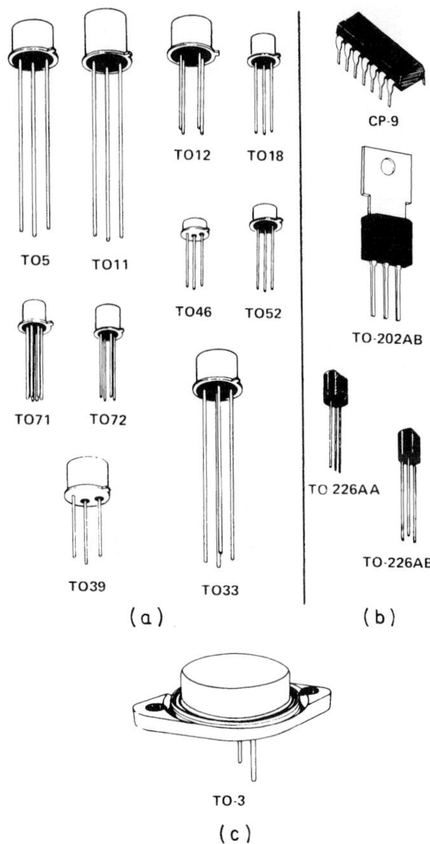

FIG. 3-2 Transistor packaging: *(a)* lead-mounded bipolar transistors in metal case, *(b)* lead-mounted bipolar transistors in plastic case, and *(c)* power transistors in metal case. *(Texas Instruments)*

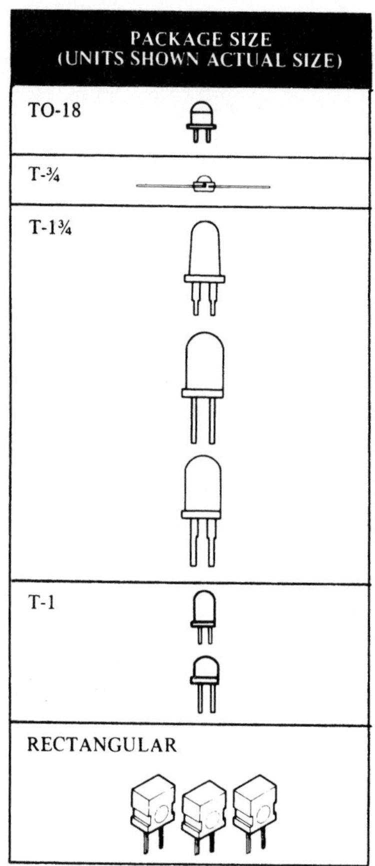

FIG. 3-3 LED packaging: TO-18, red/clear; T-¾, bright red; T-1¾, various reds; T-1, red; rectangular green, yellow, or bright red. *(General Instruments)*

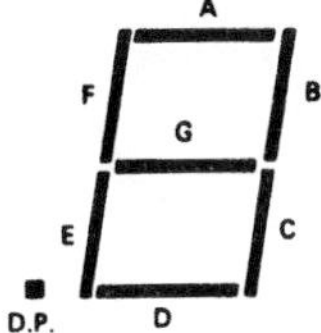

TYPICAL TRUTH TABLE

INPUT CODE				OUTPUT STATE							DISPLAY
d	c	b	a	A′	B′	C′	D′	E′	F′	G′	
0	0	0	0	0	0	0	0	0	0	1	0
0	0	0	1	1	0	0	1	1	1	1	1
0	0	1	0	0	0	1	0	0	1	0	2
0	0	1	1	0	0	0	0	1	1	0	3
0	1	0	0	1	0	0	1	1	0	0	4
0	1	0	1	0	1	0	0	1	0	0	5
0	1	1	0	1	1	0	0	0	0	0	6
0	1	1	1	0	0	0	1	1	1	1	7
1	0	0	0	0	0	0	0	0	0	0	8
1	0	0	1	0	0	0	1	1	0	0	9

FIG. 3-4 Seven-segment LED display. *(General Instruments)*

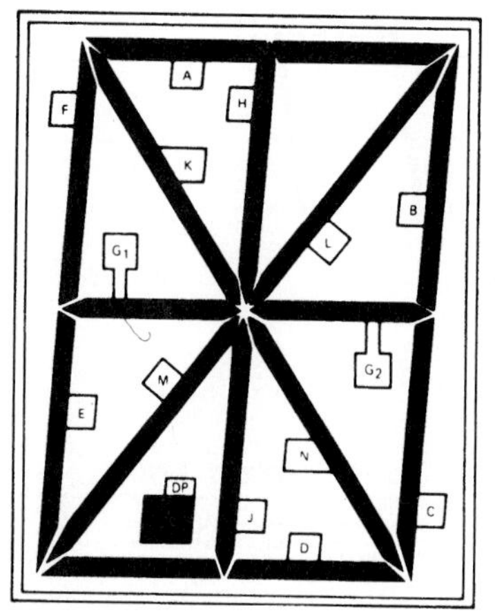

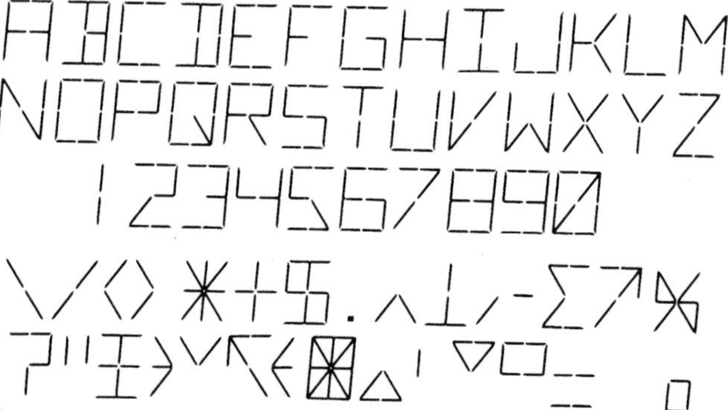

FIG. 3-5 Fourteen-segment LED display. *(General Instruments)*

3-10 INTEGRATED CIRCUITS

Pin assignments for 54/74 TTL circuits are shown in Fig. 3-6 and circuit designations and functions in Table 3-3. Semiconductor manufacturers' logos are shown in Fig. 3-7.

TABLE 3-3 Common 54/74 TTL Integrated Circuits

Designator	Function	Pin assignment (see Fig. 3-6)
00*	Quad 2-input NAND	*a*
01	Quad 2-input NAND	*b*
02*	Quad 2-input NOR	*b*
03*	Quad 2-input NAND, open collector	*a*
04*	Hex inverter	*c*
05*	Hex inverter, open collector	*c*
06	Hex inverter, buffer-driver	*c*
07	Hex noninverting buffer	*c*
08*	Quad 2-input AND	*a*
09	Quad 2-input AND, open collector	*a*
14*	Hex inverter Schmitt trigger	*b*
16	Hex inverter, buffer	*b*
17	Hex noninverting buffer	*b*
20*	Dual 4-input NAND	*d*
26	Quad 2-input NAND, high-voltage	*a*
28	Quad 2-input NOR, buffer	*c*
30*	Eight-input NAND	*e*
32*	Quad 2-input OR	*a*
33	Quad 2-input NOR, buffer	*c*
37	Quad 2-input NAND, buffer	*a*
38	Quad 2-input NAND, open collector	*a*

TABLE 3-3 Common 54/74 TTL Integrated Circuits (Continued)

Designator	Function	Pin assignment (see Fig. 3-6)
46	BCD-to-seven-segment decoder-driver, 30 V	*f*
47	BCD-to-seven-segment decoder-driver, 15 V	*f*
48	BCD-to-seven segment decoder-driver, internal pull-up	*f*
75	Quad bistable latch	*g*
76*	Dual *JK*-flip-flop	*h*
86*	Quad 2-input exclusive OR	*i*
90	Decade counter	*j*
93	Four-bit binary counter	*j*
128	Quad line driver	*c*
132*	Quad 2-input NAND, Schmitt trigger	*a*
181*	Arithmetic logic unit	*k*
190*	BCD synchronous up-down counter	*l*
191*	Binary synchronous up-down counter	*l*
192*	BCD synchronous up-down counter	*m*
193*	Binary synchronous up-down counter	*m*

*Equivalent circuit available in high-speed CMOS.

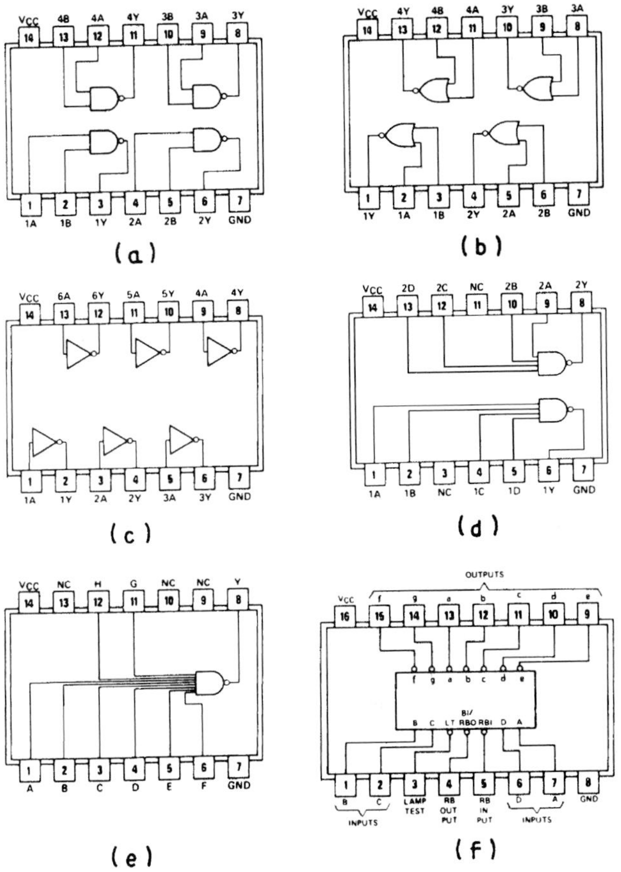

FIG. 3-6 54/74 TTL integrated circuit pin assignments (top views). *(Texas Instruments)*

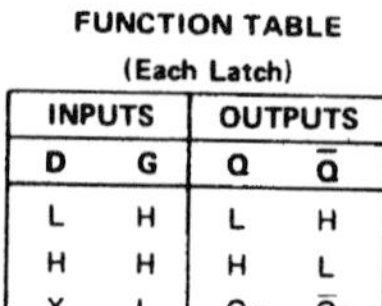

FUNCTION TABLE
(Each Latch)

INPUTS		OUTPUTS	
D	G	Q	$\bar{Q}$
L	H	L	H
H	H	H	L
X	L	Q_0	$\bar{Q}_0$

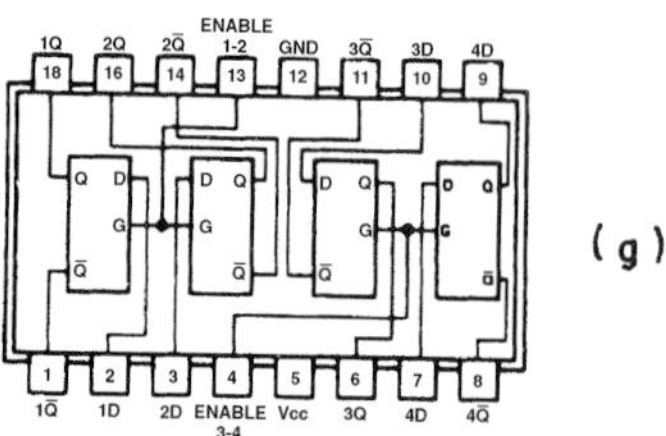

(g)

'76, 'H76
FUNCTION TABLE

INPUTS					OUTPUTS	
PRESET	CLEAR	CLOCK	J	K	Q	$\bar{Q}$
L	H	X	X	X	H	L
H	L	X	X	X	L	H
L	L	X	X	X	H*	H*
H	H	⎍	L	L	Q_0	$\bar{Q}_0$
H	H	⎍	H	L	H	L
H	H	⎍	L	H	L	H
H	H	⎍	H	H	TOGGLE	

LS76A
FUNCTION TABLE

INPUTS					OUTPUTS	
PRESET	CLEAR	CLOCK	J	K	Q	$\bar{Q}$
L	H	X	X	X	H	L
H	L	X	X	X	L	H
L	L	X	X	X	H*	H*
H	H	↓	L	L	Q_0	$\bar{Q}_0$
H	H	↓	H	L	H	L
H	H	↓	L	H	L	H
H	H	↓	H	H	TOGGLE	
H	H	H	X	X	Q_0	$\bar{Q}_0$

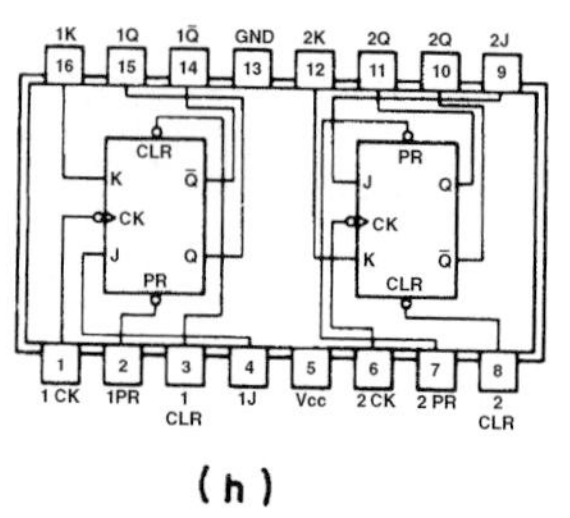

(h)

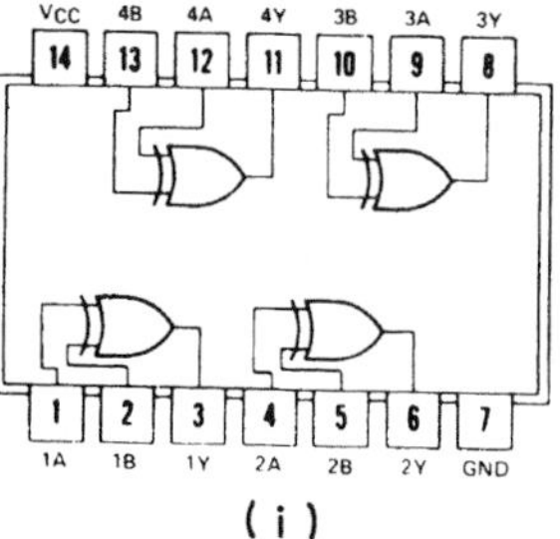

(i)

FIG. 3-6 54/74 TTL integrated circuit pin assignments (top views). *(Texas Instruments) (Continued)*

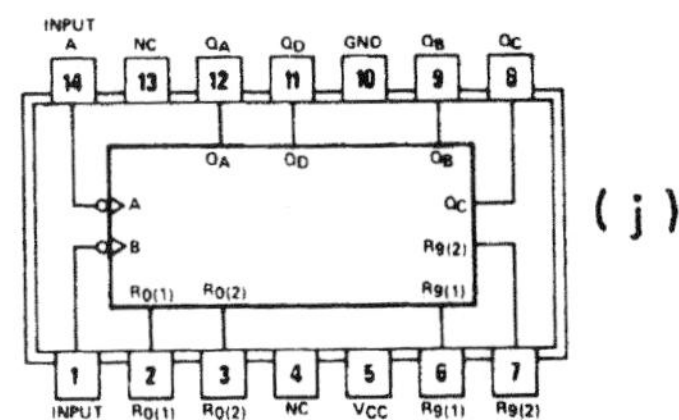

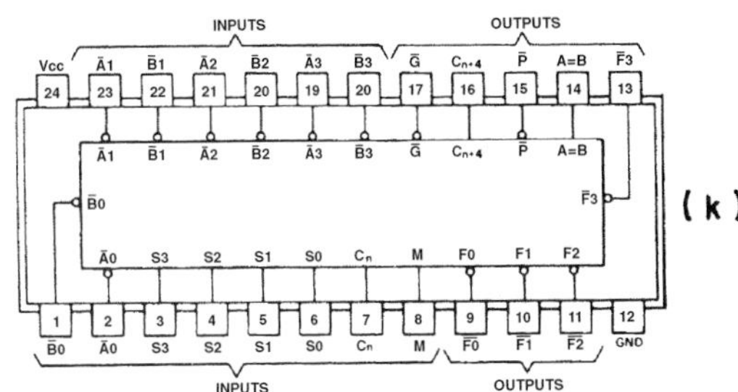

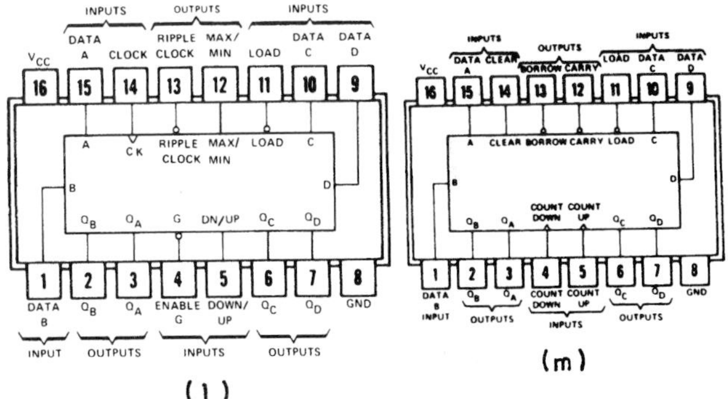

FIG. 3-6 54/74 TTL integrated circuit pin assignments (top views). *(Texas Instruments) (Continued)*

Manufacturer Contacts and World Wide Web Addresses

Manufacturer	Address and Phone Number	URL
Advanced Micro Devices	1 AMD Place P.O. Box 3453 Sunnyvale, CA 94088-3453 (408) 732-2400	www.amd.com
American Megatrends	6145-F Northbelt Parkway Norcross, GA 30071-2972 (800) 828-9264	www.aim.com
Analog Devices, Inc.	1 Technology Way P.O. Box 9106 Norwood, MA 02062-9106	www.analog.com
Fujitsu Microelectronics, Inc.	3345 No. First Street San Jose, CA 95134-1804 (800) 637-0683	www.fujitsu.com
Harris Corp.	1025 W NASA Blvd. Melbourne, FL 32919 (407) 727-9100	www.harris.com
Hitachi America Computer Division	2000 Sierra Point Parkway Brisbane, CA 94005-1835 (800) 448-2244	www.hitachi.com
Intel Corp.	2200 Mission College Dr. Santa Clara, CA 95052-8119 (408) 765-8080	www.intel.com
Mitel Corp.	350 Lagget Drive, Box 13089 Kanata, Ontario, Canada K2K 1X3 (613) 592-2122	—
Mitsubishi Electric Corp.	2-2-3 Marunouchi, 2- chrome, Chiyoda-Ku Tokyo 100JP-100, Japan 3-3218-2111	—

Manufacturer Contacts and World Wide Web Addresses (Continued)

Manufacturer	Address and Phone Number	URL
Motorola, Inc.	1303 E. Alonquin Rd. Schuamburg, IL 60196 (800) 262-8509	www.mot.com
National Semiconductor	2900 Semiconductor Dr. Santa Clara, CA 95020-8090 (408) 721-5000	www.natsemi.com
NEC Electronics, Inc.	475 Ellis St. Mountain View, CA 94039-7241 (415) 960-6000	www.nec.co.jp
Panasonic (Matsushita)	1 Panasonic Way Secaucus, NJ 07094 (201) 348-7000	—
Rockwell Semiconductor Systems	4311 Jamboree Rd., Box C Newport Beach, CA 92658-8902 (714) 833-4600	www.rockwell.com
Texas Instruments, Inc.	13500 N. Central Expressway P.O. Box 655474 Dallas, TX 75265 (214) 995-2551	www.ti.com
Toshiba America Electronic Components, Inc.	9775 Toledo Way Irvine, CA 92718 (714) 455-2000	www.toshiba.com
Zilog, Inc.	210 E. Hacienda Ave. Campbell, CA 95008-6600 (408) 370-8000	www.zilog.com

3-11 AN NOMENCLATURE

THE AN NOMENCLATURE WAS DESIGNED SO THAT A COMMON DESIGNATION COULD BE USED FOR ARMY, NAVY, AND AIR FORCE EQUIPMENT. THE SYSTEM INDICATOR AN DOES NOT MEAN THAT THE ARMY, NAVY, AND AIR FORCE USE THE EQUIPMENT, BUT MEANS THAT THE TYPE NUMBER WAS ASSIGNED IN THE AN SYSTEM.

AN NOMENCLATURE IS ASSIGNED TO COMPLETE SETS OF EQUIPMENT AND MAJOR COMPONENTS OF MILITARY DESIGN; GROUPS OF ARTICLES OF EITHER COMMERCIAL OR MILITARY DESIGN WHICH ARE GROUPED FOR MILITARY PURPOSES; MAJOR ARTICLES OF MILITARY DESIGN WHICH ARE NOT PART OF OR USED WITH A SET; AND COMMERCIAL ARTICLES WHEN NOMENCLATURE WILL NOT FACILITATE MILITARY IDENTIFICATION AND/OR PROCEDURES.

AN NOMENCLATURE IS NOT ASSIGNED TO ARTICLES CATALOGED COMMERCIALLY EXCEPT AS STATED ABOVE; MINOR COMPONENTS OF MILITARY DESIGN FOR WHICH OTHER ADEQUATE MEANS OF IDENTIFICATION ARE AVAILABLE; SMALL PARTS SUCH AS CAPACITORS AND RESISTORS; AND ARTICLES HAVING OTHER ADEQUATE IDENTIFICATION IN JOINT MILITARY SPECIFICATIONS. NOMENCLATURE ASSIGNMENTS REMAIN UNCHANGED REGARDLESS OF LATER CHANGES IN INSTALLATION AND/OR APPLICATION.

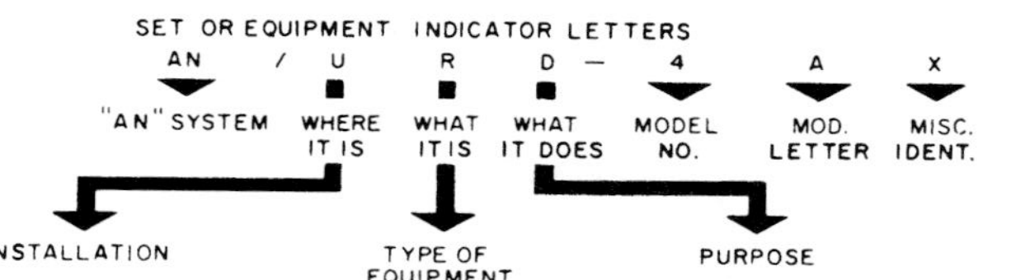

A--AIRBORNE (INSTALLED AND OPERATED IN AIRCRAFT).
B--UNDERWATER MOBILE, SUBMARINE.
C--AIR TRANSPORTABLE (INACTIVATED, DO NOT USE).
D--PILOTLESS CARRIER.
F--FIXED.
G--GROUND, GENERAL GROUND USE (INCLUDES TWO OR MORE GROUND-TYPE INSTALLATIONS).
K--AMPHIBIOUS.
M--GROUND, MOBILE (INSTALLED AS OPERATING UNIT IN A VEHICLE WHICH HAS NO FUNCTION OTHER THAN TRANSPORTING THE EQUIPMENT).
P--PACK OR PORTABLE (ANIMAL OR MAN).
S--WATER SURFACE CRAFT.
T--GROUND, TRANSPORTABLE.
U--GENERAL UTILITY (INCLUDES TWO OR MORE GENERAL INSTALLATION CLASSES, AIRBORNE, SHIPBOARD, AND GROUND).
V--GROUND, VEHICULAR (INSTALLED IN VEHICLE DESIGNED FOR FUNCTIONS OTHER THAN CARRYING ELECTRONIC EQUIPMENT, ETC., SUCH AS TANKS).
W--WATER SURFACE AND UNDERWATER.

A--INVISIBLE LIGHT, HEAT RADIATION.
B--PIGEON.
C--CARRIER.
D--RADIAC.
E--NUPAC.
F--PHOTOGRAPHIC.[1]
G--TELEGRAPH OR TELETYPE.
I--INTERPHONE AND PUBLIC ADDRESS.
J--ELECTROMECHANICAL OR INERTIAL WIRE COVERED.
K--TELEMETERING.
L--COUNTERMEASURES.
M--METEOROLOGICAL.
N--SOUND IN AIR.
P--RADAR.
Q--SONAR AND UNDERWATER SOUND.
R--RADIO
S--SPECIAL TYPES, MAGNETIC, ETC., OR COMBINATIONS OF TYPES.
T--TELEPHONE (WIRE).
V--VISUAL AND VISIBLE LIGHT.
W--ARMAMENT (PECULIAR TO ARMAMENT, NOT OTHERWISE COVERED).
X--FACSIMILE OR TELEVISION.
Y--DATA PROCESSING.

A--AUXILIARY ASSEMBLIES (NOT COMPLETE OPERATING SETS USED WITH OR PART OF TWO OR MORE SETS OR SETS SERIES).
B--BOMBING.
C--COMMUNICATIONS (RECEIVING AND TRANSMITTING).
D--DIRECTION FINDER, RECONNAISSANCE, AND/OR SURVEILLANCE.
E--EJECTION AND/OR RELEASE.
G--FIRE-CONTROL OR SEARCHLIGHT DIRECTING.
H--RECORDING AND/OR REPRODUCING (GRAPHIC METEOROLOGICAL AND SOUND).
K--COMPUTING.
L--SEARCHLIGHT CONTROL (INACTIVATED, USE G).
M--MAINTENANCE AND TEST ASSEMBLIES (INCLUDING TOOLS).
N--NAVIGATIONAL AIDS (INCLUDING ALTIMETERS, BEACONS, COMPASSES, RACONS, DEPTH SOUNDING, APPROACH, AND LANDING).

P--REPRODUCING (INACTIVATED, DO NOT USE).
Q--SPECIAL, OR COMBINATION OF PURPOSES.
R--RECEIVING, PASSIVE DETECTING.
S--DETECTING AND/OR RANGE AND BEARING, SEARCH.
T--TRANSMITTING.
W--AUTOMATIC FLIGHT OR REMOTE CONTROL.
X--IDENTIFICATION AND RECOGNITION.

[1] NOT FOR US USE EXCEPT FOR ASSIGNING SUFFIX LETTERS TO PREVIOUSLY NOMENCLATURED ITEMS.

FIG. 3-7 AN nomenclature. *(Bureau of Naval Personnel)* (Continued)

3-11 AN NOMENCLATURE

See preceding table.

3-12 DIODE COLOR CODES

The color-code convention used to identify various diodes is listed in Table 3-4 and shown in Fig. 3-8.

> **EXAMPLE:** A diode with a color stripe sequence of white—brown—yellow—red would be interpreted as 1N914B. (The prefix is normally "1N.")

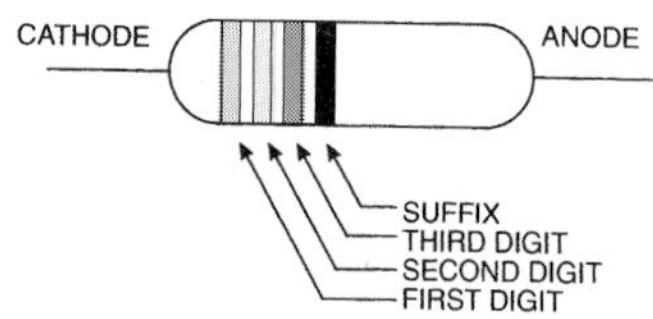

FIG. 3-8 Color code for diodes.

TABLE 3-4 Diode Color Codes

Color	Digit	Suffix
Black	0	
Brown	1	A
Red	2	B
Orange	3	C
Yellow	4	D
Green	5	E
Blue	6	F
Violet	7	G
Gray	8	H
White	9	J

3-13 SEMICONDUCTOR LETTER SYMBOLS

Letter symbols frequently used to designate semiconductor parameters are listed in Table 3-5.

TABLE 3-5 Semiconductor Letter Symbols*

Semi-conductor symbols consist of a basic letter with subscripts, either alphabetical or numerical, or both, in accordance with the following rules:

1. A capital (upper case) letter designates external circuit parameters and components, large-signal device parameters, and maximum (peak), average (d.c.), or root-mean-square values of current, voltage, and power (I, V, P, etc.)
2. Instantaneous values of current, voltage, and power, which vary with time, and small-signal values are represented by the lower case (small) letter of the proper symbol (i, v, p, i_e V_{eb}, etc.).
3. D.c. values, instantaneous total values, and large-signal values, are indicated by capital subscripts (i_C, I_C, v_{EB}, V_{EB}, P_C, etc.).
4. Alternating component values are indicated by using lower case subscripts: note the examples i_c, I_c, v_{eb}, V_{eb}, P_c, p_c.
5. When it is necessary to distinguish between maximum, average, or root-mean-square values, maximum or average values may be represented by addition of a subscript m or av; examples are i_{cm}, I_{CM}, I_{cav}, i_{CAV}.
6. For electrical quantities, the first subscript designates the electrode at which the measurement is made.

*Reprinted from *Basic Electronics*, Bureau of Naval Personnel, Dover Publications. Inc.

TABLE 3-5 Semiconductor Letter Symbols* (Continued)

7. For device parameters, the first subscript designates the element of the four-pole matrix; examples are I or i for input, O or o for output, F or f for forward transfer, and R or r for reverse transfer.

8. The second subscript normally designates the reference electrode.

9. Supply voltages are indicated by repeating the associated device electrode subscript, in which case, the reference terminal is then designated by the third subscript; note the cases V_{EE}, V_{CC}, V_{EEB}, V_{CCB}.

10. In devices having more than one terminal of the same type (say two bases), the terminal subscripts are modified by adding a number following the subscript and placed on the same line, for example, V_{B1-B2}.

11. In multiple-unit devices the terminal subscripts are modified by a number preceding the electrode subscript; note the example, V_{1B-2B}.

Semiconductor symbols change, and new symbols are developed to cover new devices as the art changes; an alphabetical list of the complex symbols is presented below for easy reference.

The list is divided into six sections. These sections are signal and rectifier diodes, zener diodes, thyristors and SCR's, transistors, unijunction transistors, and field effect transistors.

SIGNAL AND RECTIFIER DIODES

PRV	Peak Reverse Voltage
I_o	Average Rectifier Forward Current

TABLE 3-5 Semiconductor Letter Symbols* (Continued)

SIGNAL AND RECTIFIER DIODES	
I_r	Average Reverse Current
I_{surge}	Peak Surge Current
V_F	Average Forward Voltage Drop
V_R	D.c. Blocking Voltage
ZENER DIODES	
I_F	Forward current
I_Z	Zener current
I_{ZK}	Zener current near breakdown knee
I_{ZM}	Maximum D.c. zener current (limited by power dissipation)
I_{ZT}	Zener test current
V_f	Forward voltage
V_Z	Nominal zener voltage
Z_Z	Zener impedance
Z_{ZK}	Zener impedance near breakdown knee
Z_{ZT}	Zener impedance at zener test current
I_R	Reverse current
V_R	Reverse test voltage

TABLE 3-5 Semiconductor Letter Symbols* (Continued)

THYRISTORS AND SCRs

I_f	Forward current, r.m.s. value of forward anode current during the ''on'' state.
$I_{FM(pulse)}$	Repetitive pulse current. Repetitive peak forward anode current after application of gate signal for specified pulse conditions.
$I_{FM(surge)}$	Peak forward surge current. The maximum forward current having a single forward cycle in a 60 Hz single-phase resistive load system.
I_{FOM}	Peak forward blocking current, gate open. The maximum current through the thyristor when the device is in the ''off'' state for a stated anode-to-cathode voltage (anode positive) and junction temperature with the gate open.
I_{FXM}	Peak forward blocking current. Same as I_{FOM} except that the gate terminal is returned to the cathode through a stated impedance and/or bias voltage.
I_{GFM}	Peak forward gate current. The maximum instantaneous value of current which may flow between gate and cathode.
I_{GT}	Gate trigger current (continuous d.c.). The minimum d.c. gate current required to cause switching from the ''off'' state at a stated condition.

TABLE 3-5 Semiconductor Letter Symbols* (Continued)

	THYRISTORS AND SCRs
I_{HO}	Holding current. That value of forward anode current below which the controlled rectifier switches from the conducting state to the forward blocking condition with the gate open, at stated conditions.
I_{HX}	Holding current (gate connected). The value of forward anode current below which the controlled rectifier switches from the conducting state to the forward blocking condition with the gate terminal returned to the cathode terminal through specified impedance and/or bias voltage.
$P_{F(AV)}$	Average forward power. Average value of power dissipation between anode and cathode.
P_{GFM}	Peak gate power. The maximum instantaneous value of gate power dissipation permitted.
I_{ROM}	Peak reverse blocking current. The maximum current through the thyristor when the device is in the reverse blocking state (anode negative) for a stated anode-to-cathode voltage and junction temperature with the gate open.
I_{RXM}	Peak reverse blocking current. Same as I_{ROM} except that the gate terminal

TABLE 3-5 Semiconductor Letter Symbols* (Continued)

	THYRISTORS AND SCRs
	is returned to the cathode through a stated impedance and/or bias voltage.
$P_{GF(AV)}$	Average forward gate power. The value of maximum allowable gate power dissipation averaged over a full cycle.
V_F	Forward "on" voltage. The voltage measured between anode and cathode during the "on" condition for specified conditions of anode and temperature.
$V_{F(on)}$	Dynamic forward "on" voltage. The voltage measured between anode and cathode at a specified time after turn-on function has been initiated at stated conditions.
V_{FOM}	Peak forward blocking voltage, gate open. The peak repetitive forward voltage which may be applied to the thyristor between anode and cathode (anode positive) with the gate open at stated conditions.
V_{FXM}	Peak forward blocking voltage. Same as V_{FOM} except that the gate terminal is returned to the cathode through a stated impedance and/or voltage.
V_{GFM}	Peak forward gate voltage. The maximum instantaneous voltage between the gate terminal and the cathode terminal resulting from the flow of forward gate current.

TABLE 3-5 Semiconductor Letter Symbols* (Continued)

	THYRISTORS AND SCRs
V_{GRM}	Peak reverse gate voltage. The maximum instantaneous voltage which may be applied between the gate terminal and the cathode terminal when the junction between the gate region and the adjacent cathode region is reverse biased.
V_{GT}	Gate trigger voltage (continuous d.c.). The d.c. voltage between the gate and the cathode required to produce the d.c. gate trigger current.
$V_{ROM(rep)}$	Peak reverse blocking voltage, gate open. The maximum allowable value of reverse voltage (repetitive or continuous d.c.) which can be applied between anode and cathode (anode negative) with the gate open for stated conditions.
V_{RXM}	Peak reverse blocking voltage. Same as V_{ROM} except that the gate terminal is returned to the cathode through a stated impedance and/or bias voltage.
	TRANSISTORS
A_G	Available gain
A_P	Power gain
A_I	Current gain
B or b	Base electrode

TABLE 3-5 Semiconductor Letter Symbols* (Continued)

TRANSISTORS	
BV_{BCO}	D.c. base-to-collector breakdown voltage, base reverse-biased with respect to collector, emitter open.
BV_{BEO}	D.c. base-to-emitter breakdown voltage, base reverse-biased with respect to emitter, collector open.
BV_{CBO}	D.c. collector-to-base breakdown voltage, collector reverse-biased with respect to base, emitter open.
BV_{CEO}	D.c. collector-to-emitter breakdown voltage, collector reverse-biased with respect to emitter, base open.
BV_{EBO}	D.c. emitter-to-base breakdown voltage, emitter reverse-biased with respect to base, collector open.
BV_{ECO}	D.c. emitter-to-collector breakdown voltage, emitter reverse-biased with respect to collector, base open.
C or c	Collector electrode
C_c	Collector junction capacitance
C_e	Emitter junction capacitance
C_{ib}, C_{ic} C_{ie}	Input capacitance for common base, collector, and emitter, respectively.
C_{ob}, C_{oc} C_{oe}	Output terminal capacitance, a.c. input open, for common base, collector and emitter, respectively.

TABLE 3-5 Semiconductor Letter Symbols* (Continued)

TRANSISTORS	
D	Distortion
E or e	Emitter electrode
$f_{\alpha b}$, $f_{\alpha c}$, $f_{\alpha e}$	Alpha cutoff frequency for common base, collector, and emitter, respectively.
f_{co}	Cutoff frequency
f_{max}	Maximum frequency of oscillation
GC (CB), GC (CC), GE (CE)	Grounded (or common) base, collector, and emitter, respectively.
G_b, G_c, G_e	Power gain for common base, collector, and emitter, respectively.
h	Hybrid parameter
h_{fe}, h_{fb}, h_{fc}	Small signal forward current transfer ratio, a.c. output shorted, common emitter, common base, common collector, respectively.
h_{ib}	Small-signal input impedance, a.c. output shorted, common base.
h_{ob}	Small-signal output admittance, a.c. input open, common base.
I	Direct current (d.c.).

TABLE 3-5 Semiconductor Letter Symbols* (Continued)

TRANSISTORS	
I_B, I_C, I_E	D.c. current for base, collector, and emitter, respectively.
I_{CBO}	D.c. collector current, collector reverse-biased with respect to base, emitter-to-base open.
I_{CES}	D.c. collector current, collector reverse-biased with respect to emitter, base shorted to emitter.
I_{EBO}	D.c. emitter current, emitter reverse-biased with respect to base, collector-to-base open.
NF	Noise Figure
P_D	Total average power dissipation of all electrodes of a semiconductor device.
P_G	Power gain
P_{Go}	Over-all power gain
P_{in}	Input power
P_{out}	Output power
r'b	Equivalent base resistance, high frequencies
T_j	Junction temperature
T_{stg}	Storage temperature

TABLE 3-5 Semiconductor Letter Symbols* (Continued)

TRANSISTORS	
t_f	Fall time, from 90 percent to 10 percent of pulse (switching applications).
t_r	Rise time, from 10 percent to 90 percent pulse (switching applications).
t_s	Storage time (switching applications).
V_{BE}	Base-to-emitter d.c. voltage
V_{CE}	Collector-to-base d.c. voltage
V_{CE}	Collector-to-emitter d.c. voltage
V_{CEO}	D.c. collector-to-emitter voltage with collector junction reverse-biased, zero base current.
V_{CER}	Similar to V_{CEO}, except with a resistor (of value R) between base and emitter.
V_{CES}	Similar to V_{CEO}, except with base shorted to emitter.
V_{CEV}	D.c. collector-to-emitter voltage, used when only voltage bias is used.
V_{CEX}	D.c. collector-to-emitter voltage, base-emitter back biased.
V_{EB}	Emitter-to-base d.c. voltage
V_{pt}	Punch-through voltage

TABLE 3-5 Semiconductor Letter Symbols* (Continued)

UNIJUNCTION TRANSISTORS

I_E	Emitter current
I_{EO}	Emitter reverse current. Measured between emitter and base-two at a specified voltage, and base-one open-circuited.
I_p	Peak point emitter current. The maximum emitter current that can flow without allowing the UJT to go into the negative resistance region.
I_V	Valley point emitter. The current flowing in the emitter when the device is biased to the valley point.
r_{BB}	Interbase resistance. Resistance between base-two and base-one measured at a specified interbase voltage.
V_{B2B1}	Voltage between base-two and base-one. Positive at base-two.
V_p	Peak point emitter voltage. The maximum voltage seen at the emitter before the UJT goes into the negative resistance region.
V_D	Forward voltage drop of the emitter junction.
V_{EB1}	Emitter to base-one voltage

TABLE 3-5 Semiconductor Letter Symbols* (Continued)

	UNIJUNCTION TRANSISTORS
$V_{EB1(SAT)}$	Emitter saturation voltage. Forward voltage drop from emitter to base-one at a specified emitter current (larger than I_V) and specified inter-base voltage.
V_V	Valley point emitter voltage. The voltage at which the valley point occurs with a specified V_{B2B1}.
V_{OB1}	Base-one peak pulse voltage. The peak voltage measured across a resistor in series with base-one when the UJT is operated as a relaxation oscillator in a specified circuit.
α_{rBB}	Interbase resistance temperature coefficient. Variation of resistance between B2 and B1 over the specified temperature range and measured at the specific interbase voltage and temperature with emitter open circuited.
$I_{B2(mod)}$	Interbase modulation current. B2 current modulation due to firing. Measured at a specified interbase voltage, emitter and temperature.
	FIELD EFFECT TRANSISTORS
I_D	Drain current
I_{DGO}	Maximum leakage from drain to gate with source open

TABLE 3-5 Semiconductor Letter Symbols* (Continued)

FIELD EFFECT TRANSISTORS

I_{DSS}	Drain current with gate connected to source
I_G	Gate current
I_{GSS}	Maximum gate current (leakage) with drain connected to source
$V_{(BR)DGO}$	Drain to gate, source open
V_D	D.c. drain voltage
$V_{(BR)DGS}$	Drain to gate, source connected to drain
$V_{(BR)DS}$	Drain to source, gate connection not specified
$V_{(BR)DSX}$	Drain to source, gate biased to cutoff or beyond
$V_{(BR)GS}$	Gate to source, drain connection not specified
$V_{(BR)GSS}$	Gate to source, drain connected to source
$V_{(BR)GD}$	Gate to drain, source connection not specified
$V_{(BR)GDS}$	Gate to drain, source connected to drain

TABLE 3-5 Semiconductor Letter Symbols* (Continued)

FIELD EFFECT TRANSISTORS	
V_G	D.c. gate voltage
$V_{G1S(OFF)}$	Gate 1-source cutoff voltage (with gate 2 connected to source)
$V_{G2S(OFF)}$	Gate 2-source cutoff voltage (with gate 1 connected to source)
$V_{GS(OFF)}$	Cutoff

4

Linear Circuits

4-1 OPERATIONAL AMPLIFIERS (OP AMP)

The fundamental linear circuit is the op amp. Characteristics of important op-amp circuits are shown in Figs. 4-1 through 4-10. Figure 4-11 provides pin connection diagrams for some common op amps.

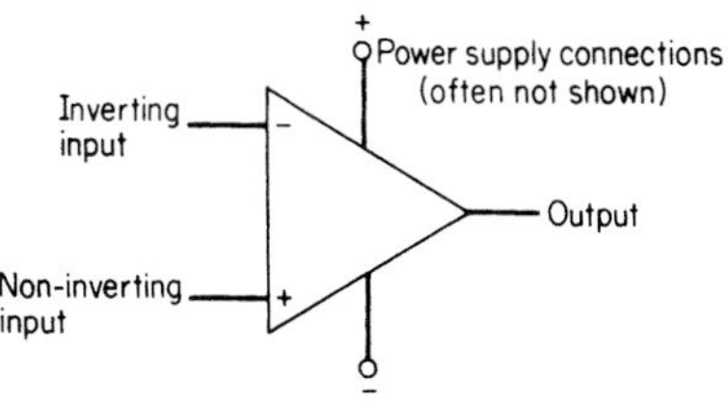

FIG. 4-1 Op-amp schematic symbol.

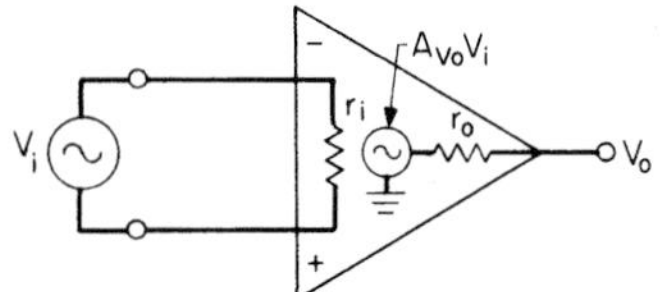

FIG. 4-2 Ideal op amp: $V_o = A_{v_o} V_i$, $A_{Vo} = \infty$, $r_I = \infty$, $r_o = 0$. Bandwidth $= \infty$, $V_o = 0$ if $V_i = 0$.

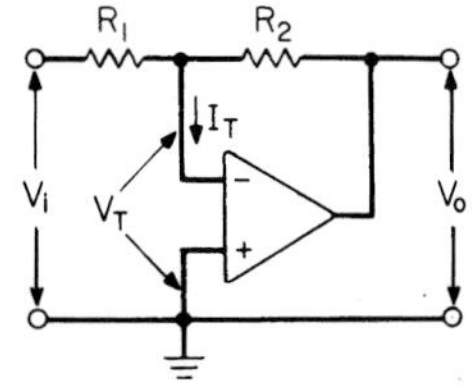

FIG. 4-3 Inverting amplifier: $I_T = 0$, $V_T = 0$, so we have a virtual ground; gain $= -R_2/R_1 = V_o/V_i$; input impedance $= R_1$.

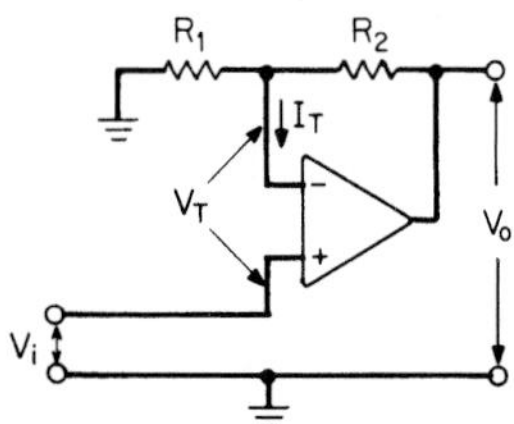

FIG. 4-4 Noninverting amplifier: $I_T = 0$, $V_T = 0$, so we have virtual ground; gain $= (R_1 + R_2)/R_1$. Input resistance $= \infty$.

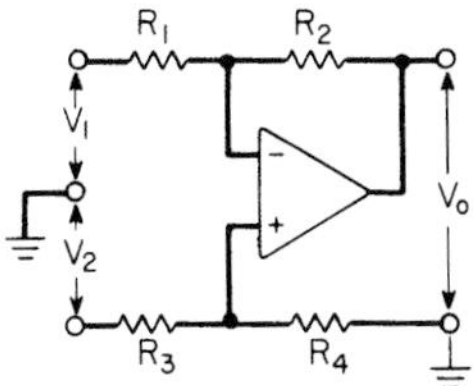

FIG. 4-5 Differential amplifier: differential-mode gain = $V_o/(V_1 + V_2) = R_4/R_3$
if $V_1 \neq V_2$
common-mode gain = 0 if $R_4/R_3 = R_2/R_1$ and if $V_1 = V_2$
otherwise = $(R_4R_1 - R_2R_3)/(R_1R_3 + R_1R_4)$
differential input resistance = $R_1 + R_3$.

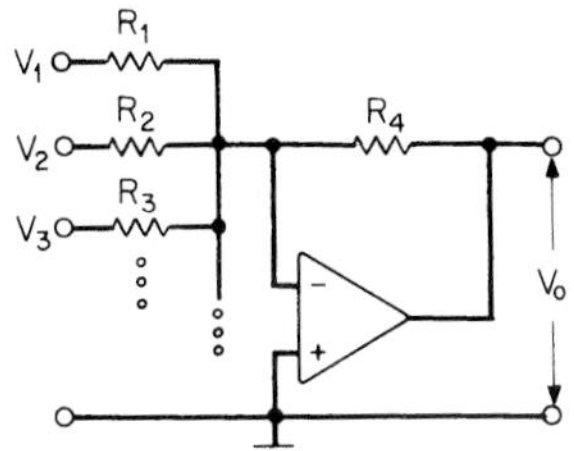

FIG. 4-6 Summing amplifier:
channel gain = $-R_4/R_1, -R_4/R_2, -R_4/R_3 \ldots$
$V_o = -[V_1(R_4/R_1) + V_2(R_4/R_2) + V_3(R_4/R_3) + \ldots]$
input impedance = $R_1, R_2, R_3 \ldots$

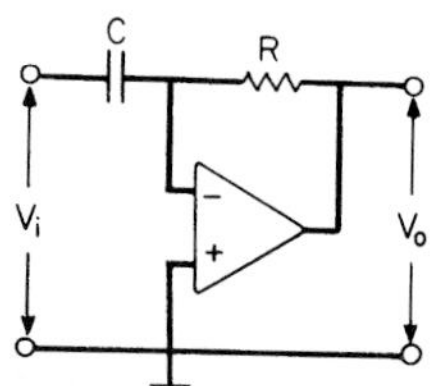

FIG. 4-7 Differentiating amplifier: $V_o = -(\Delta V_i/\Delta t)RC$.

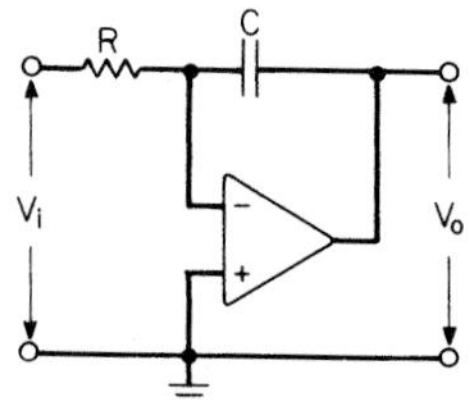

FIG. 4-8 Integrating amplifier: $\Delta V_o/\Delta t = -V_i/RC$, input resistance = R.

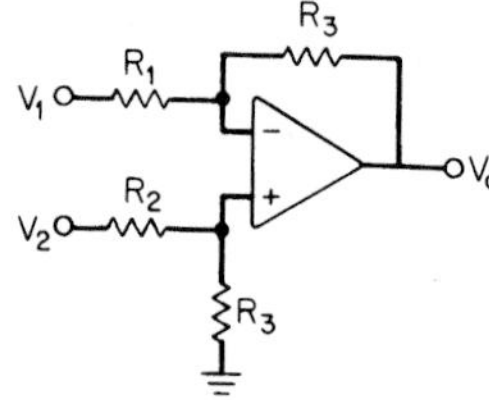

FIG. 4-9 Subtractor: $V_o = (R_3/R_1)(V_2 - V_1)$ if $R_1 = R_2$.

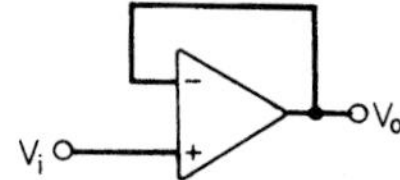

FIG. 4-10 Unity-gain amplifier: $V_o = V_i$.

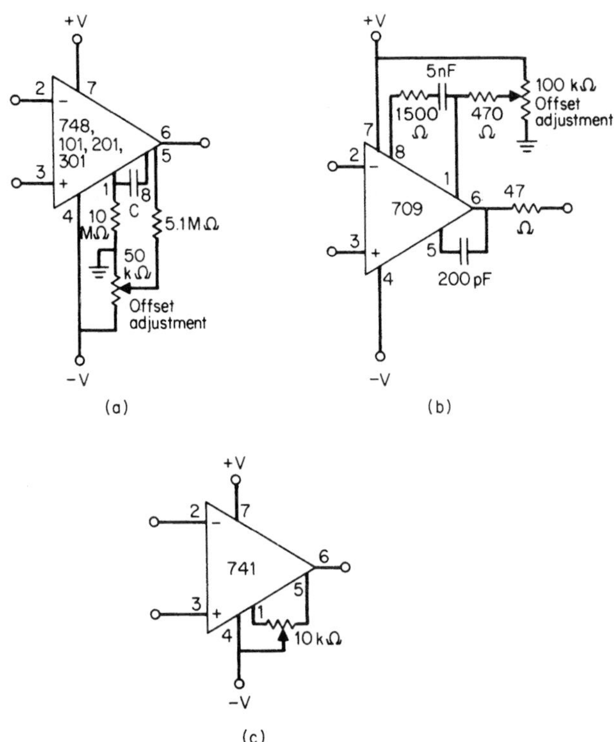

FIG. 4-11 Common op amp pin assignments and connections: *(a)* $C = 30$ pF for unity gain; *(b)* component values for unity gain; *(c)* 741 operational amplifier.

4-2 TRIGGERS AND TIMERS

The Schmitt trigger is shown in Fig. 4-12. The pin assignments for the popular 555 timer are given in Fig. 4-13.

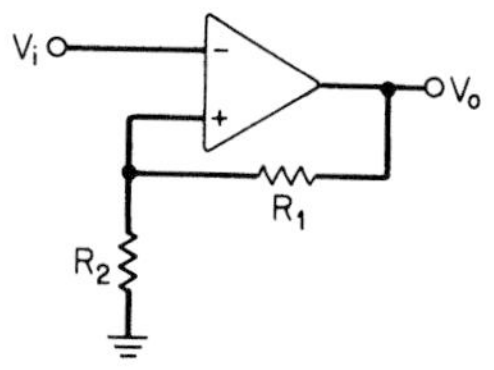

FIG. 4-12 Schmitt trigger.

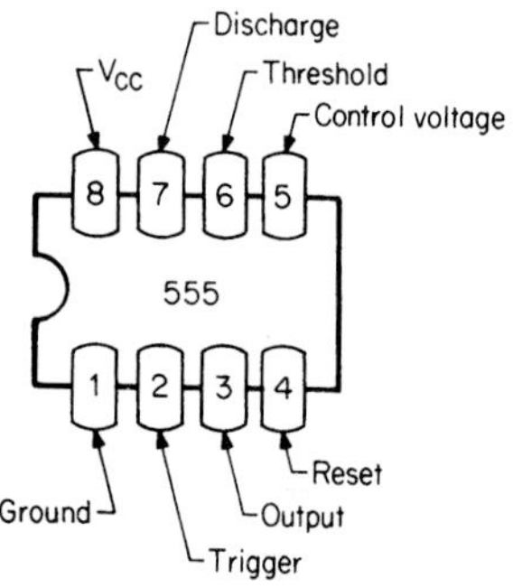

FIG. 4-13 555 pin assignments.

555 Timer*

Monostable Operation

In this mode of operation, the timer functions as a one-shot (Fig. 4-14). The external capacitor is initially held discharged by a transistor inside the timer. Upon application of a negative

*Reprinted courtesy of National Semiconductor Corporation

trigger pulse of less than ⅓ V_{CC} to pin 2, the flip-flop is set which both releases the short circuit across the capacitor and drives the output high.

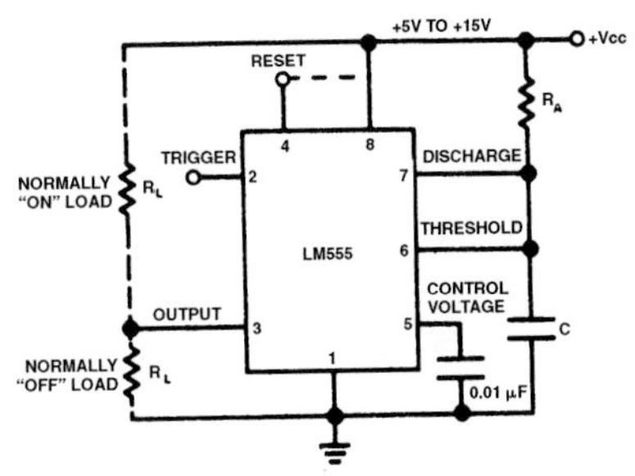

FIG. 4-14 Monostable multivibrator.

The voltage across the capacitor then increases exponentially for a period of $t = 1.1\ R_A\ C$, at the end of which time the voltage equals ⅔ V_{CC}. The comparator then resets the flip-flop which in turn discharges the capacitor and drives the output to its low state. Figure 4-15 shows the waveforms generated in this mode of operation. Since the charge and the threshold level of the comparator are both directly proportional to supply voltage, the timing internal is independent of supply.

During the timing cycle when the output is high, the further application of a trigger pulse will not effect the circuit. However, the circuit can be reset during this time by the application of a negative pulse to the reset terminal (pin 4). The output will then remain in the low state until a trigger pulse is again applied.

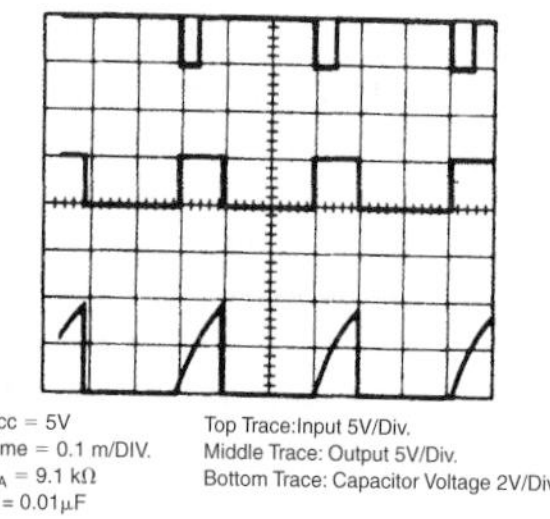

FIG. 4-15 Monostable multivibrator waveforms.

When the reset function is not in use, it is recommended that it be connected to V_{CC} to avoid any possibility of false triggering.

Figure 4-16 is a nomograph for easy determination of R, C values for various time delays.

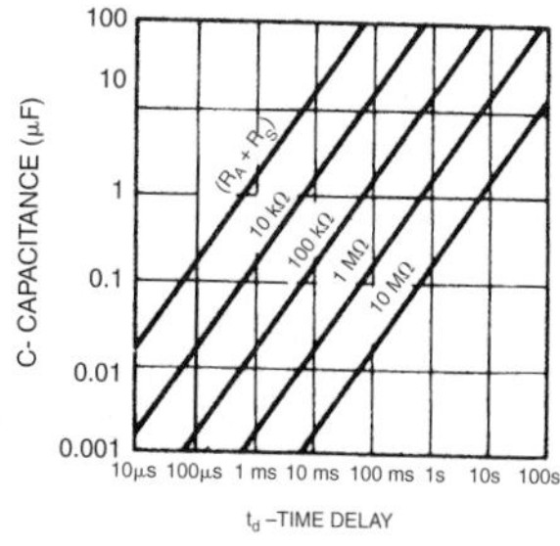

FIG. 4-16 Monostable multivibrator time delay.

NOTE: In monostable operation, the trigger should be driven high before the end of timing cycle.

Astable Operation

If the circuit is connected as shown in Fig. 4-17 (pins 2 and 6 connected) it will trigger itself and free run as a multivibrator. The external capacitor charges through $R_A + R_B$ and discharges through R_B. Thus the duty cycle may be precisely set by the ratio of these two resistors.

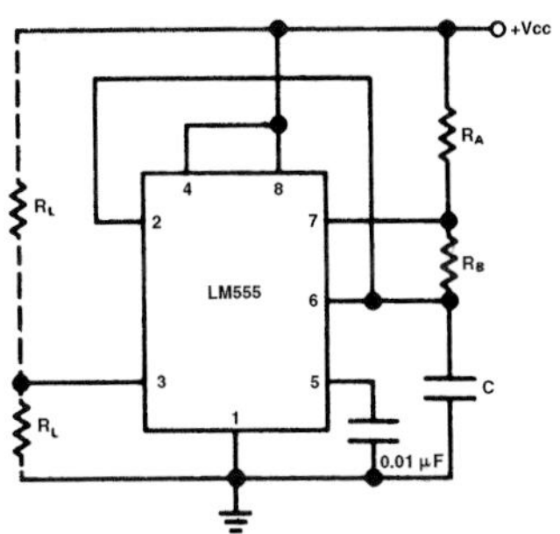

FIG. 4-17 Astable multivibrator.

In this mode of operation, the capacitor charges and discharges between ⅓ V_{CC} and ⅔ V_{CC}. As in the triggered mode, the charge and discharge times, and, therefore, the frequency are independent of the supply voltage.

Figure 4-18 shows the waveforms generated in this mode of operation.

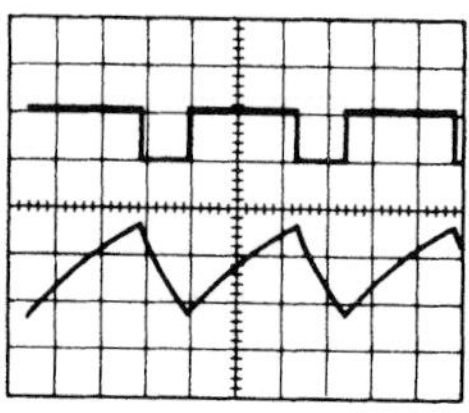

Vcc = 5V
TIME = 20μs/DIV.
R_A = 3.9 kΩ
R_B = 3 kΩ
C = 0.01 μF

Top Trace: Output 5 V/Div.
Bottom Trace: Capacitor Voltage 1V/Div.

FIG. 4-18 Astable multivibrator waveforms.

The charge time (output high) is given by:

$$t_1 = 0.693\ (R_A + R_B)\ C$$

And the discharge time (output low) by:

$$t_2 = 0.693\ (R_B)\ C$$

Thus the total period is:

$$T = t_1 + t_2 = 0.693\ (R_A + 2_{RB})\ C$$

5

Filters

5-1 *LC* FILTERS

Table 5-1, which begins on the following page, lists various *LC* filter designs—high-pass, low-pass, bandpass, notch, series, and shunt. Comments are included for each design, along with equations. The table also includes the figure numbers in which the circuit for each type of filter design is shown.

TABLE 5-1 ***LC* Filter Designs**

Filter	Comments	Circuits	Equations
High-pass, constant k	Z_0 = line impedance, f = cutoff frequency	Fig. 5-1	$L = \frac{Z_0}{4\pi f}$ $C = \frac{1}{4\pi f Z_0}$ $Z_0 = \sqrt{\frac{L}{C}}$ $f = \frac{1}{4\pi\sqrt{LC}}$
Low-pass, constant k	Z_0 = line impedance, f = cutoff frequency	Fig. 5-2	$L = \frac{Z_0}{\pi f}$ $C = \frac{1}{\pi f Z_0}$ $Z_0 = \sqrt{\frac{L}{C}}$ $f = \frac{1}{\pi\sqrt{LC}}$

TABLE 5-1 *LC* **Filter Designs (Continued)**

Bandpass, constant k	Z_0 = line impedance, f_1 = lower frequency, f_2 = upper frequency, f_0 = midfrequency	Fig. 5-3	$L = \frac{Z_0}{\pi(f_2 - f_1)}$ $L_f = \frac{f_2 - f_1}{4\pi f_1 f_2}$ $C = \frac{f_2 - f_1}{4\pi f_1 f_2 Z_0}$ $C_f = \frac{1}{\pi(f_2 - f_1)Z_0}$ $f_0 = \sqrt{f_1 f_2}$ $= \frac{1}{2\pi\sqrt{L_f C_f}}$ $= \frac{1}{2\pi\sqrt{LC}}$ $Z_0 = \sqrt{\frac{L}{C_f}} = \sqrt{\frac{L_f}{C}}$
Notch (band reject), constant k	Z_0 = lower frequency, f_1 = lower frequency, f_2 = upper frequency, f_0 = midfrequency	Fig. 5-4	$L = \frac{(f_2 - f_1)Z_0}{\pi f_1 f_2}$ $L_f = \frac{Z_0}{4\pi(f_2 - f_1)}$ $C = \frac{1}{4\pi(f_2 - f_1)Z_0}$

TABLE 5-1 *LC* **Filter Designs (Continued)**

Filter	Comments	Circuits	Equations
Notch (cont.)			$C_f = \frac{f_2 - f_1}{\pi f_1 f_2 Z_0}$ $f_0 = \sqrt{f_1 f_2}$ $= \frac{1}{2\pi\sqrt{LC}}$ $= \frac{1}{2\pi\sqrt{L_f C_f}}$ $Z_0 = \sqrt{\frac{L}{C_f}} = \sqrt{\frac{L_f}{C}}$
Series *M*-derived high-pass	Z_0 = line impedance, f = cutoff frequency, f_A = frequency of infinite attenuation, $M = \sqrt{1 - \left(\frac{f}{f_A}\right)^2}$ or $= \sqrt{1 - \left(\frac{f_A}{f}\right)^2}$	Fig. 5-5	$L_f = \frac{Z_0}{4\pi m f}$ $C = \frac{1}{4\pi m f Z_0}$ $C_f = \left(\frac{4m}{1 - m^2}\right)\left(\frac{1}{4\pi f Z_0}\right)$

TABLE 5-1 *LC* **Filter Designs (Continued)**

	(whichever is positive)		
Shunt *M*-derived high-pass	Z_0 = line impedance, f = cutoff frequency, f_A = frequency of infinite attenuation, $M = \sqrt{1 - \left(\frac{f}{f_A}\right)^2}$ or $= \sqrt{1 - \left(\frac{f_A}{f}\right)^2}$ (whichever is positive)	Fig. 5-6	$L = \left(\frac{4m}{1 - m^2}\right)\left(\frac{Z_0}{4\pi f}\right)$ $L_f = \frac{Z_0}{4\pi m f}$ $C = \frac{1}{4\pi m f Z_0}$
Series *M*-derived low-pass	Z_0 = line impedance, f = cutoff frequency, f_A = frequency of infinite attenuation, $M = \sqrt{1 - \left(\frac{f}{f_A}\right)^2}$ or	Fig. 5-7	$L = \frac{mZ_0}{2\pi f}$ $L_f = \left(\frac{1 - m^2}{4m}\right)\left(\frac{Z_0}{2\pi f}\right)$ $C_f = \frac{m}{\pi f Z_0}$

TABLE 5-1 *LC* **Filter Designs (Continued)**

Filter	Comments	Circuits	Equations
Series (cont.)	$= \sqrt{1 - \left(\frac{f_A}{f}\right)^2}$ (whichever is positive)		
Shunt *M*-derived low-pass	Z_0 = line impedance, f = cutoff frequency, f_A = frequency of infinite attenuation, $M = \sqrt{1 - \left(\frac{f}{f_A}\right)^2}$ or $= \sqrt{1 - \left(\frac{f_A}{f}\right)^2}$ (whichever is positive)	Fig. 5-8	$L = \frac{mZ_0}{\pi f}$ $C = \left(\frac{1 - m^2}{4m}\right)\left(\frac{1}{\pi f Z_0}\right)$ $C_f = \frac{m}{\pi f Z_0}$

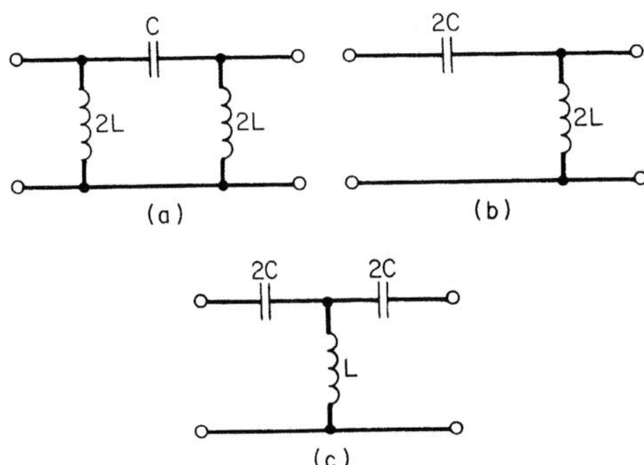

FIG. 5-1 High-pass *LC* filters: *(a)* pi section, *(b)* L section, and *(c)* T section.

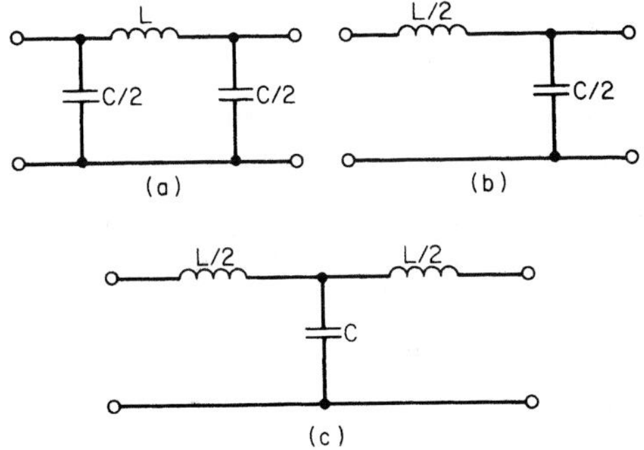

FIG. 5-2 Low-pass *LC* filters: *(a)* pi section, *(b)* L section, and *(c)* T section.

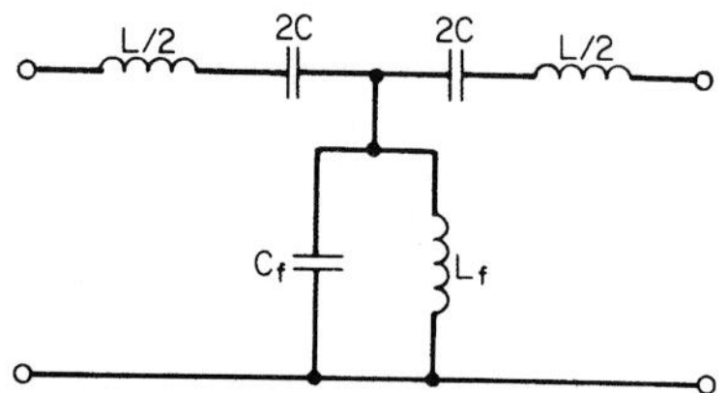

FIG. 5-3 Bandpass *LC* filter.

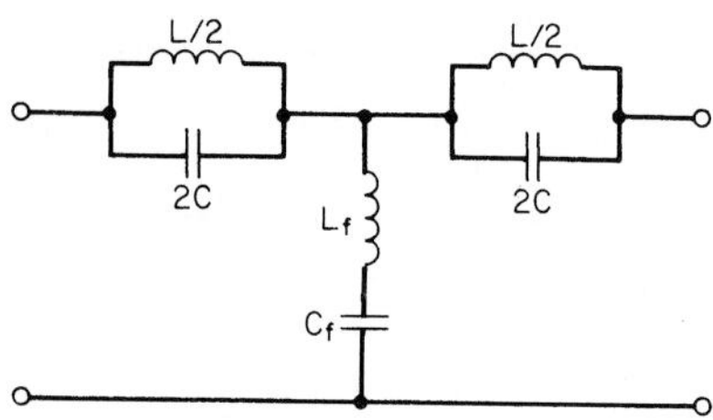

FIG. 5-4 Notch *LC* filter.

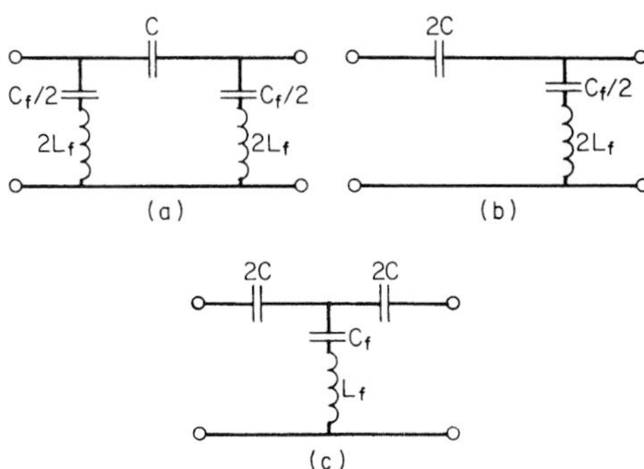

FIG. 5-5 Series high-pass *LC* filter: *(a)* pi section, *(b)* L section, and *(c)* T section.

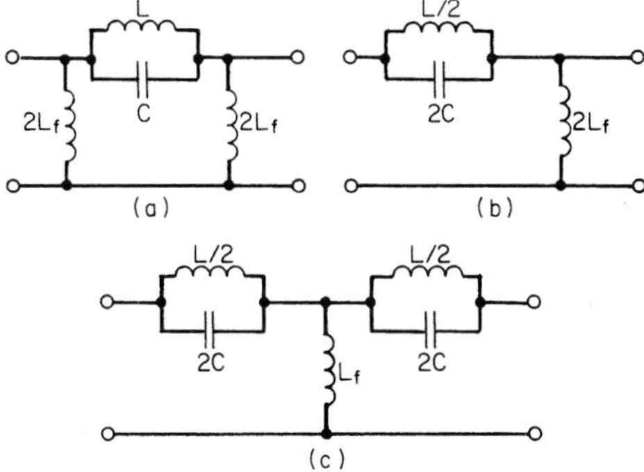

FIG. 5-6 Shunt high-pass *LC* filter: *(a)* pi section, *(b)* L section, and *(c)* T section.

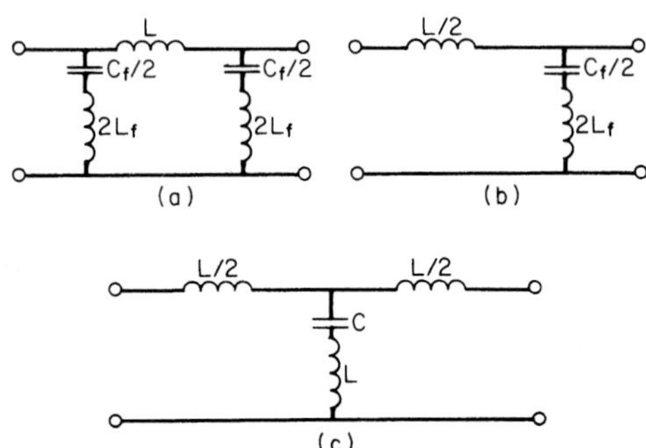

FIG. 5-7 Series low-pass *LC* filter: *(a)* pi section, *(b)* L section, and *(c)* T section.

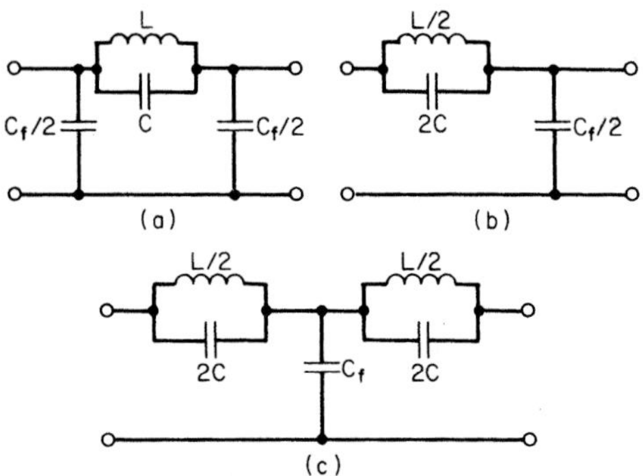

FIG. 5-8 Shunt low-pass *LC* filter: *(a)* pi section, *(b)* L section, and *(c)* T section.

5-2 RC FILTERS

Table 5-2 tabulates various RC filter designs.

TABLE 5-2 ***RC*** **Filter Designs**

Filter	Comments	Circuits	Equation
Low-pass	f = cutoff frequency	Fig. 5-9	$f = \dfrac{1}{2\pi RC}$
High-pass	f = cutoff frequency	Fig. 5-10	$f = \dfrac{1}{2\pi RC}$

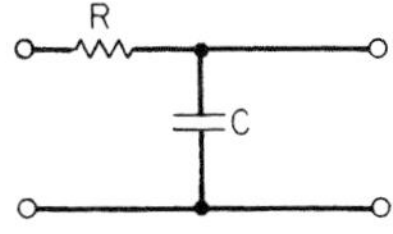

FIG. 5-9 Low-pass *RC* filter.

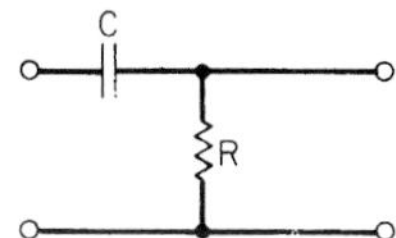

FIG. 5-10 High-pass *RC* filter.

5-3 ACTIVE FILTERS

Active filter designs are shown in Figs. 5-11 through 5-14. Equations for each filter design follow.

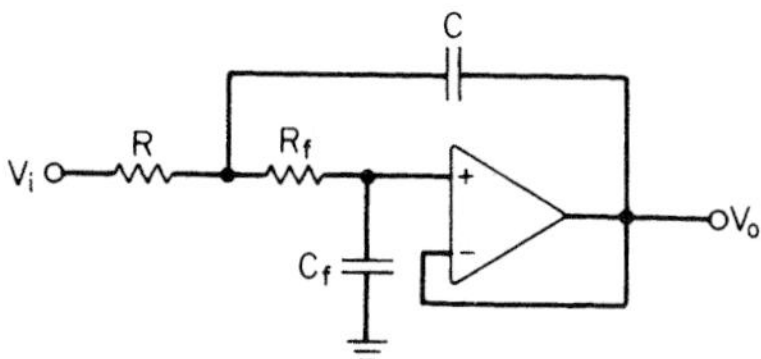

FIG. 5-11 Low-pass active filter.

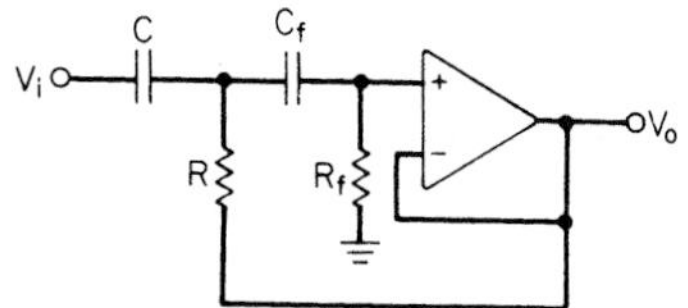

FIG. 5-12 High-pass active filter.

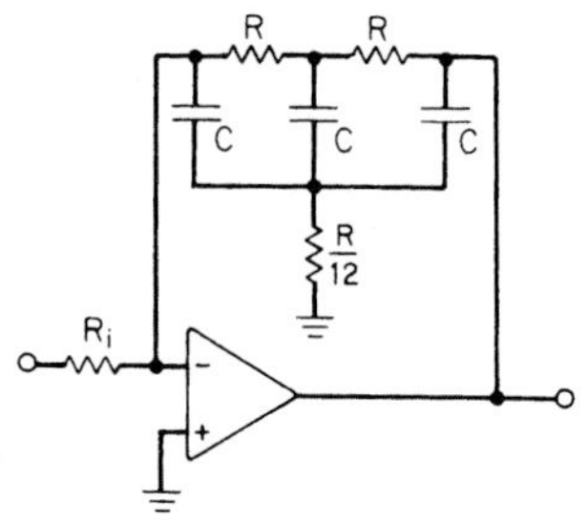

FIG. 5-13 Bandpass active filter.

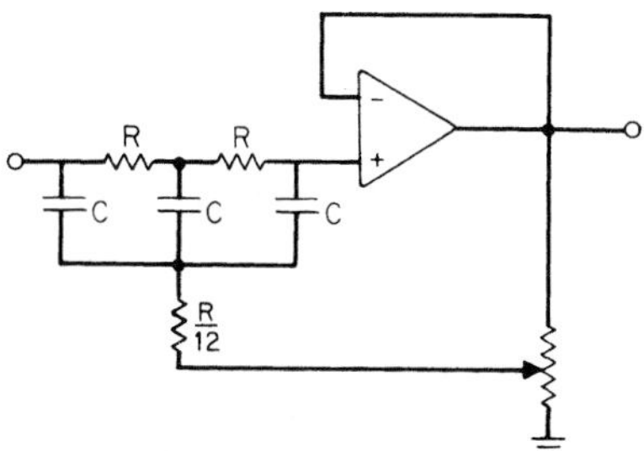

FIG. 5-14 Notch active filter.

Second-Order Low-Pass Filter (Two-Pole)

$$\frac{v_o}{v_i} = \frac{1}{RR_fCC_fs^2 + C_f(R + R_f)s + 1}$$

Second-Order High-Pass Filter (Two-Pole)

$$\frac{v_o}{v_i} = \frac{(s/\omega)^2}{(s/\omega)^2 + d(s/\omega) + 1}$$

where $\omega = \sqrt{\dfrac{1}{RCR_fC_f}}$

$$d = \frac{R(C + C_f)}{\sqrt{RCR_fC_f}}$$

Bandpass or Notch Filter

$$\omega = \frac{\sqrt{3}}{RC}$$

5-4 CHEBYSHEV FILTERS

Designs for several Chebyshev filters are described in the following section:

Type	Elements	Input/Output	Found in Table Number	Found in Figure Number
High-pass	5	Capacitive	5-3	5-15
	7	Capacitive	5-4	5-16
Low-Pass	5	Capacitive	5-5	5-17
	7	Capacitive	5-6	5-18
	5	Inductive	5-7	5-19
	7	Inductive	5-8	5-20

TABLE 5-3 50-Ω-5-Element Chebyshev High-Pass Filter Designs (Capacitive Input and Output)

Frequency, MHz			C_1,C_3	L_1,L_2,	C_2
Cutoff	3-dB	20-dB	pF	μH	pF
1.043	0.726	0.501	5100	6.447	2200
1.045	0.788	0.554	4300	5.969	2000
1.169	0.800	0.550	4700	5.851	2000
1.070	0.857	0.615	3600	5.562	1800
1.172	0.877	0.616	3900	5.358	1800
1.329	0.890	0.609	4300	5.258	1800
1.119	0.938	0.685	3000	5.195	1600
1.246	0.974	0.693	3300	4.860	1600
1.380	0.993	0.691	3600	4.714	1600
1.541	1.003	0.683	3900	4.669	1600
1.284	1.028	0.738	3000	4.635	1500
1.432	1.055	0.738	3300	4.444	1500
1.605	1.068	0.730	3600	4.380	1500
1.352	1.144	0.840	2400	4.286	1300
1.545	1.201	0.853	2700	3.935	1300
1.754	1.227	0.848	3000	3.812	1300
1.453	1.235	0.908	2200	3.985	1200
1.604	1.285	0.923	2400	3.708	1200
1.840	1.325	0.921	2700	3.536	1200
2.140	1.340	0.906	3000	3.501	1200
1.569	1.340	0.988	2000	3.686	1100
1.750	1.402	1.007	2200	3.399	1100
1.933	1.437	1.007	2400	3.267	1100
2.265	1.460	0.992	2700	3.209	1100
1.925	1.542	1.107	2000	3.090	1000
2.148	1.583	1.107	2200	2.963	1000
2.408	1.603	1.095	2400	2.920	1000
2.090	1.688	1.216	1800	2.832	910
2.357	1.739	1.217	2000	2.697	910
2.675	1.762	1.202	2200	2.656	910
2.120	1.805	1.328	1500	2.729	820
2.284	1.863	1.347	1600	2.576	820
2.612	1.930	1.350	1800	2.431	820

TABLE 5-3 50-Ω-5-Element Chebyshev High-Pass Filter Designs (Capacitive Input and Output) (Continued)

Frequency, MHz			C_1,C_3	L_1,L_2,	C_2
Cutoff	3-dB	20-dB	pF	μH	pF
3.009	1.957	1.332	2000	2.393	820
2.567	2.057	1.476	1500	2.317	750
2.762	2.097	1.479	1600	2.245	750
3.211	2.137	1.460	1800	2.190	750
2.691	2.227	1.619	1300	2.170	680
3.168	2.329	1.628	1500	2.013	680
3.443	2.352	1.616	1600	1.989	680
2.993	2.456	1.779	1200	1.959	620
3.275	2.525	1.789	1300	1.869	620
3.931	2.587	1.764	1500	1.809	620
3.371	2.736	1.974	1100	1.751	560
3.718	2.811	1.980	1200	1.673	560
4.105	2.852	1.996	1300	1.640	560
3.693	3.002	2.167	1000	1.596	510
4.113	3.091	2.174	1100	1.520	510
4.590	3.136	2.155	1200	1.491	510
3.950	3.240	2.347	910	1.485	470
4.393	3.343	2.360	1000	1.408	470
4.945	3.401	2.340	1100	1.375	470
4.244	3.517	2.559	820	1.375	430
4.772	3.650	2.580	910	1.291	430
5.358	3.714	2.560	1000	1.259	430
4.724	3.892	2.826	750	1.239	390
5.223	4.017	2.844	820	1.174	390
5.934	4.097	2.821	910	1.142	390
5.014	4.182	3.051	680	1.161	360
5.599	4.341	3.081	750	1.088	360
6.228	4.424	3.066	820	1.058	360
5.437	4.550	3.324	620	1.069	330
6.033	4.720	3.361	680	1.002	330
6.775	4.825	3.345	750	0.970	330
7.702	4.869	3.297	820	0.962	330
5.936	4.988	3.651	560	0.978	300

TABLE 5-3 50-Ω-5-Element Chebyshev High-Pass Filter Designs (Capacitive Input and Output) (Continued)

Frequency, MHz			C_1,C_3 pF	L_1,L_2, μH	C_2 pF
Cutoff	3-dB	20-dB			
6.658	5.197	3.697	620	0.910	300
7.427	5.305	3.681	680	0.882	300
8.558	5.358	3.622	750	0.875	300
6.686	5.576	4.068	510	0.870	270
7.428	5.780	4.108	560	0.817	270
8.392	5.906	4.084	620	0.792	270
7.836	6.376	4.604	470	0.752	240
8.591	6.546	4.622	510	0.719	240
9.643	6.658	4.584	560	0.702	240
8.529	6.950	5.021	430	0.690	220
9.430	7.150	5.041	470	0.658	220
10.43	7.257	5.006	510	0.644	220
9.358	7.637	5.521	390	0.628	200
10.45	7.877	5.544	430	0.596	200

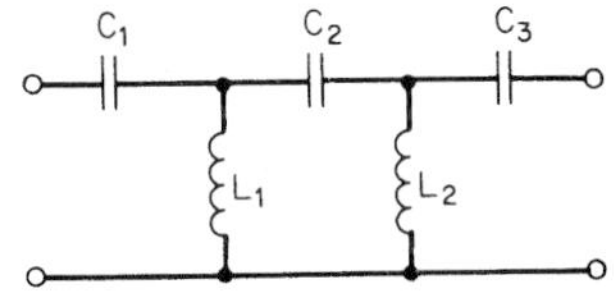

FIG. 5-15 Five-element Chebyshev high-pass filter.

TABLE 5-4 50-Ω 7-Element Chebyshev High-Pass Filter Designs (Capacitive Input and Output)

Frequency, MHz			C_1,C_4, pF	L_1,L_3, μH	C_2,C_3, pF	L_2, μH
Cutoff	3-dB	20-dB				
1.022	0.826	0.660	5100	6.162	2000	4.982
1.002	0.880	0.724	3900	5.673	1800	4.855
1.079	0.905	0.732	4300	5.554	1800	4.601
1.159	0.922	0.734	4700	5.554	1800	4.449
1.086	0.971	0.806	3300	5.153	1600	4.477
1.160	1.002	0.819	3600	4.986	1600	4.216
1.232	1.023	0.824	3900	4.930	1600	4.055
1.338	1.043	0.825	4300	4.953	1600	3.921
1.130	1.021	0.853	3000	4.919	1500	4.312
1.217	1.062	0.872	3300	4.703	1500	4.006
1.299	1.087	0.879	3600	4.626	1500	3.826
1.386	1.106	0.880	3900	4.627	1500	3.713
1.344	1.198	0.994	2700	4.171	1300	3.617
1.455	1.242	1.011	3000	4.029	1300	3.379
1.567	1.270	1.016	3300	4.004	1300	3.244
1.413	1.277	1.066	2400	3.935	1200	3.449
1.546	1.336	1.092	2700	3.739	1200	3.162
1.677	1.372	1.100	3000	3.695	1200	3.011
1.541	1.393	1.163	2200	3.607	1100	3.162
1.649	1.443	1.186	2400	3.458	1100	2.953
1.802	1.490	1.200	2700	3.388	1100	2.779
1.973	1.520	1.199	3000	3.412	1100	2.684
1.695	1.532	1.279	2000	3.279	1000	2.874
1.825	1.592	1.307	2200	3.135	1000	2.671
1.948	1.631	1.318	2400	3.084	1000	2.551
2.150	1.669	1.320	2700	3.097	1000	2.447
1.846	1.674	1.400	1800	3.007	910	2.644
2.004	1.748	1.436	2000	2.854	910	2.432
2.153	1.795	1.449	2200	2.805	910	2.314
2.312	1.827	1.451	2400	2.810	910	2.242
2.025	1.845	1.547	1600	2.737	820	2.415
2.222	1.940	1.593	1800	2.573	820	2.193
2.406	1.997	1.609	2000	2.526	820	2.077

TABLE 5-4 50-Ω 7-Element Chebyshev High-Pass Filter Designs (Capacitive Input and Output) (Continued)

Frequency, MHz			C_1,C_4, pF	L_1,L_3, μH	C_2,C_3, pF	L_2, μH
Cutoff	3-dB	20-dB				
2.606	2.034	1.610	2200	2.538	820	2.010
2.260	2.043	1.705	1500	2.459	750	2.156
2.377	2.099	1.733	1600	2.377	750	2.045
2.598	2.175	1.757	1800	2.313	750	1.913
2.834	2.221	1.760	2000	2.319	750	1.842
2.689	2.343	1.922	1500	2.130	680	1.813
2.822	2.387	1.936	1600	2.101	680	1.750
3.105	2.447	1.941	1800	2.101	680	1.673
2.660	2.429	2.040	1200	2.082	620	1.842
2.838	2.523	2.089	1300	1.980	620	1.712
3.162	2.636	2.127	1500	1.911	620	1.576
3.331	2.671	2.130	1600	1.911	620	1.538
2.982	2.711	2.270	1100	1.859	560	1.638
3.195	2.816	2.323	1200	1.772	560	1.522
3.392	2.888	2.348	1300	1.734	560	1.451
3.810	2.977	2.357	1500	1.732	560	1.373
3.269	2.974	2.491	1000	1.696	510	1.494
3.525	3.100	2.553	1100	1.610	510	1.380
3.763	3.183	2.581	1200	1.576	510	1.312
4.008	3.240	2.589	1300	1.571	510	1.270
3.510	3.205	2.691	910	1.578	470	1.396
3.786	3.347	2.764	1000	1.491	470	1.283
4.067	3.449	2.800	1100	1.453	470	1.213
4.355	3.517	2.810	1200	1.448	470	1.170
4.121	3.651	3.017	910	1.367	430	1.179
4.424	3.763	3.058	1000	1.331	430	1.114
4.768	3.846	3.071	1100	1.325	430	1.070
4.205	3.848	3.235	750	1.317	390	1.167
4.521	4.015	3.322	820	1.244	390	1.074
4.890	4.153	3.373	910	1.206	390	1.008
5.267	4.242	3.386	1000	1.202	390	0.969
4.864	4.333	3.592	750	1.153	360	0.999
5.202	4.469	3.646	820	1.118	360	0.942
5.639	4.582	3.668	910	1.108	360	0.899

TABLE 5-4 50-Ω 7-Element Chebyshev High-Pass Filter Designs (Capacitive Input and Output) (Continued)

Frequency, MHz			C_1,C_4, pF	L_1,L_3, μH	C_2,C_3, pF	L_2, μH
Cutoff	3-dB	20-dB				
5.260	4.706	3.908	680	1.063	330	0.924
5.666	4.872	3.976	750	1.026	330	0.864
6.067	4.981	4.000	820	1.016	330	0.829
5.800	5.183	4.302	620	0.965	300	0.838
6.220	5.355	4.372	680	0.933	300	0.787
6.706	5.487	4.401	750	0.923	300	0.752
7.249	5.576	4.397	820	0.931	300	0.731
6.462	5.767	4.784	560	0.867	270	0.752
6.979	5.972	4.865	620	0.837	270	0.703
7.496	6.105	4.890	680	0.831	270	0.675
6.940	6.315	5.292	470	0.798	240	0.704
7.407	6.551	5.411	510	0.762	240	0.656
7.946	6.748	5.481	560	0.742	240	0.620
8.612	6.903	5.502	620	0.740	240	0.595
7.559	6.883	5.769	430	0.733	220	0.646
8.113	7.161	5.909	470	0.697	220	0.599
8.626	7.349	5.976	510	0.681	220	0.570
9.280	7.509	6.002	560	0.677	220	0.548
8.298	7.561	6.341	390	0.667	200	0.589
8.968	7.895	6.508	430	0.632	200	0.542
9.587	8.113	6.581	470	0.618	200	0.515
10.22	8.263	6.603	510	0.616	200	0.498
9.417	8.511	7.105	360	0.590	180	0.517

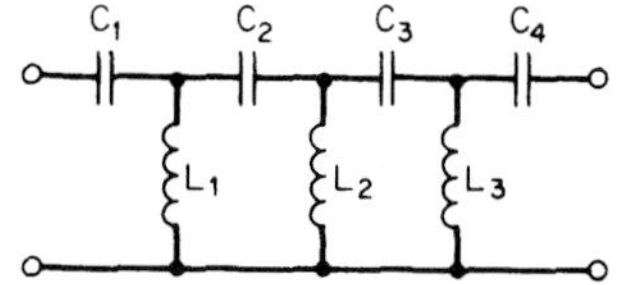

FIG. 5-16 Seven-element Chebyshev high-pass filter.

TABLE 5-5 50-Ω 5-Element Chebyshev Low-Pass Filter Designs (Capacitive Input and Output)

Frequency, MHz			C_1,C_3, pf	L_1,L_2 μH	C_2, pF
Cutoff	3-dB	20-dB			
1.016	1.209	1.652	3000	10.73	5600
1.101	1.320	1.809	2700	9.882	5100
1.039	1.371	1.944	2200	9.818	4700
1.146	1.409	1.951	2400	9.373	4700
1.127	1.496	2.125	2000	9.003	4300
1.256	1.541	2.133	2200	8.564	4300
1.054	1.619	2.379	1600	8.351	3900
1.232	1.646	2.344	1800	8.187	3900
1.388	1.701	2.353	2000	7.754	3900
1.169	1.756	2.570	1500	7.703	3600
1.275	1.771	2.547	1600	7.635	3600
1.462	1.825	2.542	1800	7.281	3600
1.430	1.939	2.773	1500	6.960	3300
1.541	1.971	2.768	1600	6.789	3300
1.315	2.101	3.108	1200	6.424	3000
1.481	2.117	3.065	1300	6.393	3000
1.754	2.190	3.050	1500	6.067	3000
1.887	2.525	3.080	1600	5.773	3000
1.506	2.337	3.440	1100	5.782	2700
1.700	2.361	3.396	1200	5.726	2700
1.868	2.403	3.383	1300	5.573	2700
1.753	2.634	3.854	1000	5.135	2400
1.985	2.671	3.810	1100	5.049	2400
2.193	2.737	3.813	1200	4.854	2400
2.402	2.838	3.865	1300	4.549	2400
1.892	2.872	4.210	910	4.709	2200
2.145	2.909	4.159	1000	4.640	2200
2.392	2.986	4.159	1100	4.449	2200
2.053	3.157	4.639	820	4.283	2000
2.362	3.201	4.575	910	4.217	2000
2.631	3.284	4.575	1000	4.045	2000
2.338	3.512	5.139	750	3.851	1800
2.628	3.557	5.083	820	3.794	1800

TABLE 5-5 50-Ω 5-Element Chebyshev Low-Pass Filter Designs (Capacitive Input and Output) (Continued)

Frequency, MHz			C_1,C_3, pf	L_1,L_2 μH	C_2, pF
Cutoff	3-dB	20-dB			
2.960	3.663	5.089	910	3.614	1800
2.705	3.959	5.763	680	3.418	1600
3.058	4.027	5.710	750	3.340	1600
3.381	4.145	5.734	820	3.182	1600
2.772	4.212	6.176	620	3.211	1500
3.135	4.265	6.101	680	3.166	1500
3.508	4.379	6.100	750	3.033	1500
3.391	4.881	7.079	560	2.772	1300
3.838	4.979	7.026	620	2.695	1300
4.259	5.147	7.080	680	2.545	1300
3.607	5.279	7.684	510	2.563	1200
4.056	5.364	7.614	560	2.509	1200
4.550	5.545	7.654	620	2.372	1200
3.963	5.762	8.376	470	2.348	1100
4.391	5.843	8.309	510	2.305	1100
4.881	6.012	8.334	560	2.198	1100
4.398	6.344	9.205	430	2.133	1000
4.907	6.448	9.135	470	2.085	1000
5.380	6.618	9.169	510	1.996	1000
4.811	6.968	10.12	390	1.942	910
5.426	7.095	10.04	430	1.894	910
5.997	7.311	10.09	470	1.799	910
4.862	7.690	11.36	330	1.756	820
5.511	7.758	11.20	360	1.743	820
6.066	7.887	11.14	390	1.702	820
6.771	8.169	11.23	430	1.602	820
5.262	8.404	12.43	300	1.606	750
6.042	8.485	12.24	330	1.594	750
6.702	8.645	12.18	360	1.550	750
7.332	8.897	12.26	390	1.475	750
6.687	9.363	13.49	300	1.444	680
7.484	9.565	13.43	330	1.398	680
8.254	9.896	13.57	360	1.317	680

TABLE 5-5 50-Ω 5-Element Chebyshev Low-Pass Filter Designs (Capacitive Input and Output) (Continued)

Frequency, MHz			C_1,C_3, pf	L_1,L_2 μH	C_2, pF
Cutoff	3-dB	20-dB			
7.213	10.25	14.82	270	1.320	620
8.181	10.48	14.73	300	1.276	620
9.109	10.88	14.90	330	1.195	620
7.818	11.32	16.45	240	1.195	560
9.021	11.59	16.31	270	1.155	560
10.16	12.09	16.52	300	1.073	560
8.659	12.44	18.04	220	1.087	510
9.636	12.65	17.91	240	1.063	510
9.224	13.48	19.61	200	1.003	470

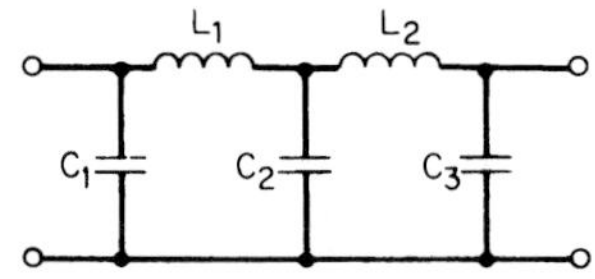

FIG. 5-17 Five-element Chebyshev low-pass filter.

TABLE 5-6 50-Ω 7-Element Chebyshev Low-Pass Filter Designs (Capacitive Input and Output)

Frequency, MHz			C_1,C_4, pF	L_1,L_3, μH	C_2,C_3, pF	L_2, μH
Cutoff	3-dB	20-dB				
1.037	1.162	1.401	2700	10.90	5600	12.57
1.047	1.229	1.511	2200	10.29	5100	12.29
1.118	1.264	1.530	2400	10.04	5100	11.66
1.033	1.299	1.633	1800	9.518	4700	11.88
1.124	1.329	1.638	2000	9.502	4700	11.40
1.208	1.368	1.658	2200	9.270	4700	10.78
1.294	1.422	1.697	2400	8.824	4700	10.01
1.101	1.412	1.785	1600	8.681	4300	10.97
1.214	1.446	1.788	1800	8.709	4300	10.51
1.314	1.492	1.810	2000	8.502	4300	9.910
1.417	1.556	1.857	2200	8.061	4300	9.138
1.250	1.566	1.967	1500	7.901	3900	9.846
1.318	1.587	1.970	1600	7.910	3900	9.617
1.440	1.641	1.993	1800	7.733	3900	9.035
1.565	1.718	2.049	2000	7.298	3900	8.268
1.445	1.726	2.135	1500	7.294	3600	8.819
1.517	1.756	2.148	1600	7.219	3600	8.537
1.660	1.837	2.201	1800	6.860	3600	7.826
1.507	1.860	2.325	1300	6.694	3300	8.265
1.682	1.929	2.350	1500	6.577	3300	7.721
1.767	1.976	2.380	1600	6.403	3300	7.370
1.556	2.020	2.560	1100	6.043	3000	7.682
1.679	2.052	2.558	1200	6.088	3000	7.472
1.786	2.092	2.570	1300	6.048	3000	7.213
1.993	2.205	2.641	1500	5.716	3000	6.522
1.746	2.248	2.844	1000	5.447	2700	6.894
1.893	2.289	2.844	1100	5.477	2700	6.677
2.022	2.341	2.863	1200	5.414	2700	6.403
2.148	2.409	2.904	1300	5.258	2700	6.064
2.006	2.539	3.198	910	4.856	2400	6.086
2.167	2.588	3.203	1000	4.863	2400	5.879
2.328	2.660	3.235	1100	4.770	2400	5.586
2.491	2.756	3.301	1200	4.573	2400	5.217

TABLE 5-6 50-Ω 7-Element Chebyshev Low-Pass Filter Designs (Capacitive Input and Output) (Continued)

Frequency, MHz			C_1,C_4, pF	L_1,L_3, μH	C_2,C_3, pF	L_2, μH
Cutoff	3-dB	20-dB				
2.155	2.762	3.490	820	4.442	2200	5.607
2.351	2.819	3.493	910	4.460	2200	5.406
2.524	2.894	3.525	1000	4.384	2200	5.147
2.717	3.006	3.601	1100	4.192	2200	4.782
2.384	3.041	3.838	750	4.041	2000	5.089
2.568	3.094	3.840	820	4.056	2000	4.933
2.778	3.184	3.878	910	3.985	2000	4.676
2.989	3.307	3.961	1000	3.811	2000	4.348
2.666	3.383	4.264	680	3.640	1800	4.570
2.889	3.451	4.271	750	3.647	1800	4.409
3.090	3.539	4.310	820	3.585	1800	4.205
3.351	3.695	4.417	910	3.404	1800	3.871
3.066	3.823	4.795	620	3.243	1600	4.029
3.300	3.902	4.811	680	3.235	1600	3.884
3.552	4.022	4.873	750	3.154	1600	3.667
3.814	4.186	4.992	820	2.995	1600	3.394
3.166	4.051	5.118	560	3.029	1500	3.821
3.445	4.133	5.122	620	3.041	1500	3.687
3.694	4.240	5.168	680	2.992	1500	3.515
3.985	4.409	5.282	750	2.858	1500	3.261
3.813	4.717	5.902	510	2.636	1300	3.260
4.103	4.819	5.927	560	2.623	1300	3.135
4.429	4.983	6.019	620	2.543	1300	2.941
4.125	5.108	6.394	470	2.433	1200	3.011
4.400	5.202	6.414	510	2.426	1200	2.913
4.719	5.354	6.491	560	2.369	1200	2.759
5.120	5.606	6.677	620	2.232	1200	2.524
4.493	5.570	6.975	430	2.230	1100	2.762
4.819	5.683	7.000	470	2.222	1100	2.663
5.123	5.827	7.073	510	2.177	1100	2.540
5.516	6.067	7.245	560	2.069	1100	2.349
4.933	6.125	7.673	390	2.027	1000	2.513
5.326	6.262	7.704	430	2.018	1000	2.413
5.694	6.442	7.801	470	1.969	1000	2.287

TABLE 5-6 50-Ω 7-Element Chebyshev Low-Pass Filter Designs (Capacitive Input and Output) (Continued)

Frequency, MHz			C_1,C_4, pF	L_1,L_3, μH	C_2,C_3, pF	L_2, μH
Cutoff	3-dB	20-dB				
6.077	6.680	7.974	510	1.879	1000	2.132
5.485	6.749	8.432	360	1.846	910	2.275
5.838	6.875	8.464	390	1.837	910	2.200
6.283	7.093	8.581	430	1.787	910	2.073
6.750	7.391	8.803	470	1.693	910	1.914
5.682	7.387	9.368	300	1.651	820	2.101
6.172	7.516	9.361	330	1.664	820	2.037
6.597	7.681	9.415	360	1.649	820	1.958

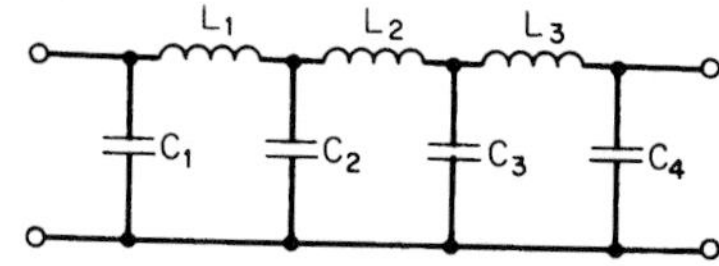

FIG. 5-18 Seven-element Chebyshev low-pass filter.

TABLE 5-7 50-Ω 5-Element Chebyshev Low-Pass Filter Designs (Capacitive Input and Output)

Frequency, MHz			L_1, L_3, μH	C_1, C_2, pF	L_2, μH
Cutoff	3-dB	20-dB			
0.74	1.15	1.69	5.6	4700	13.72
0.90	1.26	1.81	5.6	4300	12.66
1.06	1.38	1.94	5.6	3900	11.75
1.19	1.47	2.05	5.6	3600	11.15
1.32	1.58	2.17	5.6	3300	10.61
0.91	1.39	2.03	4.7	3900	11.38
1.08	1.50	2.16	4.7	3600	10.60
1.25	1.63	2.30	4.7	3300	9.92
1.42	1.77	2.46	4.7	3000	9.32
1.61	1.92	2.63	4.7	2700	8.79
1.05	1.64	2.41	3.9	3300	9.63
1.29	1.80	2.60	3.9	3000	8.83
1.54	1.99	2.80	3.9	2700	8.15
1.80	2.19	3.03	3.9	2400	7.57
1.99	2.35	3.21	3.9	2200	7.23
1.34	2.00	2.93	3.3	2700	7.89
1.68	2.25	3.20	3.3	2400	7.15
1.92	2.43	3.40	3.3	2200	6.72
2.16	2.63	3.62	3.3	2000	6.35
2.43	2.85	3.87	3.3	1800	6.02
1.66	2.46	3.59	2.7	2200	6.43
1.99	2.70	3.86	2.7	2000	5.93
2.34	2.97	4.15	2.7	1800	5.50
2.71	3.27	4.49	2.7	1600	5.13
2.92	3.44	4.68	2.7	1500	4.97
2.01	3.01	4.39	2.2	1800	5.26
2.52	3.37	4.80	2.2	1600	4.76
2.78	3.57	5.02	2.2	1500	4.55
3.34	4.02	5.52	2.2	1300	4.18
3.65	4.28	5.80	2.2	1200	4.01
2.35	3.62	5.29	1.8	1500	4.38
3.12	4.14	5.89	1.8	1300	3.88
3.51	4.45	6.23	1.8	1200	3.67

TABLE 5-7 50-Ω 5-Element Chebyshev Low-Pass Filter Designs (Capacitive Input and Output) (Continued)

Frequency, MHz			L_1, L_3, μH	C_1, C_2, pF	L_2, μH
Cutoff	3-dB	20-dB			
3.93	4.78	6.60	1.8	1100	3.48
4.37	5.16	7.01	1.8	1000	3.32
3.10	4.51	6.56	1.5	1200	3.51
3.65	4.90	6.99	1.5	1100	3.27
4.21	5.34	7.47	1.5	1000	3.06
4.75	5.77	7.96	1.5	910	2.89
5.34	6.26	8.50	1.5	820	2.74
3.53	5.41	7.94	1.2	1000	2.92
4.30	5.94	8.53	1.2	910	2.68
5.09	6.53	9.20	1.2	820	2.49
5.73	7.04	9.75	1.2	750	2.35
6.42	7.62	10.39	1.2	680	2.23
4.40	6.60	9.65	1.0	820	2.40
5.27	7.20	10.32	1.0	750	2.22
6.15	7.87	11.06	1.0	680	2.05
6.95	8.51	11.77	1.0	620	1.95
7.80	9.22	12.56	1.0	560	1.85
5.23	7.96	11.67	0.82	680	1.99
6.33	8.72	12.51	0.82	620	1.83
7.45	9.56	13.45	0.82	560	1.70
8.44	10.35	14.32	0.82	510	1.60
9.28	11.05	15.10	0.82	470	1.53
6.41	9.66	14.15	0.68	560	1.64
7.75	10.59	15.18	0.68	510	1.51
8.83	11.41	16.08	0.68	470	1.42
9.97	12.31	17.08	0.68	430	1.34

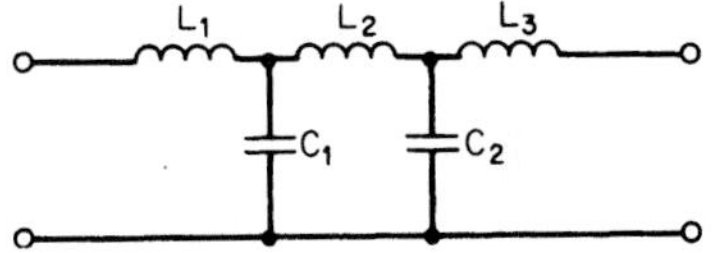

FIG. 5-19 Five-element Chebyshev low-pass filter.

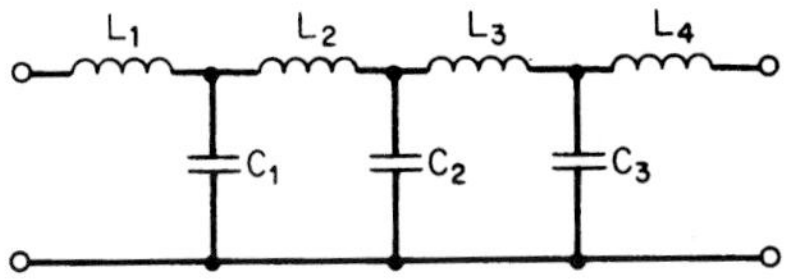

FIG. 5-20 Seven-element Chebyshev low-pass filter.

TABLE 5-8 50-Ω 7-Element Chebyshev Low-Pass Filter Designs (Capacitive Input and Output)

Frequency, MHz			L_1, L_4, μH	C_1, C_3, pF	L_2, L_3, μH	C_2, pF
Cutoff	3-dB	20-dB				
1.014	1.179	1.444	5.890	4300	13.37	5100
1.087	1.293	1.597	5.062	3900	12.04	4700
1.197	1.405	1.728	4.810	3600	11.15	4300
1.328	1.537	1.879	4.581	3300	10.29	3900
1.425	1.684	2.075	3.947	3000	9.274	3600
1.528	1.855	2.308	3.363	2700	8.316	3300
1.634	2.059	2.589	2.828	2400	7.408	3000
1.859	2.271	2.832	2.710	2200	6.775	2700
2.137	2.525	3.113	2.631	2000	6.182	2400
2.291	2.782	3.462	2.242	1800	5.544	2200
2.452	3.088	3.884	1.885	1600	4.939	2000
2.849	3.367	4.150	1.973	1500	4.637	1800
3.126	3.838	4.791	1.589	1300	4.004	1600
3.269	4.117	5.179	1.414	1200	3.704	1500
3.476	3.897	4.701	2.004	1300	4.169	1500
3.985	4.610	5.637	1.527	1100	3.429	1300
4.274	5.050	6.225	1.315	1000	3.091	1200
4.633	5.533	6.846	1.170	910	2.807	1100
5.053	6.115	7.600	1.027	820	2.525	1000
5.581	6.702	8.309	0.953	750	2.311	910
6.229	7.412	9.160	0.880	680	2.098	820
6.791	8.119	10.05	0.795	620	1.912	750
7.463	8.973	11.13	0.710	560	1.725	680
8.176	9.847	12.22	0.644	510	1.572	620
9.207	10.77	13.23	0.633	470	1.457	560
10.14	11.79	14.44	0.589	430	1.337	510
10.87	12.93	15.97	0.506	390	1.203	470

6

Power Supply and Regulation

6-1 POWER SUPPLY EQUATIONS

$$\text{Load regulation} = \frac{V_n - V_f}{} \times 100\%$$

where V_f = full-load output voltage
V_n = no-load output voltage

$$\text{Line regulation} = \frac{V_1(V_2 - V_3)}{V_3(V_4 - V_1)} \times 100\%$$

where V_1 = minimum line voltage
V_2 = maximum load voltage
V_3 = minimum load voltage
V_4 = maximum line voltage

$$\text{Percent rms ripple (sine wave)} = \frac{V_{\text{rms}}}{V_f} \times 100\%$$

$$= \frac{V_{\text{pp}}}{2.828V_f} \times 100\%$$

$$\text{Percent rms ripple (sawtooth)} = \frac{V_{\text{rms}}}{V_f} \times 100\%$$

$$= \frac{V_{\text{pp}}}{3.47V_f} \times 100\%$$

where V_{rms} = rms ripple voltage
V_{pp} = peak-to-peak ripple voltage
V_f = dc full-load voltage

Table 6-1 lists power parameters by country.

6-2 RECTIFIER CIRCUITS

Figures 6-1 through 6-5 show various rectifier circuits. The ratios of input to output current and output to input frequency for each circuit are listed in Table 6-2.

TABLE 6-1 Power Parameters by Country

Country	Frequency, Hz	Voltage, V
Australia	50	240/415*
Austria	50	220/380*
Belgium	50	220/380*
Brazil	50, 60	110, 220*
Canada	60	120/240*
Costa Rica	60	110/220
Cuba	60	115/230*
Egypt	50	110, 220*
France	50	120/240 and 220/380*
Germany	50	220/380*
Hong Kong	50	200/346
Iran	50	220/380
Iraq	50	220/380
Israel	50	230/400
Italy	50	127/220 and 220/380*
Japan	50, 60	100/200
Korea	60	100/200
Mexico	50, 60	127/220*
Panama	60	110/220*
Saudi Arabia	50, 60	120/208*
South Africa	50	220/380*
Spain	50	127/220*
Switzerland	50	220/380
Taiwan	60	100/200
United Kingdom	50	240/415*
United States	60	120/240 & 120/208

*Other voltages also.

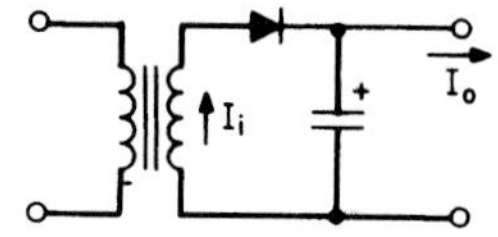

FIG. 6-1 Half-wave rectifier.

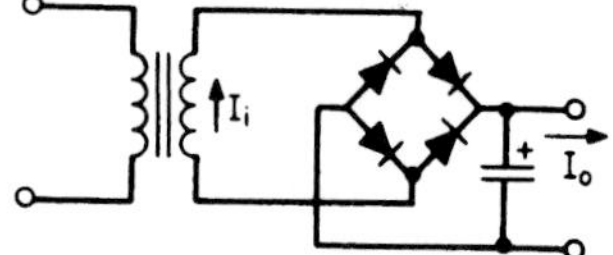

FIG. 6-2 Full-wave rectifier (bridge).

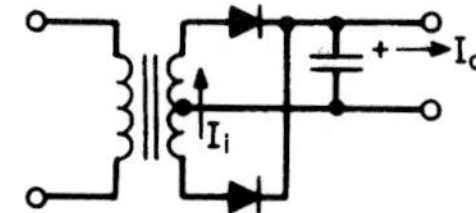

FIG. 6-3 Full-wave rectifier (center tapped).

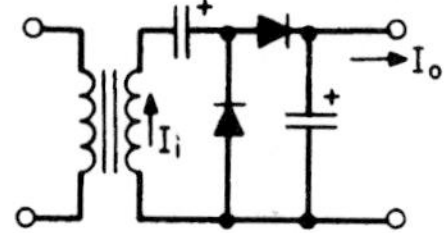

FIG. 6-4 Half-wave voltage doubler.

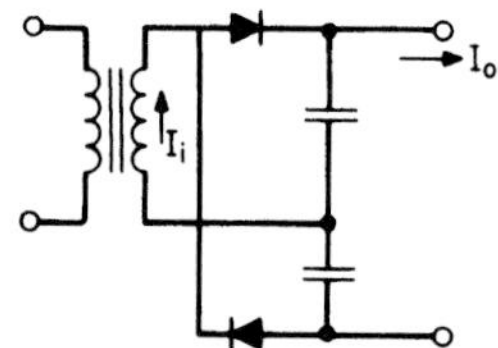

FIG. 6-5 Full-wave voltage doubler.

TABLE 6-2 Rectifier Circuit Action

Circuit	I_i/I_o	$\dfrac{\text{Output frequency}}{\text{Input frequency}}$
Half-wave	2.5	1
Full-wave (bridge)	1.7	2
Full-wave (center tapped)	0.9	2
Half-wave voltage doubler	5	1
Full-wave voltage doubler	5	2

6-3 VOLTAGE REGULATION WITH ZENER DIODES

The commonly used zener diode is shown in a regulator circuit in Fig. 6-6. Circuit parameters can be computed from the following equations:

$$R = \frac{V_i - V_z}{\quad}$$

$$V_{op} = \frac{V_{ip} R_z}{R + R_z}$$

$$\text{Line regulation} = \frac{V_i R_z}{V_z R}$$

where V_i = input voltage
V_z = zener voltage
I_z = zener current
I_o = output current
V_{op} = peak-to-peak output voltage
V_{ip} = peak-to-peak input voltage
R_z = zener resistance

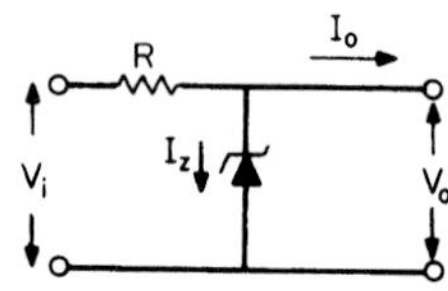

FIG. 6-6 Zener diode circuit.

6-4 THREE-TERMINAL VOLTAGE REGULATORS

Table 6-3 lists characteristics of various three-terminal regulators. Circuit diagrams are shown in Fig. 6-7.

TABLE 6-3 Regulator Characteristics

Type	Designator	Output current, A	Regulated output voltage, V
Positive	LM138/238/338	5	1.2–33
	LM123/223/323	3	5
	LM150/250/350	3	1.2–33
	LM117/217/317	1.5	1.2–37
	LM140/340	1	5, 6, 8, 10, 12, 15, 18, 24
	LM341	0.5	5, 6, 8, 10, 12, 15, 18, 24
Negative	LM145/245/345	3.0	−5, −5.2
	LM137/237/337	1.5	−1.2 to −37
	LM120/320	1.0	−5, −5.2, −6, −8, −9, −12, −15, −18, −24

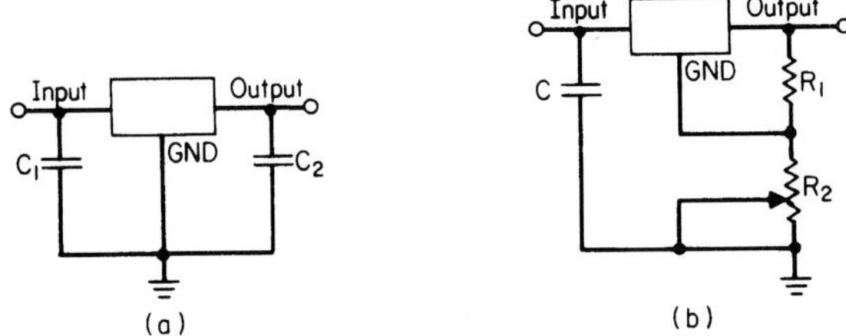

FIG. 6-7 Three-terminal regulators: *(a)* fixed regulator, *(b)* adjustable-output regulator.

7

Electronic Measurement

7-1 BRIDGE CIRCUITS

The most commonly used bridge circuits are described in Table 7-1.

7-2 MEASUREMENT ERRORS

Table 7-2 lists the number of observations necessary to obtain a specified confidence level. Confidence intervals are specified as a function of the number of observations (n) and the confidence level. In Table 7-2, the mean measurement is defined as:

$$x_m = \frac{\Sigma x_i}{n}$$

where x_i = the set of measurements $(x_1, x_2, x_3, \ldots x_n)$

TABLE 7-1 Bridge Circuits Used in Measurement

Bridge	Circuit	Quantity measured	Equations
Wheatstone	Fig. 7-1	Resistance	$R_{\text{unknown}} = \frac{R_1 R_3}{R_2}$
Wien	Fig. 7-2	Frequency	$f = \frac{1}{2\pi}\sqrt{\frac{1}{R_3 R_4 C_1 C_2}}$
		Capacitance	$\frac{C_2}{C_1} = \frac{R_2}{R_1} - \frac{R_3}{R_4}$
			$C_1^2 = \frac{R_1}{(2\pi f)^2(R_2 R_4 - R_1 R_3)R_3}$
			$C_2^2 = \frac{R_2 R_4 - R_1 R_3}{(2\pi f)^2 R_1 R_3 R_4^2}$
Scherin	Fig. 7-3	Capacitance	$C_3 = \frac{R_1 C_2}{R_2}$
		Resistance	$R_3 = \frac{C_1 R_2}{C_2}$
Hay	Fig. 7-4	Inductance	$L_1 = \frac{R_1 R_3 C_1}{1 + (2\pi f R_2 C_1)^2}$
		Resistance	$R_4 = \frac{R_1 R_2 R_3 (2\pi f C_1)^2}{1 + (2\pi f R_2 C_1)^2}$
Owen	Fig. 7-5	Inductance	$L_1 = R_1 R_3 C_2$
		Resistance	$R_4 = \frac{C_2 R_3}{C_1}$
Maxwell	Fig. 7-6	Inductance	$L_1 = R_1 R_3 C_1$
		Resistance	$R_4 = \frac{R_1 R_3}{R_2}$

TABLE 7-1 Bridge Circuits Used in Measurement (Continued)

Bridge	Circuit	Quantity measured	Equations
Resonance	Fig. 7-7	Frequency	$f = \frac{1}{2\pi}\sqrt{\frac{1}{L_1C_1}}$
		Resistance	$R_4 = \frac{R_1R_3}{R_2}$

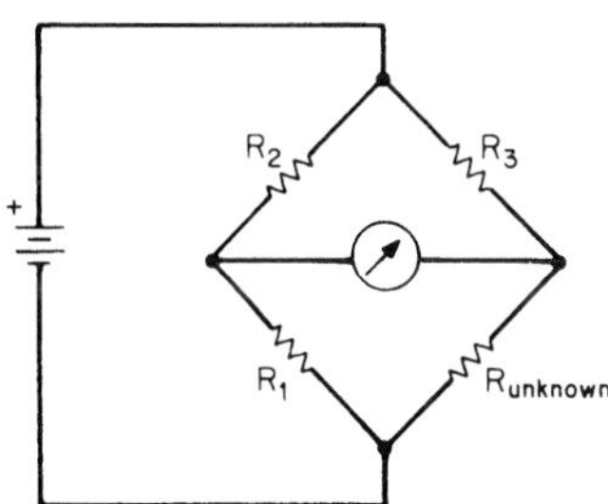

FIG. 7-1 Wheatstone bridge.

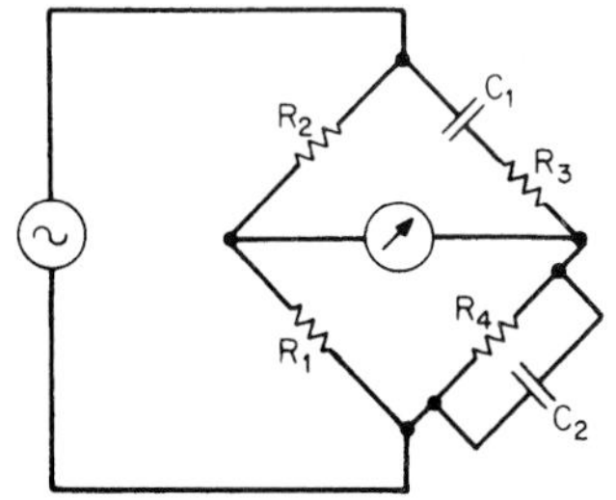

FIG. 7-2 Wien bridge.

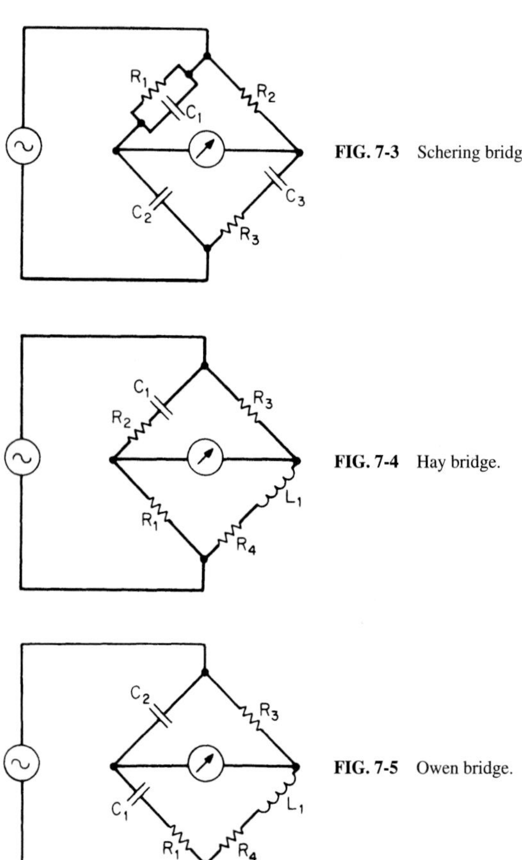

FIG. 7-3 Schering bridge.

FIG. 7-4 Hay bridge.

FIG. 7-5 Owen bridge.

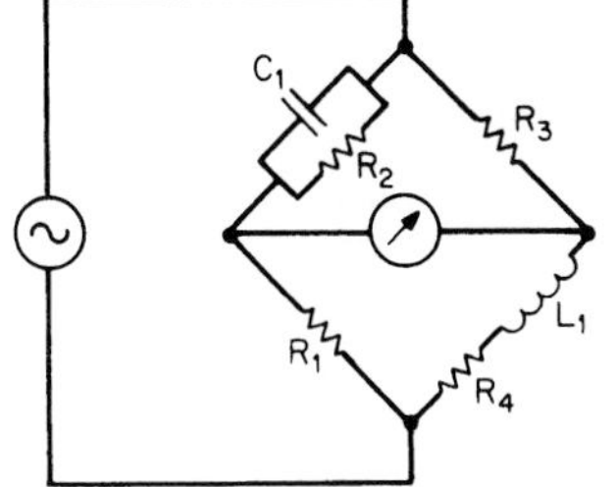

FIG. 7-6 Maxwell bridge.

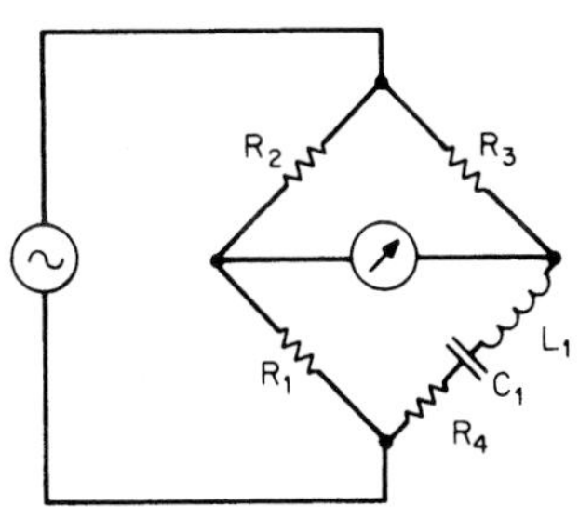

FIG. 7-7 Resonance bridge.

TABLE 7-2 Confidence Intervals

n	Confidence level			
	50%	90%	95%	99%
2	$x_m \pm 1.00$ s	$x_m \pm 6.31$ s	$x_m \pm 12.71$ s	$x_m \pm 63.66$ s
3	$x_m \pm 0.82$ s	$x_m \pm 2.92$ s	$x_m \pm 4.30$ s	$x_m \pm 9.92$ s
4	$x_m \pm 0.77$ s	$x_m \pm 2.35$ s	$x_m \pm 3.18$ s	$x_m \pm 5.84$ s
5	$x_m \pm 0.74$ s	$x_m \pm 2.13$ s	$x_m \pm 2.78$ s	$x_m \pm 4.60$ s
6	$x_m \pm 0.73$ s	$x_m \pm 2.02$ s	$x_m \pm 2.57$ s	$x_m \pm 4.03$ s
7	$x_m \pm 0.72$ s	$x_m \pm 1.94$ s	$x_m \pm 2.45$ s	$x_m \pm 3.71$ s
8	$x_m \pm 0.71$ s	$x_m \pm 1.90$ s	$x_m \pm 2.37$ s	$x_m \pm 3.50$ s
9	$x_m \pm 0.71$ s	$x_m \pm 1.86$ s	$x_m \pm 2.31$ s	$x_m \pm 3.36$ s
10	$x_m \pm 0.70$ s	$x_m \pm 1.83$ s	$x_m \pm 2.26$ s	$x_m \pm 3.25$ s

The estimate of the standard deviation is:

$$s = \sqrt{\frac{\Sigma(x_i - x_m)}{n - 1}}$$

7-3 ELECTRONIC METERS

Table 7-3 describes several common meter circuits. Electrical parameters in this table are defined below.

R = unknown resistance, Ω

R_m = meter resistance, Ω

R_r = variable resistance used to adjust meter to full scale when measuring I_o or I_s, Ω

I_o = current with probes open, A

I = current with probe connected to unknown resistor, A

I_s = current with probes short-circuited, A

N = meter multiplication factor

TABLE 7-3 Meter Circuits

Type	Circuit	Equations
Shunt ohmmeter	Fig. 7-8	Low-resistance measurement: $R = \frac{IR_m}{I_o - I}$
Series ohmmeter	Fig. 7-9	High-resistance measurement: $R = \frac{(I_s - I)(R_r + R_m)}{I}$
Shunted ammeter	Fig. 7-10	$R = \frac{R_m}{N - 1}$

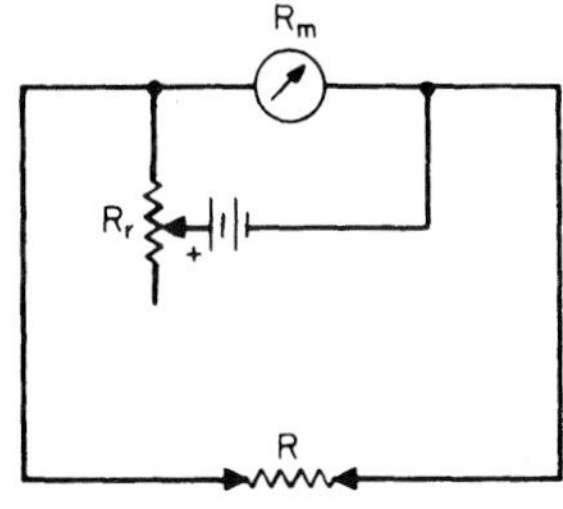

FIG. 7-8 Shunt ohmmeter.

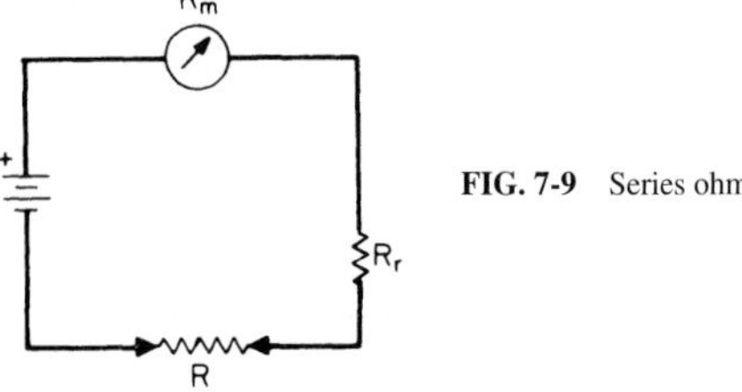

FIG. 7-9 Series ohmmeter.

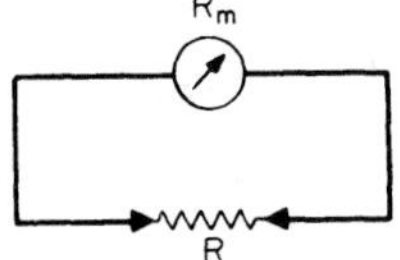

FIG. 7-10 Shunt ammeter.

8

Communications

8-1 FREQUENCY AND WAVELENGTH

Metric

$$f = \frac{300{,}000}{\lambda}$$

$$\lambda = \frac{300{,}000}{f}$$

where f = frequency, kHz
λ = wavelength, m

U. S. Customary Units

$$f = \frac{984{,}000}{\lambda}$$

$$\lambda = \frac{984{,}000}{f}$$

where f = frequency, kHz
λ = wavelength, feet

8-2 DOPPLER FREQUENCY

Sound Waves

$$f_o = \frac{v + w + v_o}{v + w - v_s} f_s$$

where f_o = observed frequency, Hz
v_o = velocity of observer, m/s
v_s = velocity of source, m/s
v = velocity of sound in the medium, m/s
w = velocity of wind in direction of sound propagation, m/s
f_s = frequency of source, Hz

Electromagnetic Waves

$$f_o = f_s \sqrt{\frac{c + v_r}{c - v_r}}$$

where f_o = observed frequency, Hz
f_s = frequency of source, Hz
v_r = velocity of source relative to observer, m/s
c = speed of light (3.0×10^8 m/s)

8-3 ANTENNA FORMULAS

Length of an Ideal Hertz Antenna

$$l = \frac{\lambda}{2}$$

where l = length
λ = wavelength

Ideal Marconi Antenna

$$l = \frac{\lambda}{4}$$

where l = length
λ = wavelength

Power Received by a Hertz Antenna

$$P = \frac{P_t G_t G_r \lambda^2}{16\pi^2 d^2} \text{ W}$$

where P_t = transmitted power, W
G_t, G_r = gain ratio of transmitting, receiving antenna relative to isotropic radiator
λ = wavelength, m
d = distance between antennas, m

EXAMPLE: Find the power received by a Hertz antenna if the 150-MHz transmitter output power is 50 W. The transmitter is 100 km distant and also uses a Hertz antenna.

The gain ratio of a Hertz antenna relative to an isotropic radiator is 1.64 (see Table 8-1). Therefore:

$$P = \frac{50 \times 1.64 \times 1.64 \times \left[\dfrac{(3 \times 10^8)}{(150 \times 10^6)}\right]}{16\pi^2(100 \times 10^3)}$$

$$= 340.6 \text{ pW}$$

Effective Radiated Power (ERP)

$$\text{ERP} = G \times P_i$$

where G = gain of transmitting antenna relative to an isotropic radiator
P_i = input power, W

Table 8-1 provides design parameters for a variety of commonly used antenna configurations.

8-4 TRANSMISSION LINES

Characteristic Impedance

$$Z_0 = \sqrt{L/C}\ \Omega$$

where L = inductance per unit length, H
C = capacitance per unit length, F

TABLE 8-1 Commonly Used Antenna Designs

Antenna Configuration	Design	*Parameters	Resistive Impedance at Resonance (Ω)	Gain Above Isotropic Antenna (dB)	†Polarization
Simple diode	Fig. 8-1	$L = \lambda/2$	High	1.74	H
Folded dipole	Fig. 8-2	$L = \lambda/2$ $L/D = 13$	6,000	1.64	H
Dipole	Fig. 8-3	$L = \lambda/2$ $L/D = 25.5$	300	2.14	H
		$L = \lambda/2$ $L/D = 276$	60	2.14	H
Cylindrical dipole	Fig. 8-4	$L = \lambda$ $L/D = 9.6$	150	3.64	H

TABLE 8-1 Commonly Used Antenna Designs (Continued)

Antenna Configuration	Design	*Parameters	Resistive Impedance at Resonance (Ω)	Gain Above Isotropic Antenna (dB)	†Polarization
Rhombic	Fig. 8-5	$L = 9\lambda$ $D = 9\ \lambda/2$	600	16.74	H
Horn	Fig. 8-6	$L = 3\lambda$ $W = 3\lambda$	50	13	H
Bicone	Fig. 8-7	$L = \lambda/2$ $\alpha = 40°$	72	2.14	H
		$L = \lambda$ $\alpha = 60°$	350	2.14	H

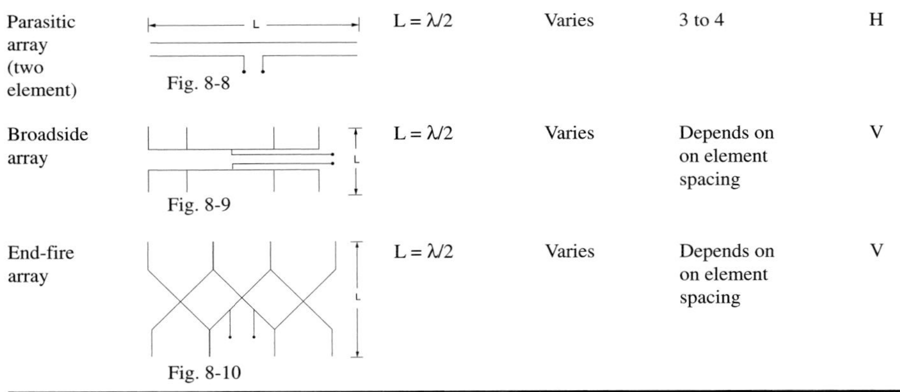

Parasitic array (two element)	Fig. 8-8	$L = \lambda/2$	Varies	3 to 4	H
Broadside array	Fig. 8-9	$L = \lambda/2$	Varies	Depends on on element spacing	V
End-fire array	Fig. 8-10	$L = \lambda/2$	Varies	Depends on on element spacing	V

* λ = wavelength

†Polarization: H = horizontal, V = vertical

For parallel-wire line,

$$Z_0 = 120 \cosh^{-1}\left(\frac{D}{d}\right)$$

$$\approx 276 \log\left(\frac{2D}{d}\right) \qquad \text{if } D >> d$$

where D = spacing between conductors (center to center)
d = diameter of one of the conductors

For coaxial line,

$$Z_0 = \frac{138}{\sqrt{\varepsilon}} \log\left(\frac{D}{d}\right)$$

where D = inner diameter of outside conductor
d = outer diameter of inner conductor
ε = relative dielectric constant of insulation

Velocity of Propagation

$$V_p = \frac{d}{\sqrt{LC}}$$

where d = distance of travel
L = inductance per unit length, H
C = capacitance per unit length, F

Standing Waves

Standing waves in open and in short-circuited lines are shown in Fig. 8-11 and are listed in Table 8-2.

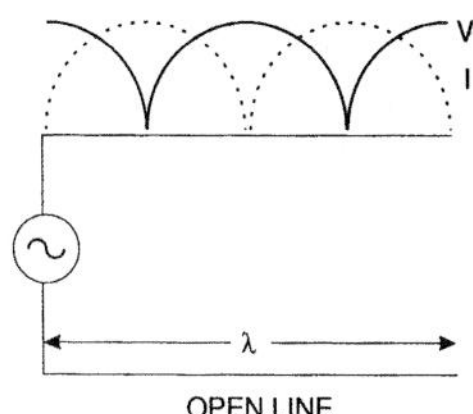

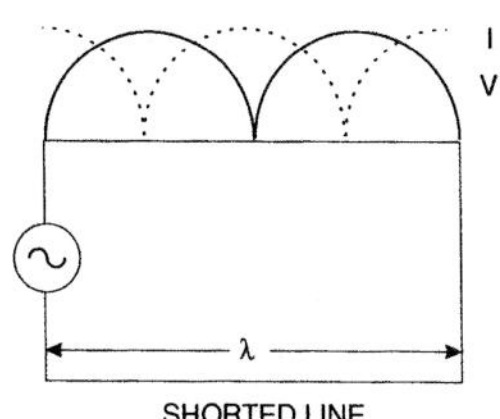

FIG. 8-11 Standing waves in open and short-circuited transmission lines.

TABLE 8-2 Voltage and Current Standing Waves

	At Open Termination	At Short-circuited Termination
Voltage	Maximum	Minimum
Current	Minimum	Maximum

Reflection Coefficient

$$\Gamma = \frac{V_r}{V_i} = \frac{Z_L - Z_0}{Z_L + Z_0}$$

where V_r = reflected voltage
V_I = incident voltage
Z_L = load impedance
Z_o = characteristic impedance

Standing Wave Ratio

$$SWR = \frac{V_{max}}{V_{min}} = \frac{I_{max}}{I_{min}} = \frac{1 + |\Gamma|}{1 - |\Gamma|}$$

where V_{max}, V_{min} = maximum, minimum voltage along line, V
I_{max}, I_{min} = maximum, minimum current along line, A
Γ = reflection coefficient

SWR is the larger of $\frac{Z_0}{R_L}$ or $\frac{R_L}{Z_0}$

where Z_0 = characteristic impedance
R_L = load resistance

Transmission Line Equivalent Circuits

Figure 8-12 provides equivalent capacitive and inductive circuits for short-circuited and open lines of lengths equal to, less than, and greater than one-quarter wavelength.

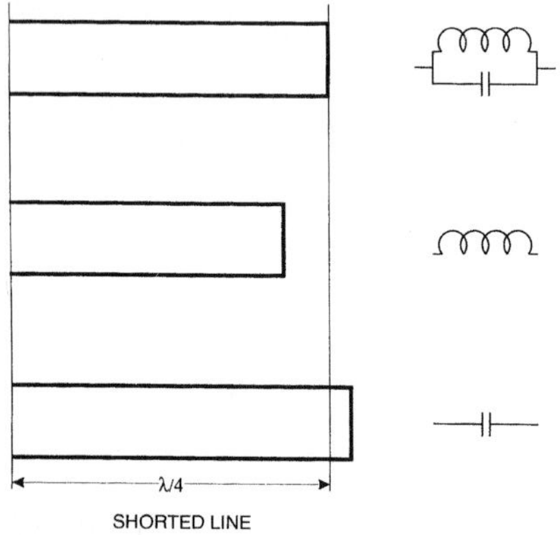

FIG. 8-12 Equivalent circuits for short-circuited and open quarter-wavelength transmission lines.

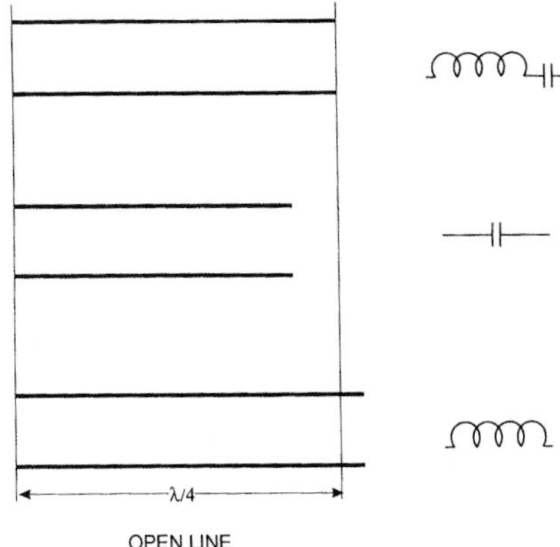

FIG. 8-12 Equivalent circuits for short-circuited and open quarter-wavelength transmission lines *(Continued)*.

8-5 AMPLITUDE MODULATION (AM)

Percent Modulation (M)

$$M = \frac{E_c - E_t}{2E_a} \times 100$$

$$= \frac{E_c - E_t}{E_c + E_t} \times 100$$

where E_c = crest amplitude of modulated carrier
E_t = trough amplitude of modulated carrier
E_a = average amplitude of modulated carrier

Sideband Power (P_s) of an AM Carrier

$$P_s = \frac{M^2 P_c}{2}$$

where M = percent modulation
P_c = carrier power, W

Total Radiated Power (P_t)

$$P_t = P_s + P_c$$

where P_s = sideband power, W
P_c = carrier power, W

8-6 FREQUENCY MODULATION (FM)

Percent Modulation (M)

$$M = \frac{\Delta f}{D} \times 100$$

where Δf = change in frequency
D = deviation (change in frequency) for 100% modulation (D = 75 kHz for commercial FM radio, 25 kHz for TV audio, and 15 kHz for two-way radio)

Modulation Index (M_i)

$$M_i = \frac{f_d}{f_a}$$

where f_d = deviation frequency, kHz
f_a = modulating audio frequency, kHz

8-7 PULSE MODULATION (PM)

See Table 8-3 for different types of pulse modulation.

TABLE 8-3 Types of Pulse Modulation

Modulation	Abbreviation	Characteristics
Pulse amplitude	PAM	Amplitude proportional to signal amplitude
Pulse time	PTM	Time of pulse parameters depends on signal
Pulse width (also known as Pulse duration and Pulse length)	PWM PDM PLM	Leading or trailing edge of pulse proportional to signal (a form of PTM)
Pulse position	PPM	Position of pulse in each time period proportional to signal (a form of PTM)
Pulse code	PCM	Signal converted to pulses by coding
Delta (also known as slope)	—	Pulses correspond to the derivative of signal amplitude

8-8 TYPES OF EMISSION

See Table 8-4 for different types of emission.

TABLE 8-4 Types of Emission

Type	Code	Characteristics	Variations
AM	A0	No modulation	
	A1	Telegraphy with on-off keying (without modulating audio frequency)	
	A2	Telegraphy by the on-off keying of an audio frequency or the modulated emission	
	A3	Telephony	Double sideband, full carrier
	A3A		Single sideband, reduced carrier
	A3B		Two independent sidebands, reduced carrier
	A3J		Single sideband, suppressed carrier
	A4	Facsimile	
	A5	Television	
	A9	Composite	
FM	F0	No modulation	
	F1	Telegraphy with FSK	
	F2	Telegraphy by the on-off keying of an audio frequency or a frequency-modulated emission	
	F3	Telephony	
	F4	Facsimile	
	F5	Television	
	F9	Composite	

TABLE 8-4 Types of Emission (Continued)

Type	Code	Characteristics	Variations
PM	P0	Absence of information-carrying modulation	
	P1	Telegraphy by on-off keying of a pulsed carrier	
	P2D	Telegraphy by on-off keying of a modulated audio frequency or a modulated pulse carrier	PAM
	P2E		PWM
	P2F		Pulse-phase (position) modulation
	P3D	Telephony	PAM
	P3E		PWM
	P3F		Pulse-phase (position) modulation
	P3G		PCM
	P9	Composite	

8-9 NATIONAL BUREAU OF STANDARDS TIME AND FREQUENCY SERVICES*

WWV and WWVH

NBS broadcasts continuous signals from its high-frequency radio stations WWV and WWVH. The radio frequencies used

*This section of chap. 8 was reprinted from *NBS Time and Frequency Dissemination Services*, Sandra Howe, Ed., National Bureau of Standards, 1979. International Time and Frequency Stations are provided in Table 8-5. NBS is now the National Institute of Standards and Technology.

are 2.5, 5, 10 and 15 MHz. WWV also broadcasts on an additional frequency of 20 MHz. All frequencies carry the same program, but because of changes in ionospheric conditions, which sometimes adversely affect the signal transmissions, most receivers are not able to pick up the signal on all frequencies at all times in all locations. Except during times of severe magnetic disturbances, however—which make all radio transmissions almost impossible—listeners should be able to receive the signal on at least one of the broadcast frequencies. As a general rule, frequencies above 10 MHz provide the best daytime reception and the lower frequencies are best for nighttime reception.

Services provided by these stations include:

Time announcements
Standard time intervals
Standard frequencies
Geophysical alerts
Marine storm warnings
Omega Navigation System status reports
UTI time corrections
BCD time code

Figure 8-13 gives the hourly broadcast schedules of these services along with station location, radiated power, and details of the modulation.

Accuracy and Stability

The time and frequency broadcasts are controlled by the primary NBS Frequency Standard in Boulder, Colorado. The frequencies as transmitted are accurate to within one part in 100 billion (1×10^{-11}) at all times. Deviations are normally less than one part in 100 billion (1×10^{-12}) from day to day. However, changes in the propagation medium (causing Doppler effect, diurnal shifts, etc.) result in fluctuations in the carrier frequencies *as received* by the user that could be very much greater than the uncertainty described above.

Radiated Power, Antennas and Modulation

Frequency, MHz	Radiated Power, kW	
	WWV	WWVH
2.5	2.5	5.0
5.0	10.0	10.0
10.0	10.0	10.0
15.0	10.0	10.0
20.0	2.5	—

The broadcasts on 5, 10, and 15 MHz from WWVH are from phased vertical half-wave dipole arrays. They are designed and oriented to radiate a cardioid pattern directing maximum gain in a westerly direction. The 2.5-MHz antenna at WWVH and all antennas at WWV are half-wave dipoles that radiate omnidirectional patterns.

At both WWV and WWVH, double sideband amplitude modulation is used with 50 percent modulation on the steady

WWV BROADCAST FORMAT

VIA TELEPHONE: (303) 499–7111
(NOT A TOLL-FREE NUMBER)

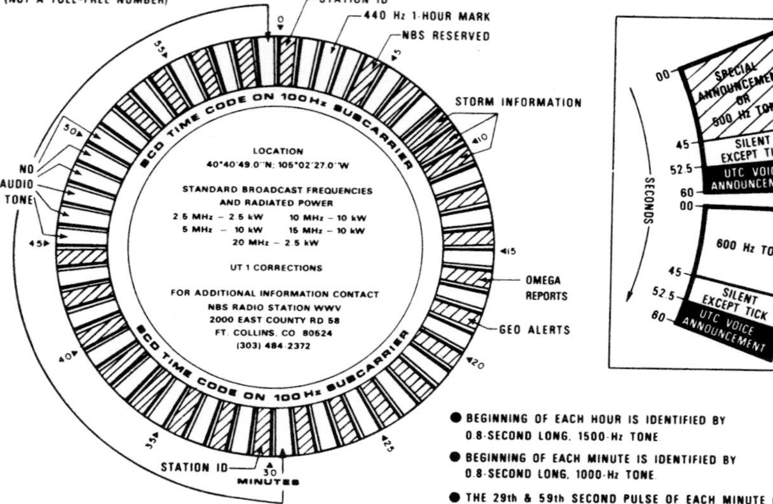

- BEGINNING OF EACH HOUR IS IDENTIFIED BY 0.8-SECOND LONG, 1500-Hz TONE.
- BEGINNING OF EACH MINUTE IS IDENTIFIED BY 0.8-SECOND LONG, 1000-Hz TONE.
- THE 29th & 59th SECOND PULSE OF EACH MINUTE IS OMITTED.

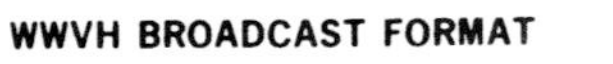

FIG. 8-13 The hourly broadcast schedule of WWV and WWVH.

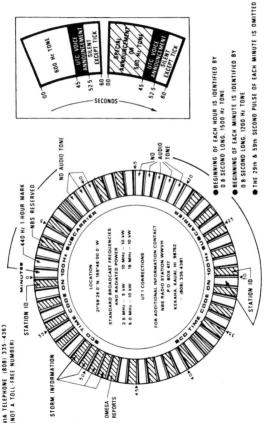

tones, 25 percent for the BCD time code, 100 percent for seconds pulses, and 75 percent for voice.

Time Announcements

Voice announcements are made from WWV and WWVH once every minute. To avoid confusion, a man's voice is used on WWV and a woman's voice on WWVH. The WWVH announcement occurs first (at 15 seconds before the minute) while the WWV announcement occurs at 7½ seconds before the minute. Although the announcements occur at different times, the tone markers referred to are transmitted simultaneously from both stations. However, they might not be received at the same time because of propagation effects.

The time referred to in the announcements is "Coordinated Universal Time" (UTC). It is coordinated through international agreements by the International Time Bureau (BIH) so that time signals broadcast from the many stations, such as WWV, throughout the world will be in close agreement.

The specific hour and minute mentioned is actually the time at the time zone centered around Greenwich, England, and can be considered generally equivalent to the more well-known "Greenwich Mean Time" (GMT). UTC time differs from your local time only by an integral number of hours. By knowing your own local time zone and using the chart of world time zones in Fig. 8-14, the appropriate number of hours to add or subtract from UTC to obtain local time can be determined. The UTC time announcements are expressed in the 24-hour clock system—i.e., the hours are numbered beginning with 00 hours at midnight through 12 hours at noon to 23 hours, 59 minutes just before the next midnight.

Standard Time Intervals

The most frequent sounds heard on WWV and WWVH are the pulses that mark the seconds of each minute, except for the 29th and 59th seconds pulses which are omitted completely. The first pulse of every *hour* is an 800-millisecond pulse of 1500 Hz. The first pulse of every *minute* is an 800-millisecond pulse of 1000 Hz at WWV and 1200 Hz at WWVH. The remaining seconds pulses are brief audio bursts (5-millisecond pulses of 1000 Hz at WWV and 1200 Hz at WWVH) that resemble the ticking of a clock. All pulses commence at the *beginning* of each second. They are given by means of double-sideband amplitude modulation.

Each seconds pulse is preceded by 10 milliseconds of silence and followed by 25 milliseconds of silence to avoid interference that might make it difficult or impossible to pick out the seconds pulses. This total 40-millisecond protected zone around each seconds pulse is illustrated in Fig. 8-15.

Standard Audio Frequencies

In alternate minutes during most of each hour, 500- or 600-Hz audio tones are broadcast. A 440-Hz tone, the musical note A above middle C, is broadcast once each hour. In addition to being a musical standard, the 440-Hz tone can be used to provide an hourly marker for chart recorders or other automated devices.

Official Announcements

Forty-five-second announcement segments (see Fig. 8-13) are available on a subscription basis to other Federal agencies to disseminate official and public service information. The accu-

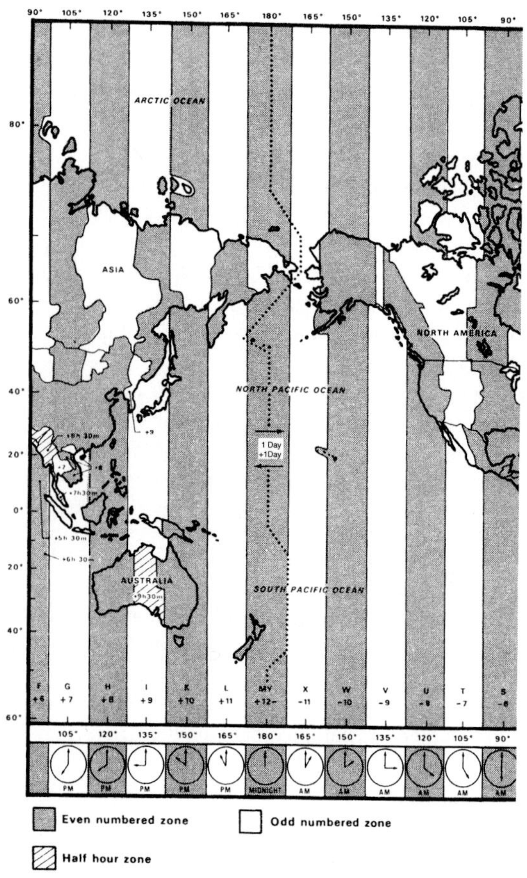

FIG. 8-14 World time zones.

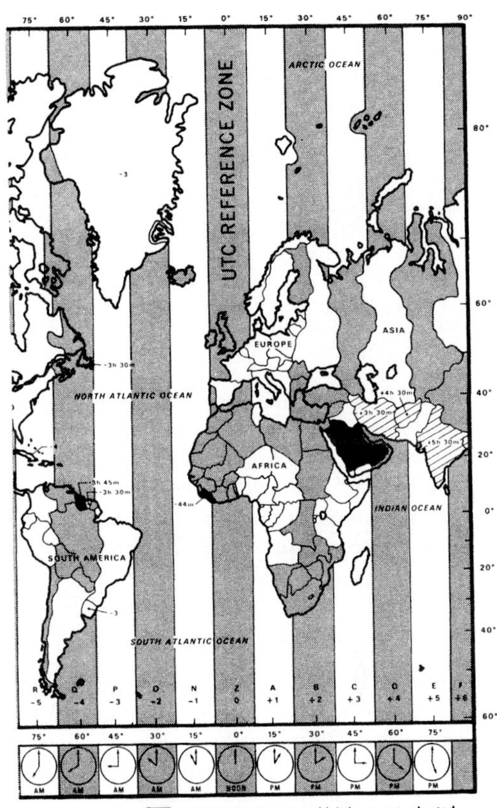

Countries and areas which have not adopted zone system, or where time differs other than half hour from neighboring zones.

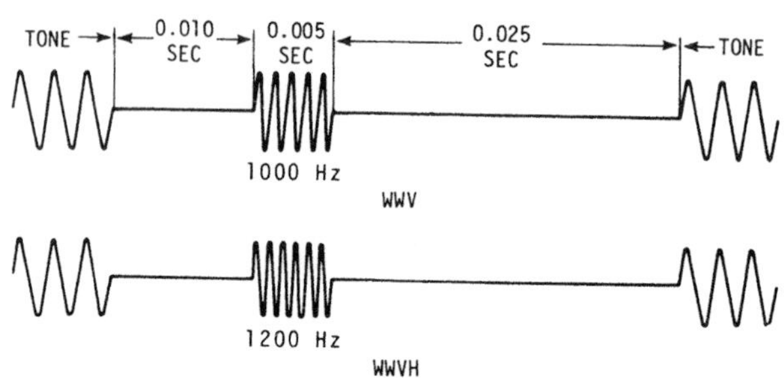

FIG. 8-15 Format of WWV and WWVH second pulses.

racy and content of these announcements are the responsibility of the originating agency, not necessarily NBS.

Most segments, except those reserved for NBS use and the semi-silent periods, are available. Arrangements for use of segments can be made through the *Time and Frequency Services Group, 524.06, National Bureau of Standards, Boulder, CO 80303*.

Omega Navigation System Status Reports

Omega Navigation System status reports are broadcast in voice from WWV at 16 minutes after the hour and from WWVH at 47 minutes after the hour. The international Omega Navigation System is a very low frequency (VLF) radio navigation aid operating in the 10- to 14-kHz frequency band. Eight stations are in operation around the world. Omega, like other radio navigation systems, is subject to signal degradation caused by ionospheric disturbances at high latitudes. The Omega announcements on WWV and WWVH are given to provide users with immediate notification of such events and other information on the status of the Omega system.

Geophysical Alerts

Current geophysical alerts (Geoalerts) are broadcast in voice from WWV at 18 minutes after each hour. The messages are changed approximately every three hours at 0000, 0300, 0600, 0900, 1200, 1500, 1800, and 2100 UTC. Part A of the message gives the solar-terrestrial indices for the day; namely, the 1700Z solar flux from Ottawa, Canada at 2800 MHz, the estimated A value for Fredericksburg, Virginia, and the current, Boulder, Colorado, K-index. Part B gives the solar-terrestrial conditions for the previous 24 hours and Part C gives the forecast for the next 24 hours. If stratwarm conditions exist, a brief advice is given at the end of the message.

1. Solar activity is classified as:
   - VERY LOW = usually only quiet regions on the solar disk and no more than five of these; fewer than 10 class-C subflares without centimetric radio burst or SID (sudden ionospheric disturbance) observed or expected.
   - LOW = usually more than five but less than 10 quiet regions on the solar disk; only class-C subflares without centimetric radio bursts or SID observed or expected.
   - MODERATE = eruptive regions on the solar disk; fewer than five class-M X-ray events with centimetric radio bursts and SID observed or expected.

HIGH = active regions on solar disk; several class-M X-ray events with centimetric radio bursts and strong SID; and/or one to two importance-2 chromospheric flares or class-X X-ray events observed or expected.

VERY HIGH = region capable of producing protons on the sun; one or more chromospheric flares of importance-2 or greater; with outstanding centimetric radio bursts (500 flux units or greater), class-X X-ray bursts, and major SID observed or expected.

2. The geomagnetic field is classified as:

   QUIET = $A \leq 7$, usually no K-indices > 2.

   UNSETTLED = $7 < A < 15$, usually no K-indices > 3.

   ACTIVE = $15 \leq A < 30$, a few K-indices of 4.

3. The geomagnetic storms are classified as:

   MINOR = $30 \leq A < 50$, K-indices mostly 4 and 5.

   MAJOR = $A \geq 50$, some K-indices 6 or greater.

   SSC = Sudden commencements indicated by the beginning time given to the nearest minute. Gradual commencements indicated by the beginning time to the nearest hour.

4. The rest of the report is as follows:

   SOLAR FLARES

   CLASS C = any solar X-ray burst with a peak flux at 1-8Å of less than 10^{-6} watts m^{-2}.

   CLASS M = a solar X-ray burst with a peak flux at 1-8Å greater than or equal to 10^{-5} but less than 10^{-4} watts m^{-2}.

CLASS X =	a solar X-ray burst with peak flux at 1-8Å greater than or equal to 10^{-4} watts m^{-2}.
MAJOR SOLAR FLARE =	a flare of optical importance ≥ 2B (Bright) with a centimetric radio outburst of 500 flux units or more; or an X-ray event of Class-X intensity of duration ≥ 180 minutes regardless of optical flare importance.
PROTON FLARE =	protons by satellite detectors (or polar cap absorption by riometer) have been observed in time association with the Hα flare.
SATELLITE-LEVEL PROTON EVENT =	a proton enhancement detected by Earth orbiting satellites with measured particle flux of at least 10 protons cm^{-2} sec^{-1} $ster^{-1}$ at ≥ 10 MeV.
POLAR CAP ABSORPTION =	proton-induced absorption ≥ 2 dB day time, 0.5 dB night, as measured by a 30 MHz riometer located within the polar cap.
STRATWARM =	reports of stratospheric warmings in the high-latitude regions of winter hemisphere of the earth associated with gross distortions of the normal circulation associated with the winter season.

Inquiries regarding these messages should be addressed to *NOAA, Space Environment Services Center R432 Boulder, Colorado 80303*. These messages are also available by dialing (303) 499-8129.

Propagation Forecasts

The radio propagation forecasts broadcast on WWV were discontinued on September 30, 1976. Some of the information previously contained in these forecasts is now included in the Geophysical Alert announcements at 18 minutes after each hour. However, neither NBS nor the Space Environment Services Center, which sponsors the Geophysical Alerts, make radio propagation predictions, nor do they maintain a literature file on the subject. Users interested in further reading material on the effect of solar and geophysical activity on radio propagation should consult the latest edition of the *Amateur Radio Handbook*, published by the American Radio Relay League.

Marine Storm Warnings

Weather information about major storms in the eastern and central North Pacific are given from WWVH at 48, 49, 50, and 51 minutes after the hour. Similar storm warnings covering the eastern and central North Pacific are given from WWVH at 48, 49, and 50 minutes after each hour. An additional segment (at 11 minutes after the hour on WWV and at 51 minutes on WWVH) can be used when there are unusually widespread storm conditions. The brief messages are designed to tell mariners of storm threats in their areas. If there are no warnings in the designated areas, the broadcasts will so indicate. The ocean areas involved are those for which the U.S. has warning responsibility under international agreement. The regular times of issue by the National Weather Service are 0500, 1100, 1700, and 2300 UTC for WWV and 0000, 0600, 1200, and 1800 UTC for WWVH. These broadcasts are updated, effective with the next scheduled announcement following the time of issue.

Mariners might expect to receive a broadcast similar to the following:

> North Atlantic weather west of 35 West at 1700 UTC: Hurricane Donna, intensifying, 24 North, 60 West, moving northwest, 20 knots, winds 75 knots; storm, 65 North, 35 West, moving east, 10 knots; winds 50 knots, seas 15 feet.

Information regarding these announcements can be obtained from the *Director, National Weather Service, Silver Spring, MD 20910.*

"Silent" Periods

These are periods with no tone modulation. However, the carrier frequency, seconds pulses, time announcements, and 100-Hz BCD time code continue. The main silent periods extend from 45 to 52 minutes after the hour on WWV and from 15 to 20 minutes after the hour on WWVH. An additional 3-minute period from 8 to 11 minutes after the hour is silent on WWVH. Also, minutes 29 and 59 on WWV and minutes 00 and 30 on WWVH are silent.

BCD Time Code

A binary coded decimal (BCD) time code is transmitted continuously by WWV and WWVH on a 100-Hz subcarrier. The 100-Hz subcarrier is synchronous with the code pulses so that 10-millisecond resolution is attained. The time code provides a standard timing base for scientific observations made simultaneously at different locations. It has application, for example, where signals telemetered from a satellite are recorded along with the time code pulses. Data analysis is then aided by having accurate, unambiguous time markers superimposed directly on the recording.

The WWV/WWVH time code format presents UTC information in serial fashion at a rate of one pulse per second. Groups of pulses can be decoded to ascertain the current minute, hour, and day of year. Although the 100-Hz subcarrier is not considered one of the standard audio frequencies, the code does contain the 100-Hz frequency and can be used as a standard with the same accuracy as the audio frequencies. A description of the time code is contained in the Appendix.

UT1 Time Corrections

The UTC time scale broadcast by WWV and WWVH runs at a rate that is almost perfectly constant because it is based on ultra-stable atomic clocks. This time scale meets the needs of most users. Somewhat surprisingly, however, some users of time signals need time, which is not this stable. In such applications as very precise navigation and satellite tracking, which must be referenced to the rotating earth, a time scale that speeds up and slows down with the earth's rotation rate must be used. The particular time scale needed is known as *UT1* and is inferred from astronomical observations.

To be responsive to these users, information needed to obtain UT1 time is included in the UTC broadcasts. This occurs at two different levels of accuracy. First, for those users needing to know UT1 only to within about one second (this includes nearly all boaters/navigators), occasional corrections of exactly one second—called "leap" seconds—are inserted into the UTC time scale whenever needed to keep the UTC time signals within ±0.9 second of UT1 at all times. These leap seconds can be either positive or negative and are coordinated under international agreement by the International Time Bureau (BIH) in Paris. Ordinarily, a positive leap second must be added about once per year (usually on June 30 or December 31), depending on how the earth's rotation rate is behaving in each particular year. Information on how to assign dates to events that occur near the time of a leap second insertion is given in the Appendix.

The second level of correction is included in the UTC broadcasts for the very small number of users who need UT1 time to better than one second. These corrections, in units of 0.1 second, are encoded into the broadcasts by using double ticks or pulses after the start of each minute. The amount of correction is determined by counting the number of successive double ticks heard each minute. The 1st through the 8th seconds ticks indicate a "plus" correction, and the 9th through the 16th, a "minus" correction. For example, if the 1st, 2nd, and 3rd ticks are doubled, the correction is "plus" 0.3 second: UT1 = UTC + 0.3 second, or if UTC is 8:45:17, then UTI is 8:45:17.3. If the 9th, 10th, 11th, and 12th ticks are doubled, the correction is "minus" 0.4 second, or as in the previous example, UT1 = 8:45:16.6.

WWVB

WWVB transmits continuously on a standard radio carrier frequency of 60 kHz. Standard time signals, time intervals; daylight savings time and leap second indicators, and UTI corrections are provided by means of a BCD time code. The station is located on the same site as WWV. Effective coverage area is the continental U.S.

Accuracy and Stability

The frequency of WWVB is normally within its prescribed value to better than 1 part in 100 billion (1×10^{-11}). Deviations from day to day are less than 5 parts in 1,000 billion (5×10^{-12}). Effects of the propagation medium on received signals are relatively minor at low frequencies; therefore, frequency comparisons to better than 1 part in 10^{11} are possible using appropriate receiving and averaging techniques.

Station Identification

WWVB identifies itself by advancing its carrier phase 45° at 10 minutes after every hour and returning to normal phase at 15 minutes after the hour. WWVB can also be identified by its unique time code.

Radiated Power, Antenna, and Coverage

The effective radiated power from WWVB is 13 kW. The antenna is a 122-meter, top-loaded vertical installed over a radial ground screen. Some measured field intensity contours are shown in Fig. 8-16.

BCD Time Code

WWVB broadcasts time information in the form of a BCD time code. The time code is synchronized with the 60-kHz carrier and is broadcast continuously at a rate of one pulse per second. Each pulse is generated by reducing the carrier power 10 dB at the beginning of the second, so the leading edge of every negative-going pulse is on time. Details of the WWVB time code are presented in the Appendix.

Summary of Broadcast Services

The services provided by the NBS radio stations are summarized in the following chart. Coordinates for the stations are also listed.

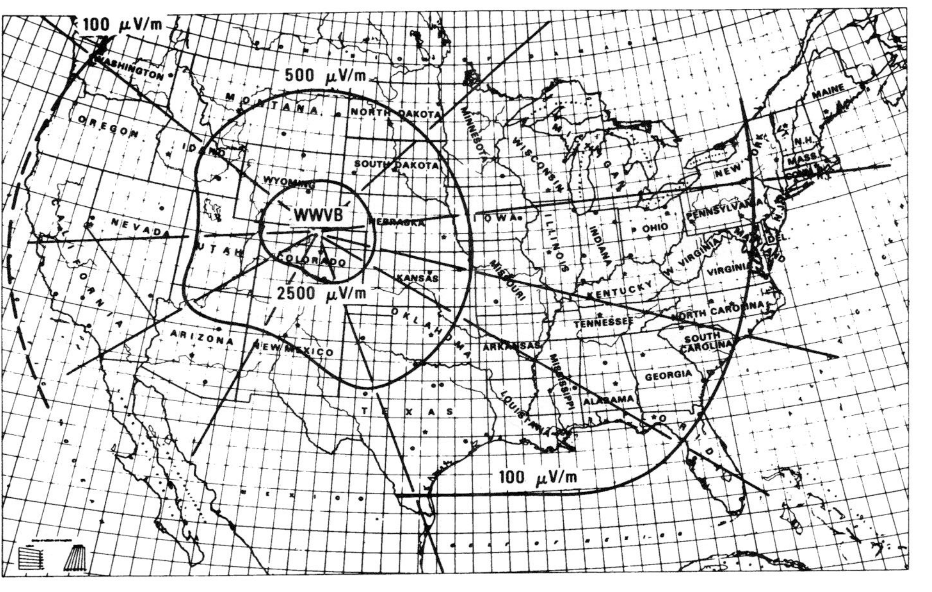

FIG. 8-16 Measured field-intensity contours of WWVB at 13-kW ERP.

STATION	DATE SERVICE BEGAN	RADIO FREQUENCIES	AUDIO FREQUENCIES	MUSICAL PITCH	TIME INTERVALS	TIME SIGNALS	UT1 CORRECTIONS	OFFICIAL ANNOUNCEMENTS
WWV	1923	X	X	X	X	X	X	X
WWVH	1948	X	X	X	X	X	X	X
WWVB	1956	X			X	X	X	

COORDINATES:		
WWV	40°40'49.0''N	105°02'27.0''W
WWVB	40°40'28.3''N	105°02'39.5''W
WWVH	21°59'26.0''N	159°46'00.0''W

How NBS Controls the Transmitted Frequencies

A simplified diagram of the NBS frequency control system is shown in Fig. 8-17. The entire system depends upon the reference shown in this diagram as the NBS Primary Time and Frequency Standard. This standard is comprised of a number of commercial cesium beam clocks, up to two primary cesium beam frequency and time standards, and computer-aided measurement and computation methods which combine all of the clock data to generate an accurate and uniform time scale, TA (NBS). Another scale, UTC (NBS), is also generated by adding leap seconds and small corrections to TA (NBS) as needed to keep UTC (NBS) synchronized with the internationally coordinated time scale, UTC, which is maintained by the BIH.

Utilizing the Line-10 horizontal synchronizing pulses from a local television station, the Fort Collins master clock is compared on a regular basis with the UTC (NBS) time scale. All other clocks and time-code generators at the Fort Collins site are then compared with the Fort Collins master clock. Frequency corrections of the WWVB controlling oscillators are based on their phase relative to the UTC (NBS) time scale.

The transmissions from WWV and WWVH are controlled by three commercial cesium standards located at each site. To ensure accurate time transmission from each station, the time-code generators are compared with the stations' master clock several times each day.

Control of the signals transmitted from WWVH is based not only upon the cesium standards, but upon signals from WWVB as received by phase-lock receivers. The cesium standards controlling the transmitted frequencies and time signals are continuously compared with the received signals.

NBS Time via Satellite

As a complement to its other time and frequency services, NBS is now sponsoring a satellite-disseminated time code using the GOES (Geostationary Operational Environmental Satellite) satellites of the National Oceanic and Atmospheric Administration (NOAA). The time code is referenced to the NBS time

NATIONAL BUREAU OF STANDARDS
FREQUENCY AND TIME FACILITIES

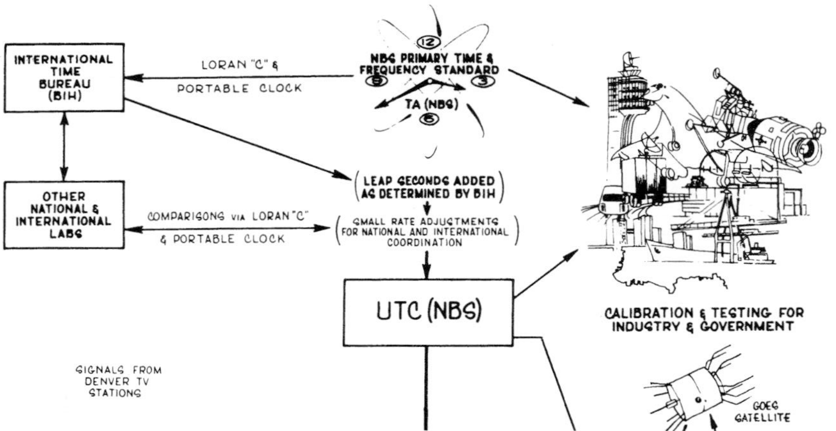

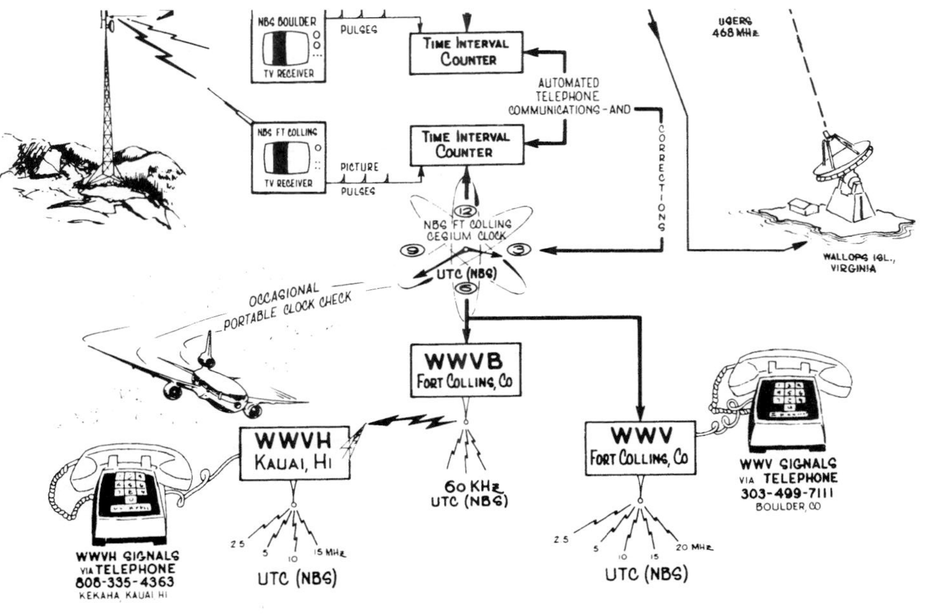

FIG. 8-17 The NBS frequency-control system.

scale and gives Coordinated Universal Time (UTC). Although the time code was designed to provide a means of dating environmental data collected by the GOES satellites, it can also be used as a general-purpose time reference for many other applications. The time code is available to the entire Western Hemisphere from two satellites on a near full-time basis.

There are three GOES satellites in orbit, two in operational status with a third serving as an in-orbit spare. The western satellite operates at 468.825 MHz and is located at 135° West Longitude. The eastern satellite is received on 468.8375 MHz and is positioned at 75° West Longitude. The spare is at 105° West Longitude. Coverages of the two operational satellites are shown in Fig. 8-18.

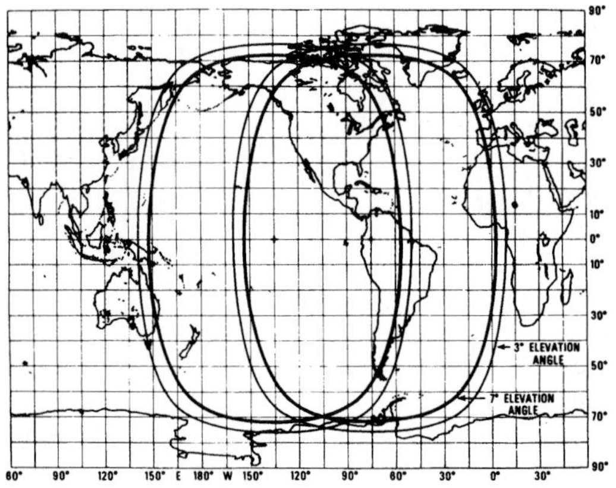

FIG. 8-18 Coverage of the GOES satellites.

The GOES satellites collect environmental data from remote sensors. The time code is part of the interrogation channel which is used to communicate with these sensors. The interrogation messages and time code are prepared and sent to the GOES satellites from Wallops Island, Virginia. NBS maintains atomic clocks, referenced to UTC(NBS), at this site to generate the time code. The time code includes a sync word, a time-of-year message (including day of year, hour, minute, and second), UTI correction, and satellite position. A description of the time code is given in the Appendix.

Performance

The GOES time code can be used at three levels of performance: uncorrected for path delay, corrected for mean path delay only, and fully corrected.

Uncorrected: The path delay from point of origin (Wallops Island, Virginia) to the earth via the satellite is approximately 260,000 microseconds. Since the signals are advanced in time by this amount before transmission from Wallops Island, they arrive at the earth's surface on time to within 16 milliseconds.

Corrected for Mean Path Delay: Accounting for the mean path delay to any point on the earth's surface, but ignoring the cyclic (24-hour) delay variation, generally guarantees the signal arrival time to ±0.5 millisecond.

Fully Corrected: The cyclic delay variation is a result of the satellite orbit or path around the earth not being perfectly circular and not in the plane of the equator. The orbit is actually an ellipse and has a small inclination—usually less than 1°. To compensate for these and other effects, the satellite position is included with the time message for correction of path delay by the user. This correction provides path delays accurate to ±50 microseconds. The ultimate accuracy of the recovered time depends upon knowledge of user equipment delays and noise levels as well as path delay.

Precautions

Because the GOES time code is transmitted outside the spectrum reserved exclusively for time and frequency broadcasts, it *cannot* be considered an NBS service in the same sense that the radio broadcasts and television methods are services. The "land-mobile" services and the GOES interrogation channels use the same frequency allocations (468.825 and 468.8375 MHz), which means that the time code might suffer interference from land-mobile transmissions. This is particularly true in urban areas, where there is a high density of land-mobile activity. The satellite frequency allocations are secondary to the land-mobile services. Therefore, any interference must be accepted by the time-signal users. Complaints to the FCC will not result in any adjustments in favor of time code users.

Because of the spacing of frequency assignments to the land-mobile users, there is far less interference to the eastern satellite than to the western satellite. Therefore, the eastern satellite should be used by those users situated in large urban areas.

Outages

Although the GOES satellites transmit continuously, there may be interruptions during the periods of solar eclipses. The GOES satellites undergo spring and autumn eclipses during a 46-day interval at the vernal and autumnal equinoxes. The eclipses vary from approximately 10 minutes at the beginning and end of eclipse periods to a maximum of approximately 72 minutes at the equinox. The eclipses begin 23 days prior to equinox and end 23 days after equinox (i.e., March 1 to April 15 and September 1 to October 15). The outages occur during local midnight for the satellite's mean meridian.

There will also be shutdowns for periodic maintenance at the Wallops Island ground station.

Continuity

NBS cannot give an absolute guarantee to the long-term continuance of the GOES time code because the satellites belong to NOAA. However, NBS and NOAA have agreed to include the time code in the transmissions to the maximum extent possible.

For further information on the GOES time code or commercial equipment availability, write the *Time and Frequency Services Group, 524.06, NBS, Boulder, CO 80303.*

Appendix

Dating of Events in the Vicinity of Leap Seconds
WWV/WWVH Time Code
WWVB Time Code
GOES Satellite Time Code

Dating Events in the Vicinity of Leap Seconds When leap-second adjustments are necessary to keep the broadcast time signals (UTC) within ±0.9 second of the earth-related UT1 time scale, the addition or deletion of exactly 1 second occurs at the end of the UTC month. By international agreement, first preference is given to December 31 or June 30, second preference to March 31 or September 30, and third preference to any other month.

When a positive leap second is required (that is, when UT1 is slow, relative to UTC), an additional second is inserted beginning at 23h 59m 60s of the last day of the month and ending at 0h 0m 0s of the first day of the following month. In this case, the last minute of the month in which there is a leap second contains 61 seconds. To assign dates to events that occur around this extra second, refer to Fig. 8-19.

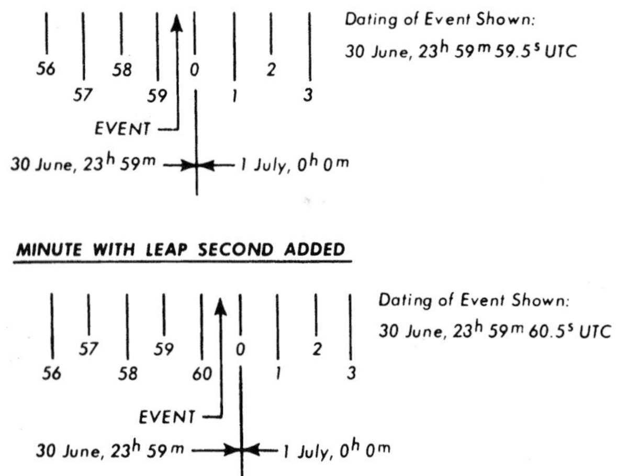

FIG. 8-19 Dating events in vicinity of a leap second.

Assuming that unexpected large changes do not occur in the earth's rotation rate in the future, it is likely that positive leap seconds will continue to be needed about once per year. If, however, the earth should speed up significantly at some future time, so that UT1 runs at a *faster* rate than UTC, then provision is also made for negative leap seconds in the UTC time scale. In this case, exactly one second would be *deleted* at the end of some UTC month, and the last minute would contain only 59 seconds.

Positive leap seconds were inserted in all NBS broadcasts at the end of June 30, 1972, and December 31, 1972 through 1978.

WWV/WWVH Time Code The WWV/WWVH time code is a modified version of the IRIG-H format. Data is broadcast on a 100-Hz subcarrier at a rate of one pulse per second. Certain pulses in succession comprise binary-coded groups representing decimal numbers. The binary-to-decimal weighting scheme is 1-2-4-8 with the least-significant binary digit always transmitted first. The binary groups and their basic decimal equivalents are shown in the following table:

	BINARY GROUP	DECIMAL
Weight:	1 2 4 8	EQUIVALENT
	0 0 0 0	0
	1 0 0 0	1
	0 1 0 0	2
	1 1 0 0	3
	0 0 1 0	4
	1 0 1 0	5
	0 1 1 0	6
	1 1 1 0	7
	0 0 0 1	8
	1 0 0 1	9

In every case, the decimal equivalent of a BCD group is derived by multiplying each binary digit times the weight factor of its respective column and then adding the four products together. For instance, the binary sequence 1010 in the 1-2-4-8 scheme means $(1 \times 1) + (0 \times 2) + (1 \times 4) + (0 \times 8) = 1 + 0 + 4 + 0 = 5$, as shown in the table. If fewer than nine decimal digits are needed, one or more of the binary columns can be omitted.

In the standard IRIG-H code, a binary 0 pulse consists of exactly 20 cycles of 100-Hz amplitude modulation (200 milliseconds duration), whereas a binary 1 consists of 50 cycles of 100 Hz (500 milliseconds duration). In the WWV/WWVH broadcast format, however, all tones are suppressed briefly while the seconds pulses are transmitted.

Because the tone suppression applies also to the 100-Hz subcarrier frequency, it has the effect of deleting the first 30-millisecond portion of each binary pulse in the time code. Thus, a binary 0 contains only 17 cycles of 100-Hz amplitude modulation (170 milliseconds duration) and a binary 1 contains 47 cycles of 100 Hz (470 milliseconds duration). The leading edge of every pulse coincides with a positive-going zero crossing of the 100-Hz subcarrier, but it occurs 30 milliseconds after the beginning of the second.

Within a time frame of one minute, enough pulses are transmitted to convey in BCD language the current minute, hour, and day of year. Two BCD groups are needed to express the hour (00 through 23); and three groups are needed to express the day of year (001 through 366). When representing units, tens, or hundreds, the basic 1-2-4-8 weights are simply multiplied by 1, 10, or 100, as appropriate. The coded information always refers to time at the beginning of the one-minute frame. Seconds can be determined by counting pulses within the frame.

Each frame commences with a unique spacing of pulses to mark the beginning of a new minute. No pulse is transmitted during the first second of the minute. Instead, a one-second space or hole occurs in the pulse train at that time. Because all pulses in the time code are 30 milliseconds late with respect to UTC, each minute actually begins 1030 milliseconds (or 1.03 seconds) prior to the leading edge of the first pulse in the new frame.

For synchronization purposes, every 10 seconds a so-called position identifier pulse is transmitted. Unlike the BCD data pulses, the position identifiers consist of 77 cycles of 100 Hz (770 milliseconds duration).

UTI corrections to the nearest 0.1 second are broadcast via BCD pulses during the final 10 seconds of each frame. The coded pulses that occur between the 50th and 59th seconds of each frame are called *control functions*. Control function #1, which occurs at 50 seconds, tells whether the UT1 correction is negative or positive. If control function #1 is a binary 0, the correction is negative; if it is a binary 1, the correction is positive. Control functions #7, #8, and #9, which occur respectively at 56, 57, and 58 seconds, specify the amount of UT1 correction. Because the UT1 corrections are expressed in tenths of a second, the basic binary-to-decimal weights are multiplied by 0.1 when applied to these control functions.

Control function #6, which occurs at 55 seconds, is programmed as a binary 1 throughout those weeks when Daylight Savings Time is in effect and as a binary 0 when Standard Time is in effect. The setting of this function is changed at 0000 UTC on the date of change. Throughout the U.S. mainland, this schedule allows several hours for the function to be received before the change becomes effective locally—i.e., at 2:00 A.M. local time. Thus, control function #6 allows clocks or digital recorders operating on local time to be programmed to make an automatic one-hour adjustment in changing from Daylight Saving Time to Standard Time and vice versa.

Figure 8-20 depicts one frame of the time code as it might appear after being rectified, filtered, and recorded. In this example, the leading edge of each pulse is considered to be the positive-

FORMAT H, SIGNAL H001, IS COMPOSED OF THE FOLLOWING:

1) 1 ppm FRAME REFERENCE MARKER R = (P_0 AND 1.03 SECOND "HOLE")
2) BINARY CODED DECIMAL TIME-OF-YEAR CODE WORD (23 DIGITS)
3) CONTROL FUNCTIONS (9 DIGITS) USED FOR UT_1 CORRECTIONS, ETC.
4) 6 ppm POSITION IDENTIFIERS (P_0 THROUGH P_5)
5) 1 pps INDEX MARKERS

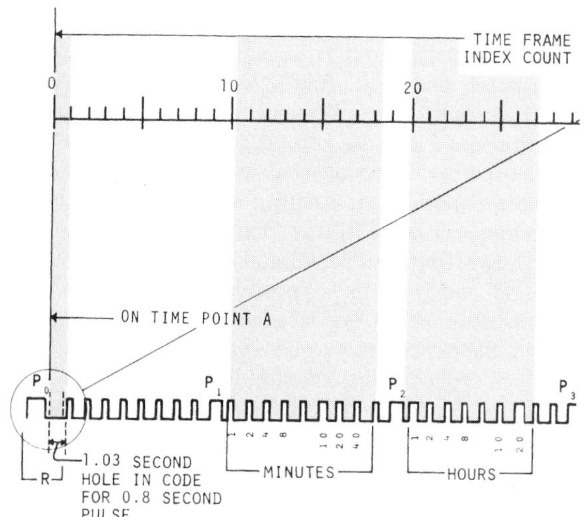

P_0-P_5 POSITION IDENTIFIERS (0.770 SECOND DURATION)

W WEIGHTED CODE DIGIT (0.470 SECOND DURATION)

C WEIGHTED CONTROL ELEMENT (0.470 SECOND DURATION)

DURATION OF INDEX MARKERS, UNWEIGHTED CODE, AND UNWEIGHTED CONTROL ELEMENTS = 0.170 SECONDS

NOTE: <u>BEGINNING</u> OF PULSE IS REPRESENTED BY POSITIVE-GOING EDGE.

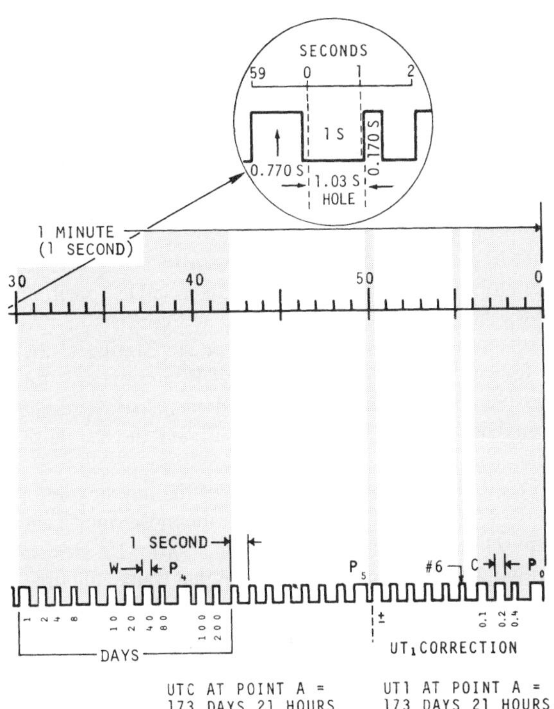

FIG. 8-20 WWV and WWVH time-code format.

going excursion. The pulse train in the figure is annotated to show the characteristic features of the time-code format. The six position identifiers are denoted by symbols P_1, P_2, P_3, P_4, P_5, and P_0. The minutes, hours, days, and UT1 sets are marked by brackets, and the applicable weighting factors are printed beneath the coded pulses in each BCD group. With the exception of the position identifiers, all uncoded pulses are set permanently to binary 0.

The first 10 seconds of every frame always include the 1.03-second hole followed by eight uncoded pulses and the position identifier P_1. The minutes set follows P_1 and consists of two BCD groups separated by an uncoded pulse. Similarly, the hours set follows P_2. The days set follows P_3 and extends for two pulses beyond P_4 to allow enough elements to represent three decimal digits. The UT1 set follows P_5, and the last pulse in the frame is always P_0.

In Fig. 8-20, the least-significant digit of the minutes set is $(0 \times 1) + (0 \times 2) + (0 \times 4) + (0 \times 8) = 0$; the most-significant digit of that set is $(1 \times 10) + (0 \times 20) + (0 \times 40) = 10$. Hence, at the beginning of the 1.03-second hole in that frame, the time was exactly 10 minutes past the hour. By decoding the hours set and the days set, it is seen that the time of day is in the 21st hour on the 173rd day of the year. The UT1 correction is +0.3 second. Therefore, at point A, the correct time on the UT1 scale is 173 days, 21 hours, 10 minutes, 0.3 second.

WWVB Time Code The WWVB time code is generated by shifting the power of the 60-kHz carrier. The carrier power is reduced 10 db at the beginning of each second and restored to full power 200 milliseconds later for a binary zero, 500 milliseconds later for a binary one, and 800 milliseconds later for a reference marker or position identifier. Certain groups of pulses are encoded to represent decimal numbers, which identify the minute, hour, and day of year.

The binary-to-decimal weighting scheme is 8-4-2-1 with the most-significant binary digit transmitted first. Note that this weighting sequence is the reverse of the WWV/WWVH code. The BCD groups and their basic decimal equivalents are tabulated below:

	BINARY GROUP	DECIMAL
Weight:	8 4 2 1	EQUIVALENT
	0 0 0 0	0
	0 0 0 1	1
	0 0 1 0	2
	0 0 1 1	3
	0 1 0 0	4
	0 1 0 1	5
	0 1 1 0	6
	0 1 1 1	7
	1 0 0 0	8
	1 0 0 1	9

The decimal equivalent of each group is derived by multiplying the individual binary digits by the weight factor of their respective columns and then adding the four products together. For example, the binary sequence 1001 in 8-4-2-1 code is equivalent to $(1 \times 8) + (0 \times 4) + (0 \times 2) + (1 \times 1) = 8 + 0 + 0 + 1 = 9$, as shown in the table. If fewer than nine decimal digits are required, one or more of the high-order binary digits may be dispensed with.

Once every minute, in serial fashion, the code format presents BCD numbers corresponding to the current minute, hour, and day on the UTC scale. Two BCD groups identify the minute (00 through 59); two groups identify the hour (00 through 23); and three groups identify the day of year (001 through 366). When representing units, tens, or hundreds, the basic 8-4-2-1 weights are multiplied by 1, 10, or 100, respectively. The coded information refers to the time at the beginning of the one-minute frame. Within each frame, the seconds can be determined by counting pulses.

Every new minute commences with a frame-reference pulse, which lasts for 0.8 second. Also, every 10-second interval within the minute is marked by a position-identifier pulse of 0.8-second duration.

UT1 corrections to the nearest 0.1 second are transmitted at seconds 36 through 44 of each frame. Coded pulses at 36, 37, and 38 seconds indicate the positive or negative relationship of UT1 with respect to UTC. Pulses at 36 and 38 seconds are transmitted as binary ones only if UT1 is *early* with respect to UTC, in which case the correction to be added to the UTC signals to obtain UT1 is a positive. The pulse transmitted at 37 seconds is a binary one if UT1 is *late* with respect to UTC, in which case the required UT1 correction is negative. The magnitude of the UT1 correction is transmitted as a BCD group at 40, 41, 42, and 43 seconds. Because UT1 corrections are expressed in tenths of seconds, the basic 8-4-2-1 weight of that particular binary group is multiplied by 0.1 to obtain its proper decimal equivalent.

Bit 55 is a leap-year indicator. It will be changed from "0" to "1" during each leap year sometime *after* January 1, but before February 29. Bit 55 would then remain set to "1" *through* January 1 of the year following the leap year. This procedure is designed to allow receivers to automatically convert day of year properly to month and day during leap years and also to allow receivers that convert UTC to local time to "back up" properly to either day 365 or 366, as appropriate upon the change of year.

Bit #57 is a Daylight Savings Time (DST) indicator. It will be set to "0" during periods of Standard Time (October-April, under present law) and to "1" during periods when DST is in effect (April–October, under present law). The bit will be changed at WWVB during the 24-hour period preceding 2:00 A.M. Eastern Time on the appropriate days.

Figure 8-21 shows a sample frame of the time code in rectified or dc form. The negative-going edge of each pulse coincides with the beginning of a second. Position identifiers are labeled P_1, P_2, P_3, P_4, P_5, and P_0. Brackets show the demarcation of the minutes, hours, days, and UT1 sets. The applicable weight factor is printed beneath the coded pulses in each BCD group. Except for the position identifiers and the frame-reference marker, all uncoded pulses are binary zeros.

In Fig. 8-21, the most significant digit of the minutes set is $(1 \times 40) + (0 \times 20) + (0 \times 10) = 40$; the least-significant digit of that set is $(0 \times 8) + (0 \times 4) + (1 \times 2) + (0 \times 1) = 2$. Thus, at the beginning of the frame, UTC was precisely 42 minutes past the hour. The sets for hours and days reveal further that it is the 18th hour of the 258th day of the year. The UT1 correction is -0.7 second, so at the beginning of the frame, the correct time on the UT1 scale was 258 days, 18 hours, 41 minutes, and 59.3 seconds.

GOES Satellite Time Code The GOES time code is part of the interrogation channel which is used to communicate with remote data sensors that send information to GOES. Interrogation messages are continuously relayed through the GOES satellites. The format of the messages is shown in Fig. 8-22.

As shown, an interrogation message contains more than timing information. A complete message consists of four bits representing a BCD time code word followed by a maximum length sequence (MLS) 15 bits in length for message synchronization, and ends with 31 bits as an address for a particular remote weather-data sensor.

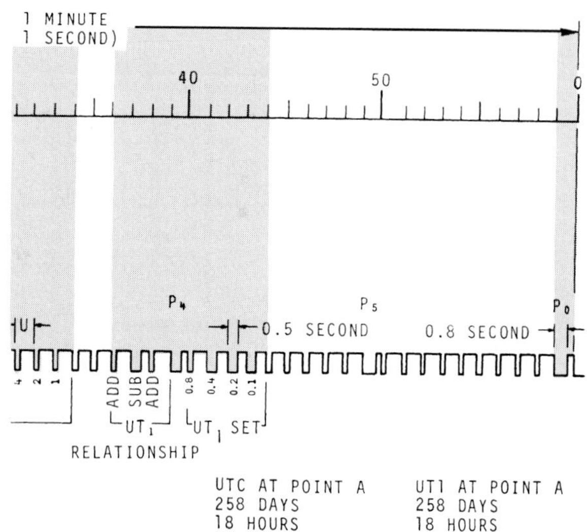
1 MINUTE
1 SECOND)
40
50
0
P4
P5
P0
U
0.5 SECOND
0.8 SECOND
4 2 1
ADD
SUB
ADD
0.8 0.4 0.2 0.1
UT1
UT1 SET
RELATIONSHIP
UTC AT POINT A
258 DAYS
18 HOURS
42 MINUTES
UT1 AT POINT A
258 DAYS
18 HOURS
41 MINUTES
59.3 SECONDS

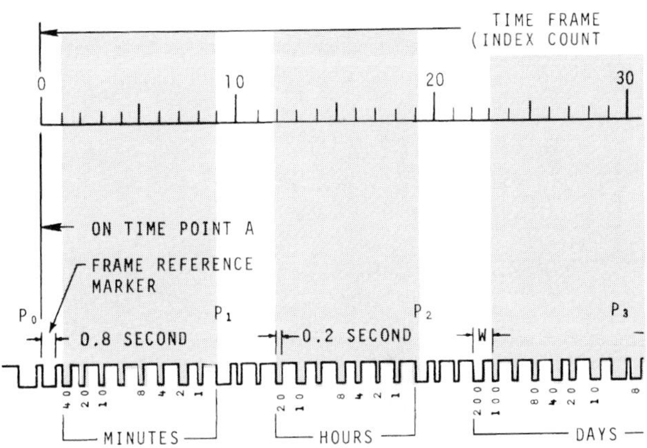

BINARY CODED DECIMAL TIME-OF-YEAR CODE WORD
(23 DIGITS)
CONTROL FUNCTIONS (15 DIGITS) USED FOR UT_1 CORRECTIONS
6 PPM POSITION IDENTIFIER MARKERS AND PULSES
(P_0 THRU P_5)
(REDUCED CARRIER 0.8 SECOND DURATION PLUS 0.2 SECOND DURATION PULSE)
W - WEIGHTED CODE DIGIT (CARRIER RESTORED IN 0.5 SECOND - BINARY ONE)
U - UNWEIGHTED CODE DIGIT (CARRIER RESTORED IN 0.2 SECOND - BINARY ZERO)

NOTE: BEGINNING OF PULSE IS REPRESENTED BY NEGATIVE - GOING EDGE.

FIG. 8-21 WWVB time-code format.

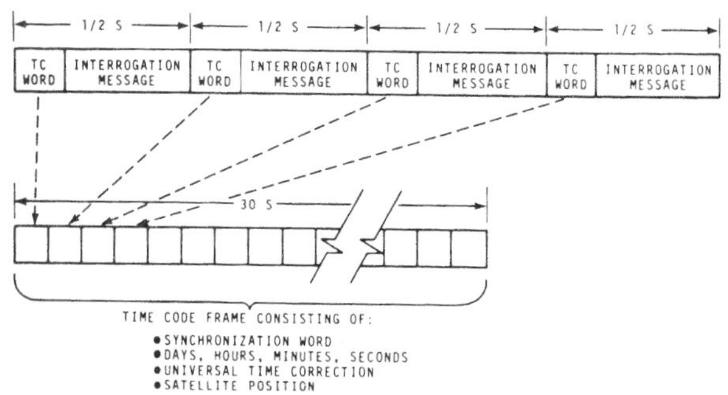

FIG. 8-22 GOES interrogation channel format.

Each interrogation message is one-half second in length or 50 bits. The data rate is 100 bits per second. The time-code frame begins on the one-half minute and takes 30 seconds to complete (see Fig. 8-23). Sixty interrogation messages are required to send the 60 BCD time-code words constituting a time-code frame.

The time-code frame contains a synchronization word, a time-of-year word (UTC) (including day of year, hour, minute,

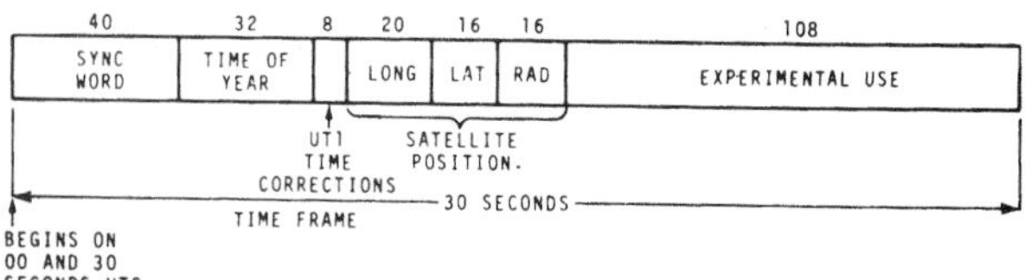

FIG. 8-23 GOES time-code format.

and second), the UT1 correction, and the satellites position in terms of its latitude, longitude, and height above the earth's surface minus a bias of 119,300 microseconds. The position information is presently updated on the half-hour.

8-10 INTERNATIONAL TIME AND FREQUENCY STANDARDS

Table 8-5 provides a listing of time-and-frequency broadcasts from around the world. Each entry in the table provides the identifier for the station, location, frequencies, and times of broadcasts.

8-11 TIME STANDARD FORMATS

Table 8-6 lists formats used for timing in a variety of applications.

8-12 ELECTROMAGNETIC SPECTRUM

The range of electromagnetic signals is shown in the spectrum of Fig. 8-24. Of more practical application is the allocation of the radio-frequency portion of the spectrum. For ease in discussing various portions of the spectrum, bands of frequencies have been given unique designations, as listed in Table 8-7. By international agreement, broadcast allocation in the frequency bands from 9 kHz to 400 GHz were revised in the World Administrative Radio Conference in 1979. The bands were designated as either exclusive or shared. Only appropriate broadcasts should use these bands, but exceptions are allowed. Shared bands provide equal or secondary rights to particular types of applications. Tables 8-8 through 8-12 describe and list the prescribed uses for these bands.

TABLE 8-5 Worldwide Time and Frequency Broadcasts

Country	Station	Location	Frequency (kHz)	Schedule (UT)
Argentina	LOL	Buenos Aires	4,856, 8,030, 17,180	0100, 1300, 2100
	LQB9	Buenos Aires	8,167.5	2200-2205, 2345-2350
	LQC20	Buenos Aires	17,500	1000-1005, 1145-1150
Australia	VNG	Lyndhurst	4,500 7,500 12,000	0945-2130 2245-2230 2145-0930
Brazil	PPE	Rio de Janeiro	8,721	0025-0030, 1125-1130, 1325-1330, 1825-1830, 2025-2030, 2325-2330
	PPR	Rio de Janeiro	435, 4,244, 8634, 12,687, 16,984, 17,194.4, 22,352.5, 22,420	0125-0130, 1425-1430, 2125-2130
Canada	CHU	Ottawa	3,330, 7,335, 14,670	0000-2359
Chile	CCV	Valparaiso	148.125, 4,298, 8,558	0055-0100, 1155-1200, 1555-1600, 1955-2000
China	BPV	Shanghai	5,000 10,000 15,000	1600-0100 0100-1600 0100-1600

TABLE 8-5 Worldwide Time and Frequency Broadcasts (Continued)

Country	Station	Location	Frequency (kHz)	Schedule (UT)
	VPS	Hong Kong	500	Even hours
	VPS8		3,842	Odd hours 1100-2100
	VPS35		8,539	Odd hours
	VPS60		13,020	Odd hours 0100-1500
	VPS80		17,096	Odd hours 2100-1300
	VPS22		22,536	Odd hours 0100-0900
Ecuador	HD210A	Guayaquil	1,510 3,810 5,000 7,600	0000-2359 0000-1200 1200-1300 1300-0000
England	MSF	Rugby	60, 2,500, 5,000, 10,000	0000-2359
	GBR		16,000	0255-0300, 0855-0900, 1455-1500, 2055-2100
France	FFH	Ste. Assise	2,500	0000-2359
	FTH42		7,428	0900, 2100
	FTK77		10,775	0800, 2000
	FTN87		13,873	0930, 1300, 2230

TABLE 8-5 Worldwide Time and Frequency Broadcasts (Continued)

Country	Station	Location	Frequency (kHz)	Schedule (UT)
Germany	DCF77	Braunschweig	77.5	0000-2359
	DAN	Norddeich	2,614	1155-1206, 2355-0006
	DAM	Elmshorn	8638.5, 16,980.4	1155-1206
	DAO	Kiel	2,775	1155-1206, 2355-0006
Guam	NPN	Barrigada	4,955, 8,150, 13,380, 27,760	0555-0600, 1155-1200, 1755-1800, 2355-0000
India	ATA	New Delhi	5,000, 10,000, 15,000	0330-1430 (Sun. and holidays 0430-0830)
	VWC	Calcutta	434	0825-0830, 1625-1630
			4,286	1625-1630
			12,745	0825-0830
Indonesia	PKI	Djakarta	8,542	0045-0055
	PLC		11,440	0045-0055
Italy	IAM	Rome	5,000	0730-0830, 1030-1130 (1 hour earlier in summer)

TABLE 8-5 Worldwide Time and Frequency Broadcasts (Continued)

Country	Station	Location	Frequency (kHz)	Schedule (UT)
	IBF	Turin	5,000	0645-0700, 0845-0900, 0945-1000, 1145-1200, 1245-1300, 1345-1400, 1445-1500, 1545-1600, 1645-1700 1745-1800 (1 hour earlier in summer)
Japan	JJY	Tokyo	2,500, 5,000, 8,000, 10,000, 15,000	0000-2359
New Zealand	ZLF	Wellington	2,500	0100-0400 Wednesday
Peru	OBC	Callao	8,560, 12,307	0155-0200, 1555-1600, 1855-1900
Russia	RBU	Moscow	66.66	2100-0807, 0900-1307, 1700-2100
	RWM		4,996, 9,996, 14,996	0000-2359
	UTR3	Gorky	20.0, 23.0, 25.0, 25.1	0536-0617, 1436-1517, 1836-1917

TABLE 8-5 Worldwide Time and Frequency Broadcasts (Continued)

Country	Station	Location	Frequency (kHz)	Schedule (UT)
	UQC3	Khabarovsk	20.0, 23.0, 25.0, 25.1	0036-0117, 0336-0417, 0636-1717, 1736-1817
	RID	Irkutsk	5,004, 10,004, 15,004	0000-2359
	RTZ		50	0000-2359
	RW-166		200	2200-2100
	RTA	Novosibirsk	10,000	0200-0500, 1400-1700, 1800-0130
			15,000	0630-0930, 1000-1330
	RCH	Tashkent	2,500	0530-0400
			5,000	0200-0400, 1400-1730, 1800-0130
			10,000	0530-0930, 1000-1330
South Africa	ZUO	Pretoria	2,500	1800-0400
			5,000	0000-2359
			100,000	0000-2359
	ZSC	Capetown	418, 4,291, 8,461, 12,724, 17,018, 22,245	0755-0800, 1655-1700
Spain	EBC	Cadiz	6,840	1029-1055
			12,008	0959-1025
Sri Lanka	4PB	Colombo	482, 8,473	0555-0600, 1325-1330

TABLE 8-5 Worldwide Time and Frequency Broadcasts (Continued)

Country	Station	Location	Frequency (kHz)	Schedule (UT)
Switzerland	HBG	Neuchâtel	75	0000-2359
Taiwan	BSF	Chung-Li	5,000, 15,000	0000-2359
Venezuela	YVTO	Caracas	6,100	0000-2359

TABLE 8-6 Time Standard Formats

Standard	Time Frame	Resolution*
IRIG format A	0.1 s	U:1 ms M: 0.1 ms
IRIG format B	1.0 s	U: 10 ms M: 1 ms
IRIG format C†	1 min	U: 0.5 s M: 0.01 s/0.001s
IRIG format D	1 h	1 min
IRIG format E	10 s	U: 100 ms M: 0.01 s/0.001 s
IRIG format G	0.001 s	U: 0.10 ms M: 0.01 ms
IRIG format H	1 min	U: 1 s M: 0.01 s/0.001 s
NASA 36-bit	1 s	U: 10 ms M: 1 ms
NASA 28-bit	1 min	U: 0.5 s M: 0.01 s/0.001 s
NASA 20-bit	1 h	U: 1 min M: 10 ms/1 ms
FAA NAFEC	10 s to 50 ms	100 ms to 500 μs

*M—modulated; U—unmodulated.
†Obsolete standard (for historical information only).

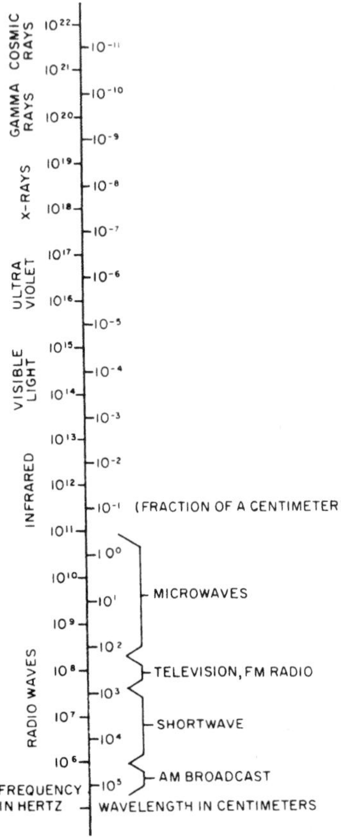

FIG. 8-24 Frequency spectrum. *(Bureau of Naval Personnel)*

TABLE 8-7 International Telecommunications Union Band Designators

Acronym (Description)	Descriptive Name (Wave Length)	Band Number	Frequency (MHz)
VLF (very low frequency)	--------	4	0.003–0.03
LF (low frequency)	--------	5	0.3–0.3
MF (medium frequency)	--------	6	0.3–3
HF (high frequency)	--------	7	3–30
VHF (very high frequency)	Metric	8	30–300
UHF (ultra high frequency)	Decimetric	9	300–3,000
SHF (super high frequency)	Centrimetric	10	3,000–30,000
EHF (extremely high frequency)	Millimetric	11	30,000–300,000
--------	--------	12	300,000–3,000,000

TABLE 8-8 Band Designators

Frequency bands are often identified by alphabetical or numerical designators in place of their numerical frequency limits. Several such band identifier systems are in use. Probably one of the oldest alphabetical systems is applied primarily to microwave and radar frequencies within the U.S. The military alphabetical designators use some of the same letters for different frequency ranges, so be certain to determine which of the two alphabetical designator systems is being used.

Microwave and Radar Bands

Alphabetical Band Designator	Frequency Range (MHz)
I	100–150
G	150–225
P	225–390
L	390–1550
S*	1,550–3,900
C	3,900–6,200
X*	6,200–10,900
K	10,900–36,000
K_u	15,250–17,250
K_a	33,000–36,000
Q	36,000–46,000
V	46,000–56,000
W	56,000–100,000

*The S band is sometimes extended from 1,550 to 5,200 MHz and the X band is sometimes extended from 5,200 to 10,900 MHz.

TABLE 8-8 Band Designators (Continued)

Military Band Designators

Alphabetical Band Designator	Frequency Range (MHz)
A	0–250
B	250–500
C	500–1,000
E	1,000–2,000
E	2,000–3,000
F	3,000–4,000
G	4,000–6,000
H	6,000–8,000
I	8,000–10,000
J	10,000–20,000
K	20,000–40,000
L	40,000–60,000
M	60,000–100,000

TABLE 8-9 Broadcast Bands Below 2 MHz

Frequency, kHz	Primary use	Shared use
148.5 – 283.5	European broadcast	255 – 283.5 kHz, radio navigation and mobile service
526.5 – 1606.5	European, Asian, Near Eastern, USSR, African, and Pacific broadcast	526.5 – 535 kHz, mobile service
525 - 1605	North American, Central American, and South American broadcast	525 – 535 kHz, aeronautical navigation

TABLE 8-10 Frequency Bands, 2.3–3.0 MHz

Frequency, kHz	Tropical zone* broadcast	Fixed service	Mobile service	Time/ frequency standard	Amateur	Broadcast†
2,300–2,498	X	X	X		X	
2,498–2,502				X		
2,502–3,200		X	X			
3,200–3,400	X	X	X			
3,400–3,900		X	X		X	
3,900–4,000		X	X			1,3
4,000–4,750		X	X			
4,750–5,060	X	X	X			
5,060–5,950		X	X			
5,950–6,200						X
6,200–7,000		X	X			
7,000–7,100					X	
7,100–7,300					X	1,3
7,300–9,500		X	X			
9,500–9,750						X
9,750–9,900		X				X
9,900–11,650		X	X	X		
11,650–11,700		X				X
11,700–11,975						X
11,975–12,050		X				X
12,050–13,600		X	X			
13,600–13,800		X				X
13,800–15,100		X	X	X	X	
15,100–15,450						X
15,450–15,600		X				X
15,600–17,550		X	X			
17,550–17,700		X				X
17,700–17,900						X
17,900–21,450		X	X	X	X	
21,450–21,750						X
21,750–21,850		X				X
21,850–25,670		X	X	X	X	
25,670–26,100						X
26,100–30,000		X	X	X		X

* Area between 30°N and 30°S latitude.

† Regions indicated by numbers: (1) Europe, Africa, USSR, Turkey, Arabia, (2) the Americas, (3) other parts of the world.

TABLE 8-11 Frequency Bands above 30 MHz

Region 1*	Region 2†	Region 3‡
47–68 MHz, TV	54–68 MHz, fixed and mobile service	47–50, 54–68 MHz, fixed and mobile service
87.5–108 MHz, FM	88–108 MHz, FM	87–108 MHz, fixed and mobile service
174–230 MHz, TV (223–230 shared with fixed and mobile service)	174–216 MHz, fixed and mobile service	174–230 MHz, fixed and mobile service
470–790 MHz TV, 790–960 MHz TV (shared with fixed and mobile service); 862–960 MHz for Africa only; 620–790 MHz satellite TV	470–512 MHz, 614–806 MHz, fixed and mobile service, 512–608 MHz TV; 620–790 MHz satellite TV	470–960 MHz, TV (shared with fixed and mobile service), 620–790 MHz satellite TV
2500–2690 MHz broadcast satellite	2500–2690 MHz broadcast satellite	2500–2690 MHz broadcast satellite
11.7–12.5 GHz TV (shared with fixed services)	12.1–12.7 GHz TV (shared with fixed and mobile services)	11.7–12.75 GHz TV (shared with fixed and mobile services), 12.2–12.5 GHz terrestrial broadcast only

TABLE 8-11 Frequency Bands above 30 MHz (Continued)

Region 1*	Region 2†	Region 3‡
	22.5–23 GHz TV (shared with fixed and mobile satellite)	22.5–23 GHz TV (shared with fixed and mobile satellite)
84–86 GHz broadcast (shared with fixed and mobile services)	84–86 GHz broadcast (shared with fixed and mobile services)	84–86 GHz broadcast (shared with fixed and mobile services)

* Europe, Africa, USSR, Turkey, Arabia
† The Americas
‡ Other parts of the world

TABLE 8-12 High-Frequency Bands

Frequency, kHz	MHz band	Meter band
2,300–2,495	2	120
3,200–3,400	3	90
3,900–4,000	4	75
4,750–5,060	5	60
5,950–6,200	6	49
7,100–7,300	7	41
9,500–9,775	9	31
11,700–11,975	11	25
15,100–15,450	15	19
17,700–17,900	17	16
21,450–21,750	21	13
25,600–26,100	26	11

8-13 TV STANDARDS

Standard U.S. TV channel frequencies are listed in Table 8-13. World television standards are described in Table 8-14. Picture-line amplifier standard output signals and sync waveforms for RS-170 monochrome TV are shown in Figs. 8-25 and 8-26. Figure 8-27 shows the composite video waveform for RS-343 TV.

TABLE 8-13 TV Channel Frequency Assignments

Channel	Limits, MHz	Picture carrier, MHz	Sound carrier, MHz
2	54–60	55.25	59.75
3	60–66	61.25	65.75
4	66–72	67.25	71.75
5	76–82	77.25	81.75
6	82–88	83.25	87.75
7	174–180	175.25	179.75
8	180–186	181.25	185.75
9	186–192	187.25	191.75
10	192–198	193.25	197.75
11	198–204	199.25	203.75
12	204–210	205.25	209.75
13	210–216	211.25	215.75
14	470–476	471.25	475.75
15	476–482	477.25	481.75
16	482–488	483.25	487.75
17	488–494	489.25	493.75
18	494–500	495.25	499.75
19	500–506	501.25	505.75
20	506–512	507.25	511.75
21	512–518	513.25	517.75
22	518–524	519.25	523.75
23	524–530	525.25	529.75
24	530–536	531.25	535.75
25	536–542	537.25	541.75
26	542–548	543.25	547.75

TABLE 8-13 TV Channel Frequency Assignments *(Continued)*

Channel	Limits, MHz	Picture carrier, MHz	Sound carrier, MHz
27	548–554	549.25	553.75
28	554–560	555.25	559.75
29	560–566	561.25	565.75
30	566–572	567.25	571.75
31	572–578	573.25	577.75
32	578–584	579.25	583.75
33	584–590	585.25	589.75
34	590–596	591.25	595.75
35	596–602	597.25	601.75
36	602–608	603.25	607.75
37	608–614	609.25	613.75
38	614–620	615.25	619.75
39	620–626	621.25	625.75
40	626–632	627.25	631.75
41	632–638	633.25	637.75
42	638–644	639.25	643.75
43	644–650	645.25	649.75
44	650–656	651.25	655.75
45	656–662	657.25	661.75
46	662–668	663.25	667.75
47	668–674	669.25	673.75
48	674–680	675.25	679.75
49	680–686	681.25	685.75
50	686–692	687.25	691.75
51	692–698	693.25	697.75
52	698–704	699.25	703.75
53	704–710	705.25	709.75
54	710–716	711.25	715.75
55	716–722	717.25	721.75
56	722–728	723.25	727.75
57	728–734	729.25	733.75
58	734–740	735.25	739.75
59	740–746	741.25	745.75
60	746–752	747.25	751.75
61	752–758	753.25	757.75
62	758–764	759.25	763.75

TABLE 8-13 TV Channel Frequency Assignments *(Continued)*

Channel	Limits, MHz	Picture carrier, MHz	Sound carrier, MHz
63	764–770	765.25	769.75
64	770–776	771.25	775.75
65	776–782	777.25	781.75
66	782–788	783.25	787.75
67	788–794	789.25	793.75
68	794–800	795.25	799.75
69	800–806	801.25	805.75
70	806–812	807.25	811.75
71	812–818	813.25	817.75
72	818–824	819.25	823.75
73	824–830	825.25	829.75
74	830–836	831.25	835.75
75	836–842	837.25	841.75
76	824–848	843.25	847.75
77	848–854	849.25	853.75
78	854–860	855.25	859.75
79	860–866	861.25	865.75
80	866–872	867.25	871.75
81	872–878	873.25	877.75
82	878–884	879.25	883.75
83	884–890	885.25	889.75

TABLE 8-14 World TV Standards

Number of lines	Channel width, MHz	Video bandwidth, MHz	Video/audio separation, MHz	Vestigial sideband, MHz	Video modulation	Audio modulation	Country
405	5	3	−3.5	0.75	Pos	AM	Ireland, United Kingdom
625	7	5	+5.5	0.75	Neg	FM	Continental Europe, Italy, Morocco, Australia, New Zealand
625	8	6	+6.5	0.75	Neg	FM	USSR, China
819	14	10	+11.15	2	Pos	AM	France, Monaco
625	8	5.5	+6	1.25	Neg	FM	Ireland, South Africa
625	8	6	+6.5	1.25	Neg	FM	French overseas territories
525	6	4.2	+4.5	0.75	Neg	AM	USA, Japan

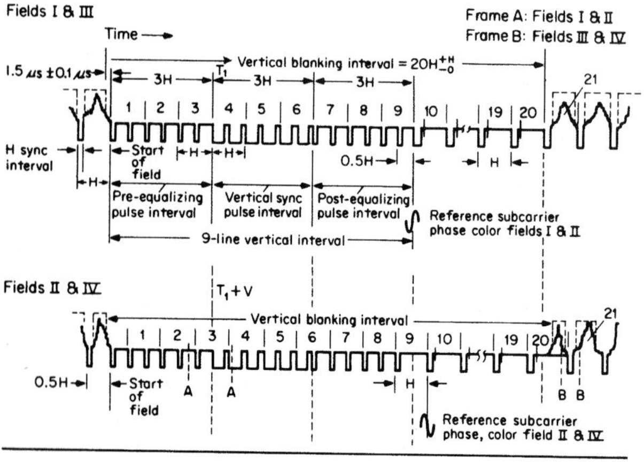

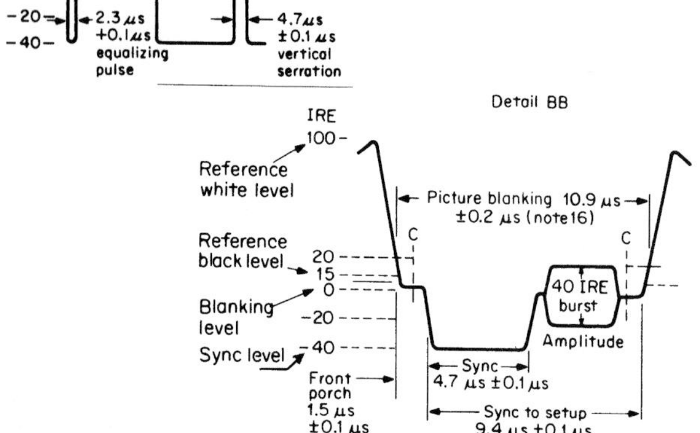
Fields I & III
Time
Frame A: Fields I & II
Frame B: Fields III & IV
1.5 μs ±0.1 μs
Vertical blanking interval = 20H +H −0
T1
3H
3H
3H
21
1 2 3 4 5 6 7 8 9 10 19 20
H sync interval
Start of field
H
H
H
0.5H
H
Pre-equalizing pulse interval
Vertical sync pulse interval
Post-equalizing pulse interval
Reference subcarrier phase color fields I & II
9-line vertical interval
Fields II & IV
T1+V
Vertical blanking interval
21
1 2 3 4 5 6 7 8 9 10 19 20
0.5H
Start of field
A
A
H
B B
Reference subcarrier phase, color field II & IV
Detail AA 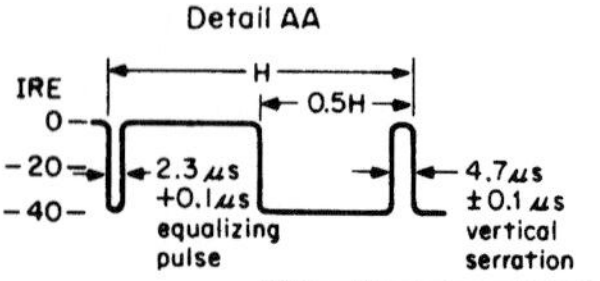
IRE
0
−20
−40
H
0.5H
2.3 μs +0.1 μs equalizing pulse
4.7 μs ±0.1 μs vertical serration
Detail BB
IRE
100
Reference white level
Picture blanking 10.9 μs ±0.2 μs (note 16)
C
C
Reference black level
20
15
0
Blanking level
−20
−40
Sync level
40 IRE burst
Amplitude
Front porch 1.5 μs ±0.1 μs
Sync 4.7 μs ±0.1 μs
Sync to setup 9.4 μs ±0.1 μs

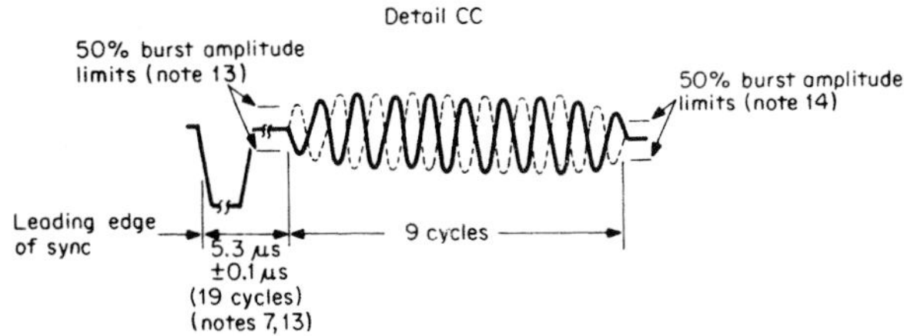

FIG. 8-25 RS-170 composite display waveforms: (1) tolerances and limits are for long time variations; (2) applicable to local facilities—common-carrier transmitter characteristics are not included; (3) burst frequency is 3.579545 MHz ± 10 Hz; (4) horizontal scanning frequency is 2/455 times the burst frequency, scan period (H) is 63.556 μs; (5) vertical scanning frequency is 2/525 times horizontal scanning frequency, scan period (V) is 16.683 μs; (6) fields I and III begin with a whole line between the first equalizing pulse and the proceeding H sync pulse, fields II and IV begin with a half line between the first equalizing pulse and the proceeding H pulse, color field I begins with the positive-going zero-crossing of the reference subcarrier most coincident with the 50-percent amplitude point of the leading edges of even-numbered horizontal sync pulses; (7) the zero-crossing of the reference subcarrier will nominally be coincident with the 50 percent point of the leading edges of all horizontal sync pulses; (8) unspecified rise and fall times are 0.14 μs ± 0.02 μs measured from the 10 and 90 percent points—all pulse widths are measured from the 50 percent amplitude points; (9) tolerances on sync level, reference black level; and peak-to-peak burst amplitude are ± 2 IRE units; (10) interval from line 17 through line 20 can be used for test, cue, and control signals; (11) extraneous synchronous signals during blanking intervals, including subcarrier, cannot exceed 1 IRE unit, extraneous nonsynchronous signals cannot exceed 0.5 IRE units, special signals added to vertical blanking interval are excepted, and overshoot on all pulses cannot exceed 2 IRE units; (12) the burst envelope rise time is 0.3 μs + 0.2 μs, − 0.1 μs; (13) the burst starts with the zero-crossing that precedes the first half cycle of the subcarrier that is 50 percent or more of the burst amplitude; (14) the burst ends with the zero-crossing that follows the last half cycle of the subcarrier that is 50 percent or more of the burst amplitude; (15) for monochrome signals the burst is omitted—fields I and III and II and IV are identical; and (16) occasionally picture blanking at 20 IRE units is impossible because of image content on a monitor. *(RS-343A, Electrical Performance Standard for High-Resolution Monochrome Closed-Circuit Television Camera, Copyright Registration A179358, September 1970, Electronic Industries Association, 2001 Eye St. NW, Washington, DC 20006 (202) 457-4900)*

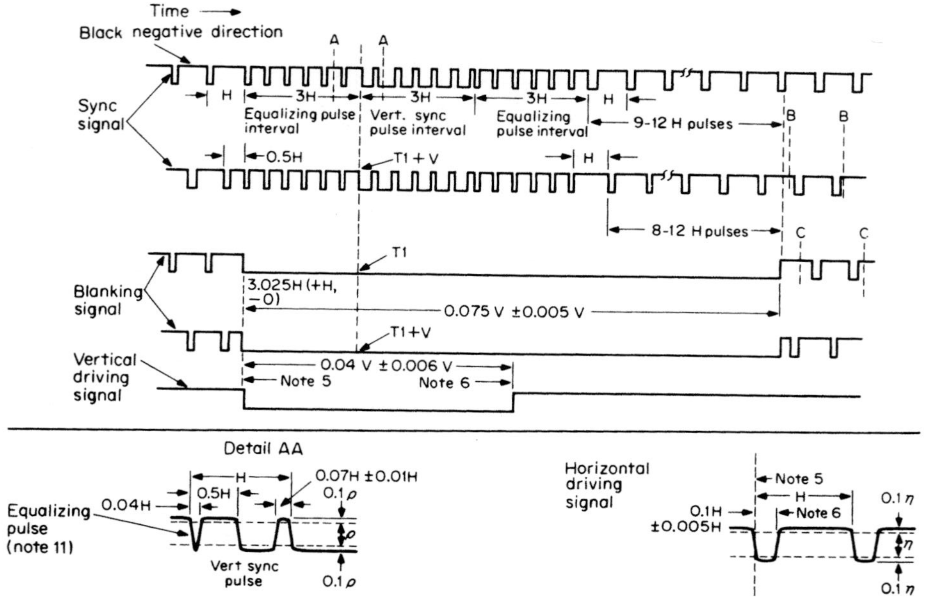

Time
Black negative direction
A
A
H
3H
3H
3H
H
Equalizing pulse interval
Vert. sync pulse interval
Equalizing pulse interval
9-12 H pulses
B
B
Sync signal
0.5H
T1 + V
H
8-12 H pulses
C
C
T1
Blanking signal
3.025H (+H, −0)
0.075 V ±0.005 V
T1+V
Vertical driving signal
0.04 V ±0.006 V
Note 5
Note 6
Detail AA
H
0.07H ±0.01H
0.5H
0.1 ρ
Equalizing pulse (note 11)
0.04H
ρ
Vert sync pulse
0.1 ρ
Horizontal driving signal
Note 5
H
0.1 η
0.1H ±0.005H
Note 6
η
0.1 η

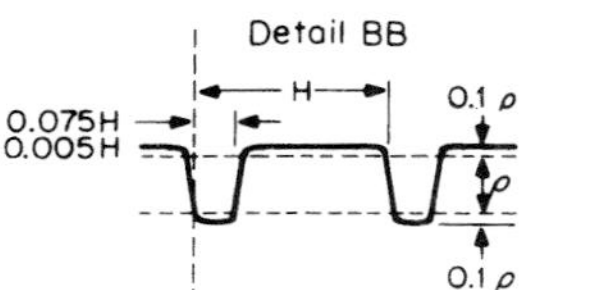

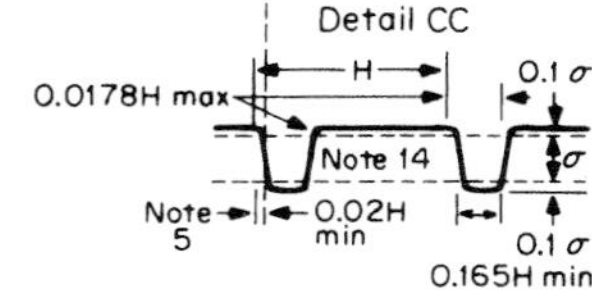

FIG. 8-26 RS-170 sync generator waveforms: (1) H is the time from start of one line to the next; (2) V is the time from start to one field to the next (3) leading and trailing edges of vertical driving and blanking signals shall complete in $0.1H$; (4) tolerances and limits are for long time variations; (5) signal amplitude shall be adjustable over 3.5 to $4.5V$ with load impedance of 75 Ω ± 5 percent; (6) vertical driving pulse durtaion is $0.04V \pm 0.006V$, horizontal driving pulse duration is $0.1H \pm 0.005H$; (7) time relationships and waveform of the blanking and sync signals must result in a standard RETMA signal; (8) standard RETMA values of frequency and rate of change of frequency for horizontal components of the output signal of the recommended sync generator; (9) rise and decay times are measured between the 10 and 90 percent amplitudes; (10) time of occurrence of the leading edge of any horizontal pulse N of any group of 20 horizontal pulses appearing on any output signal of a standard sync generator will not differ from NH by more than $0.0008H$ where H is the average interval between the leading edges of the pulses as determined by an averaging process carried out over 20 to 100 lines; (11) equalizing pulses are 0.45 to 0.5 of the area of a horizontal sync pulse; (12) overshoot of any pulse shall not exceed 5 percent; (13) the output level of the blanking signal and sync signal will not vary by more than ± 3 percent under conditions where (*a*) the ac voltage supplied to the sync generator is within 110 to 120 V and does not vary by more than 5 percent during the test, (*b*) a 5-h period of continuous operation is adequate for the measurement (after suitable warmup), (*c*) ambient temperature is 20 to 40°C and does not change by more than 10°C during the test; (14) adjustments between minimum and maximum limits will be possible so the aspect ratio can be set to the normal value. [*RS-343A, Electrical Performance Standard for High-Resolution Monochrome Closed-Circuit Television Camera, Copyright Registration A179358, September 1970, Electronic Industries Association, 2001 Eye St. NW, Washington, DC 20006 (202) 457–4900.*]

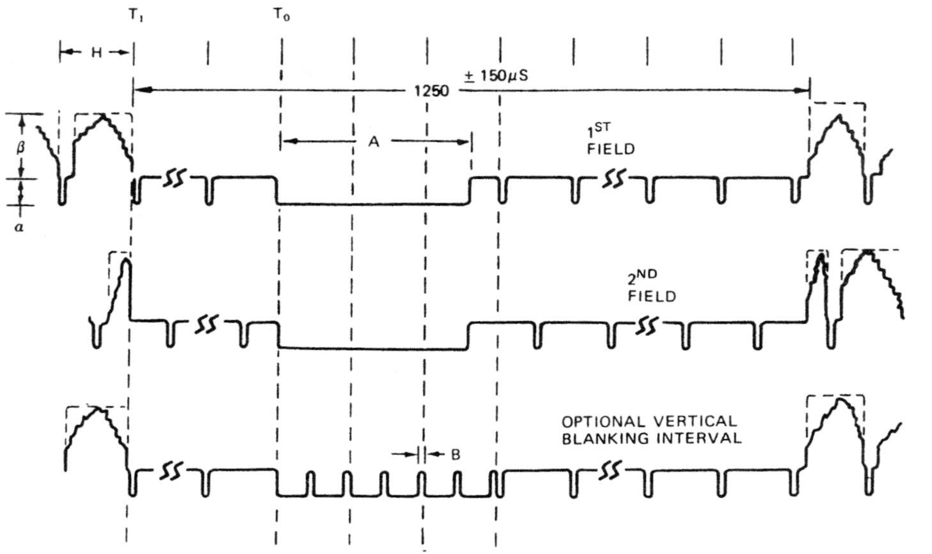
T1
T0
H
1250 ± 150μS
A
1ST FIELD
β
α
2ND FIELD
OPTIONAL VERTICAL BLANKING INTERVAL
B

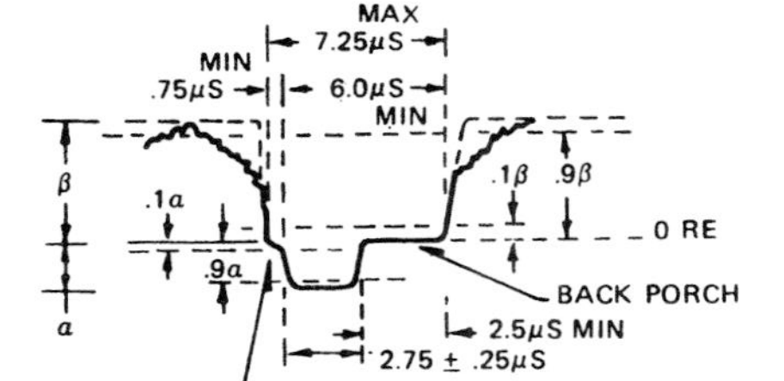

NOTES:

1. $\beta = 0.714 \pm 0.1$ volts (100 IRE Units).
2. $\alpha = 0.286$ (40 IRE Units) nominal.
3. Sync to total signal ratio $(\frac{\alpha}{\beta + \alpha}) = 28.6 \pm 5\%$.
4. Blanking = 7.5 ± 5 IRE Units (2.5% to 12.5% of β).
5. Horizontal Rise Times measured from 10% to 90% amplitudes shall be less than 0.1 μ/s.
6. Overshoot on horizontal blanking signal shall not exceed 0.02β at beginning of front porch and 0.05β at end of back porch.
7. Overshoot on sync signal shall not exceed 0.05β.
8. T_0 = start of vertical sync pulse.
9. T_1 = start of vertical blanking.
10. $T_1 = T_0 {}^{+0}_{-250}$ μ/s
11. A - vertical sync pulse = 125 ± 50 μ/s measured between 90% amplitude points.
12. Rise and fall times of vertical blanking and vertical sync pulse, measured from 10% to 90% amplitudes, shall be less than 5 μ/s.
13. Tilt on vertical sync pulse shall be less than 0.1α.
14. If horizontal information is provided during the vertical sync pulse it must be at 2H frequency and as shown in the optional vertical blanking interval waveform.
15. B - vertical serration = 2 ± .5 μ/s measured between the 90% amplitude points. Rise time measured from 10% to 90% amplitudes shall be less than 0.1 μ/s.
16. If equalizing pulses are used in the vertical blanking interval waveforms they shall be 6 in number preceding and following the vertifcal sync pulse, be at 2H frequency and 1/2 the width of H sync pulse.
17. It is recommended that for proper interlace the time duration between the leading edge of vertical sync and the leading edge of horizontal sync be a multiple of H/2.

FIG. 8-27 Composite video waveform high-resolution monochrome television camera. [*RS-343A, Electrical Performance Standard for High-Resolution Monochrome Closed-Circuit Television Camera, Copyright Registration A179358, September 1970, Electronic Industries Association, 2001 Eye St. NW, Washington, DC 20006 (202) 457–4900.*]

8-14 AUDIO SPECTRUM

The audio spectrum is normally considered to extend from 15 Hz to 20 kHz, as shown in Fig. 8-28. The frequencies of an 88-note piano are listed in Table 8-15.

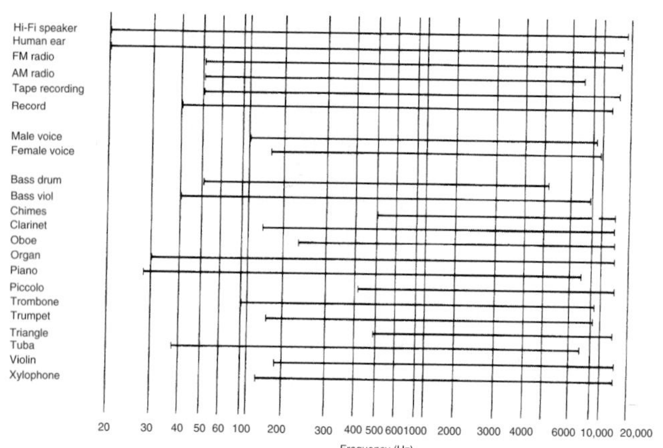

FIG. 8-28 Audio spectrum.

TABLE 8-15 Piano Keyboard Scale

Note	Frequency, Hz							
A	27.5	55.0	110.0	220.0	440.0	880.0	1760	3520
A#	29.1	58.3	116.5	233.1	466.2	932.3	1865	3729
B	30.9	61.7	123.5	246.9	493.9	987.8	1976	3951
C	32.7	65.4	130.8	261.6	523.3	1047	2093	4186
C#	34.6	69.3	138.6	277.2	554.4	1109	2217	
D	36.7	73.4	146.8	293.7	587.3	1175	2349	
D#	38.9	77.8	155.6	311.1	622.3	1245	2489	
E	41.2	82.4	164.8	329.6	659.3	1319	2637	
F	43.7	87.3	174.6	349.2	698.5	1397	2794	
F#	46.2	92.5	185.0	370.0	740.0	1480	2960	
G	49.0	98.0	196.0	392.0	784.0	1568	3136	
G#	51.9	103.8	207.7	415.3	830.6	1661	3322	

8-15 CODES

Codes are frequently used in communications. The most common codes are listed in Tables 8-16 through 8-18.

TABLE 8-16 International Morse Code

Character	Code	Character	Code
A	•—	U	••—
B	—•••	V	•••—
C	—•—•	W	•——
D	—••	X	—••—
E	•	Y	—•——
F	••—•	Z	——••
G	——•	1	•————
H	••••	2	••———
I	••	3	•••——
J	•———	4	••••—
K	—•—	5	•••••
L	•—••	6	—••••
M	——	7	——•••
N	—•	8	———••
O	———	9	————•
P	•——•	0	—————
Q	——•—	Period	•—•—•—
R	•—•	Comma	——••——
S	•••	Question Mark	••——••
T	—		

TABLE 8-17 American Standard Code for Information Interchange (ASCII)

			Column →	Ø	1	2	3	4	5	6	7
				128-Symbol printing set							
b7				0	0	0	0	1	1	1	1
	b6			0	0	1	1	0	0	1	1
		b5		0	1	0	1	0	1	0	1
b4	b3	b2	b1	Nonprinting		96-Symbol printing subset					
0	0	0	0	NUL	DLE	SP	Ø	@	P	`	p
0	0	0	1	SOH	DC1	!	1	A	Q	a	q
0	0	1	0	STX	DC2	"	2	B	R	b	r
0	0	1	1	ETX	DC3	#	3	C	S	c	s
0	1	0	0	EOT	DC4	$	4	D	T	d	t
0	1	0	1	ENQ	NAK	%	5	E	U	e	u
0	1	1	0	ACK	SYN	&	6	F	V	f	v
0	1	1	1	BEL	ETB	'	7	G	W	g	w
1	0	0	0	BS	CAN	(	8	H	X	h	x
1	0	0	1	HT	EM	)	9	I	Y	i	y
1	0	1	0	LF	SUB	*	:	J	Z	j	z
1	0	1	1	VT	ESC	+	;	K	[	k	{
1	1	0	0	FF	FS	,	<	L	\	l	|
1	1	0	1	CR	GS	-	=	M	]	m	}
1	1	1	0	SO	RS	.	>	N	^	n	~
1	1	1	1	SI	US	/	?	O	_	o	DEL

TABLE 8-18 Extended Binary-Coded Decimal Interchange Code (EBCDIC)

0	F0	c	83	o	96	A	C1	M	D4	Y	E8	)	5D		70
1	F1	d	84	p	97	B	C2	N	D5	Z	E9	;	5E	=	7E
2	F2	e	85	q	98	C	C3	O	D6	¢	4A	–	60	"	7F
3	F3	f	86	r	99	D	C4	P	D7	.	4B	/	61	b̸ *	40
4	F4	g	87	s	A2	E	C5	Q	D8	<	4C	,	6B	LF†	25
5	F5	h	88	t	A3	F	C6	R	D9	(	4D	%	6C		
6	F6	i	89	u	A4	G	C7	S	E2	+	4E	—	6D		
7	F7	j	91	v	A5	H	C8	T	E3	|	4F	>	6E		
8	F8	k	92	w	A6	I	C9	U	E4	&	50	?	6F		
9	F9	l	93	x	A7	J	D1	V	E5	!	5A	:	7A		
a	81	m	94	y	A8	K	D2	W	E6	$	5B	#	7B		
b	82	n	95	z	A9	L	D3	X	E7	*	5C	@	7C		

b̸ * = blank space.
† LF = line feed.

8-16 ERROR DETECTION AND CORRECTION

Parity Codes

Parity codes detect single-bit errors.

Odd parity: sum of bits = odd number

Even parity: sum of bits = even number

EXAMPLE:

0110 1101 sum of bits = 5, odd parity

0110 0101 sum of bits = 4, even parity

Longitudinal Redundancy Check

Longitudinal redundancy checks detect double-bit errors. A common method uses two-dimensional parity checking of each data block. The vertical odd-parity word follows its respective data block when the data is transmitted.

EXAMPLE:

horizontal odd-parity bit
↓

0110 1000	
1001 0100	data block
0110 1110	
1111 0111	
1001 1010	vertical odd-parity word for each column

Hamming Codes

Hamming codes detect double-bit errors and correct single-bit errors. The message is divided into blocks of n bits.

$$n = 2^c - 1$$

where c = number of check bits

Each check bit provides parity over $(n + 1)/2$ bits.

8-17 TOUCH-TONE TELEPHONE CODING

A touch-tone keypad comprises three columns by four rows of the digits 1 through 9 and the special characters * and #. The keys are encoded using two tones at different frequencies. Table 8-19 shows the two-tone codes for each key.

TABLE 8-19 Touch-tone Frequencies

Frequency, Hz	1209	1336	1477
697	1	2	3
770	4	5	6
852	7	8	9
941	*	0	#

8-18 NATIONAL OCEANIC AND ATMOSPHERIC ADMINISTRATION WEATHER FREQUENCIES

The following three frequencies are used by NOAA to broadcast National Weather Service information:

162.40 MHz
162.475 MHz
162.55 MHz

9

Digital Circuits

9-1 BOOLEAN ALGEBRA

The fundamental Boolean functions are the NOT, AND, and OR operations shown in Figs. 9-1 through 9-5. In each case, the Venn diagram truth table, equivalent switching circuit, and logic diagram are shown. The derived NAND, NOR, and EXCLUSIVE OR functions are similarly illustrated in Figs. 9-6 through 9-8.

The fundamental Boolean algebra laws and axioms are listed and shown in Figs. 9-9 through 9-18. See also Section 3-10.

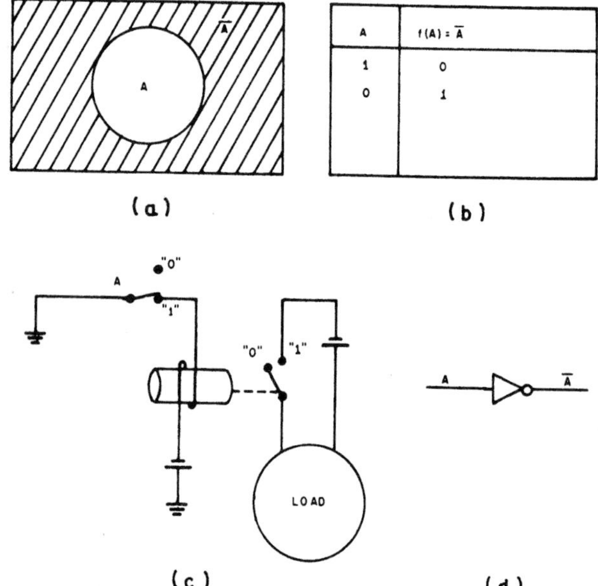

FIG. 9-1 NOT operation: *(a)* Venn diagram, *(b)* truth table, *(c)* NOT switching circuit, *(d)* logic diagram—mechanization of $f(A) = \overline{A}$. *(Bureau of Naval Personnel)*

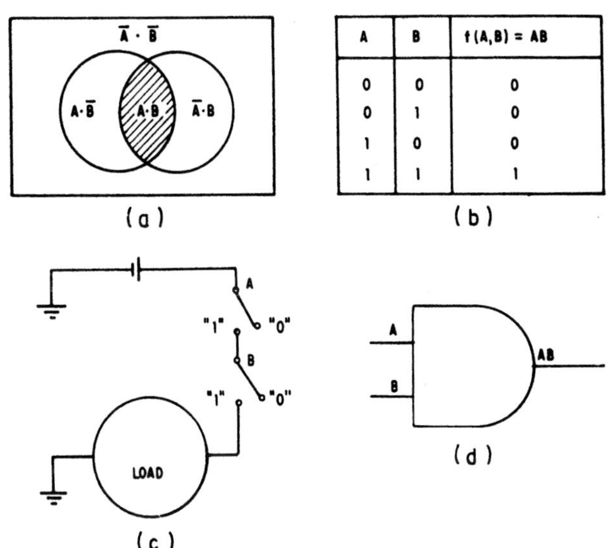

A	B	f(A,B) = AB
0	0	0
0	1	0
1	0	0
1	1	1

FIG. 9-2 AND operation: *(a)* Venn diagram, *(b)* truth table, *(c)* AND switching circuit, *(d)* logic diagram—mechanization of $f(A,B) = AB$. *(Bureau of Naval Personnel)*

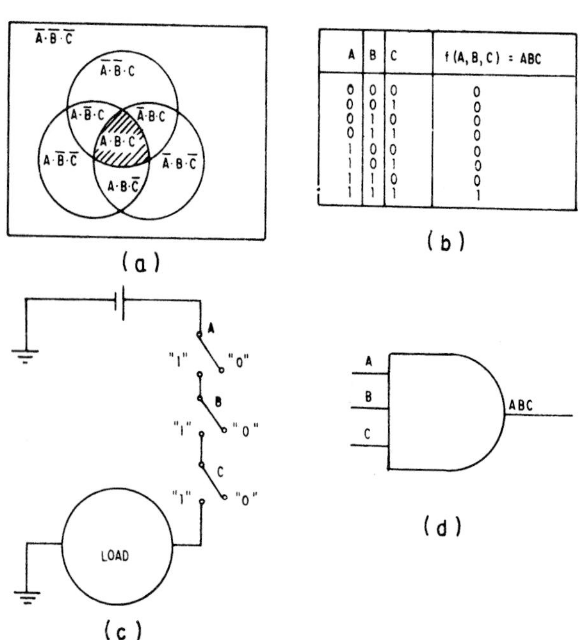

A	B	C	f (A, B, C) = ABC
0	0	0	0
0	0	1	0
0	1	0	0
0	1	1	0
1	0	0	0
1	0	1	0
1	1	0	0
1	1	1	1

FIG. 9-3 Three-input AND operation: *(a)* Venn diagram, *(b)* truth table, *(c)* AND switching circuit, and *(d)* logic diagram—mechanization of $f(A,B,C) = ABC$. *(Bureau of Naval Personnel)*

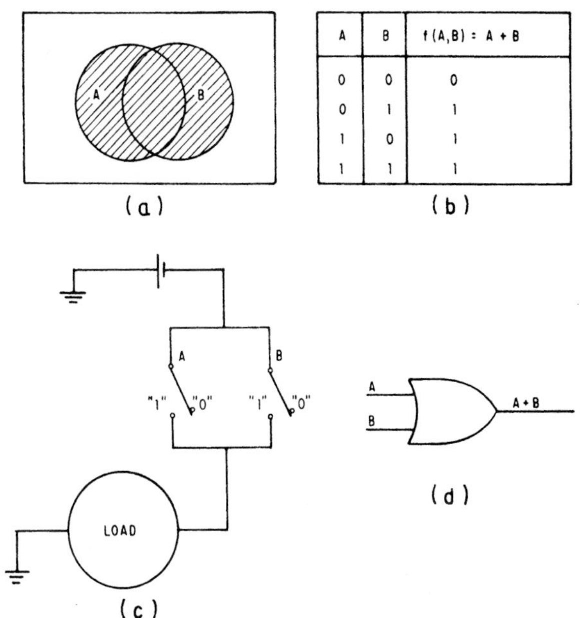

A	B	f(A,B) = A + B
0	0	0
0	1	1
1	0	1
1	1	1

FIG. 9-4 OR operation: *(a)* Venn diagram, *(b)* truth table, *(c)* OR switching circuit, and *(d)* logic diagram—mechanization of $f(A,B) = A + B$. *(Bureau of Naval Personnel)*

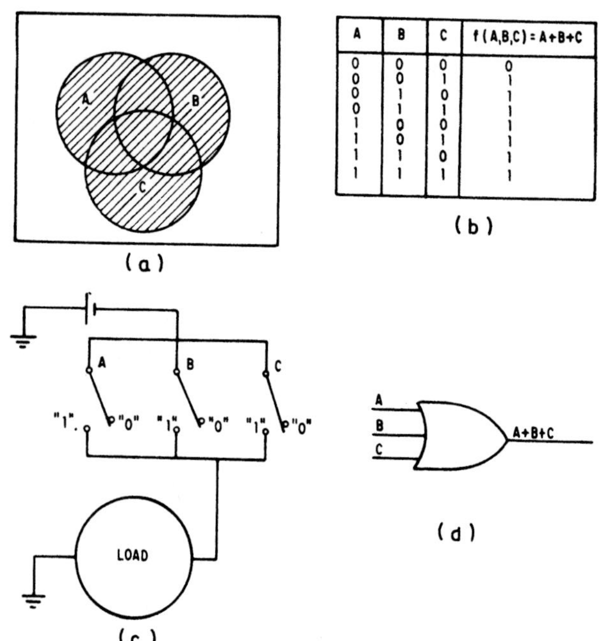

A	B	C	f(A,B,C) = A+B+C
0	0	0	0
0	0	1	1
0	1	0	1
0	1	1	1
1	0	0	1
1	0	1	1
1	1	0	1
1	1	1	1

FIG. 9-5 Three-input OR operation: *(a)* Venn diagram, *(b)* truth table, *(c)* OR switching circuit, and *(d)* logic diagram—mechanization of $f(A,B,C) = A+B+C$. *(Bureau of Naval Personnel)*

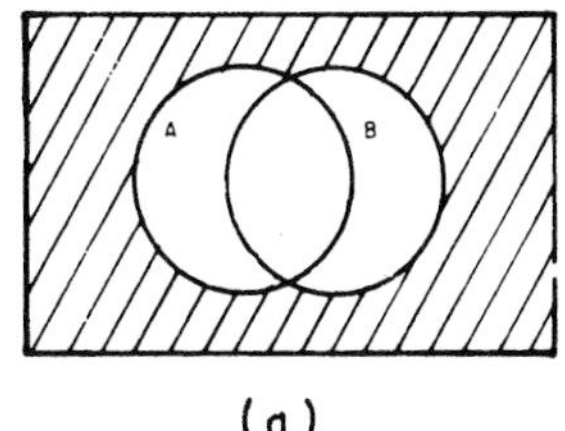

A	B	AB	$f(A,B) = \overline{AB}$
0	0	0	1
0	1	0	1
1	0	0	1
1	1	1	0

(b)

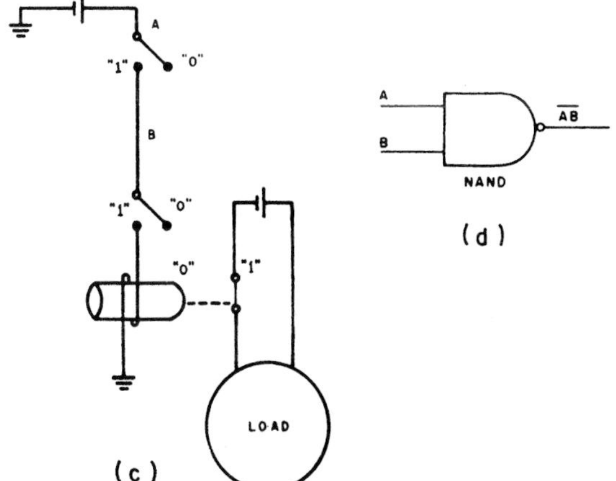

FIG. 9-6 NAND operation: *(a)* Venn diagram, *(b)* truth table, *(c)* NAND switching circuit, and (D) logic diagram—mechanization of $f(A,B) = \overline{AB}$. *(Bureau of Naval Personnel)*

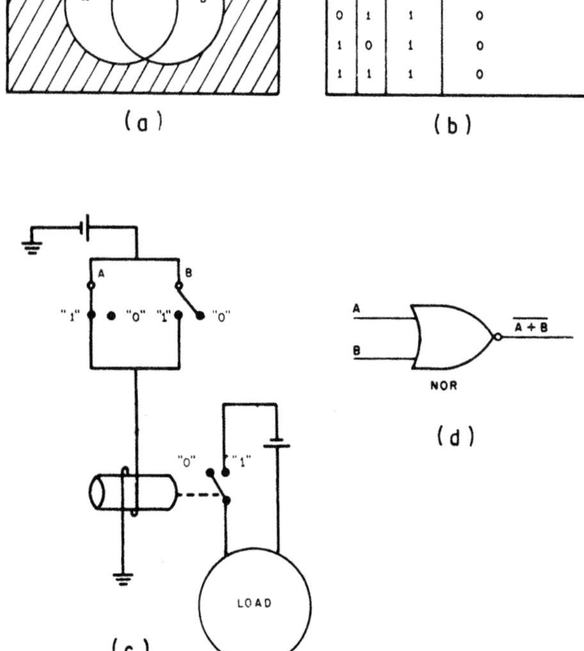

A	B	A + B	$f(A,B) = \overline{A+B}$
0	0	0	1
0	1	1	0
1	0	1	0
1	1	1	0

FIG. 9-7 NOR operation: *(a)* Venn diagram, *(b)* truth table, *(c)* NOR switching circuit, and *(d)* logic diagram—mechanization of $f(A,B) = \overline{A + B}$. *(Bureau of Naval Personnel)*

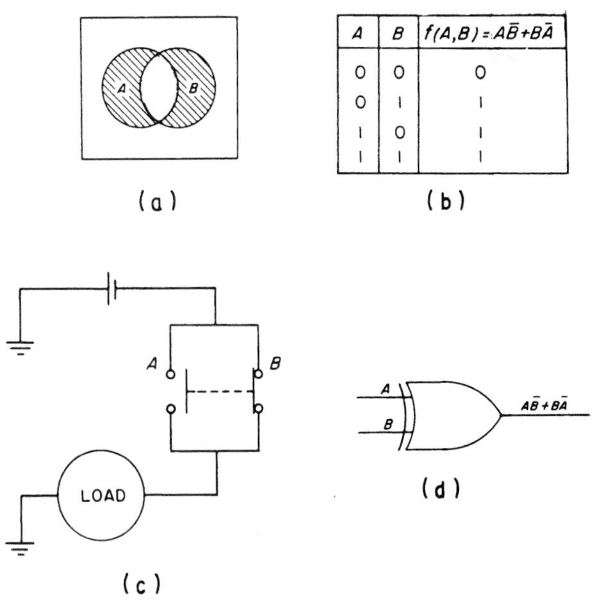

A	B	$f(A,B) = A\bar{B} + B\bar{A}$
0	0	0
0	1	1
1	0	1
1	1	1

FIG. 9-8 EXCLUSIVE OR operation: *(a)* Venn diagram, *(b)* truth table, *(c)* EXCLUSIVE OR switching circuit, and *(d)* logic diagram—EXCLUSIVE OR function. *(Bureau of Naval Personnel)*

Identity Law $$A = A$$

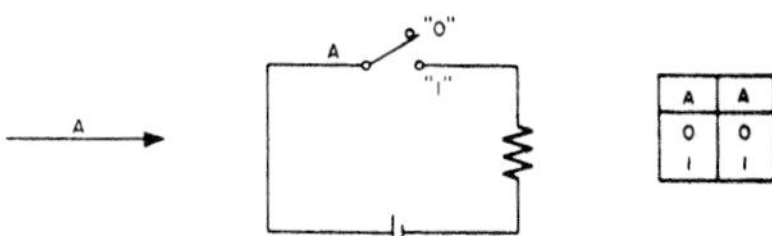

FIG. 9-9 Identity law. *(Bureau of Naval Personnel)*

Complementary Law $$A\overline{A} = 0$$

$$A + \overline{A} = 1$$

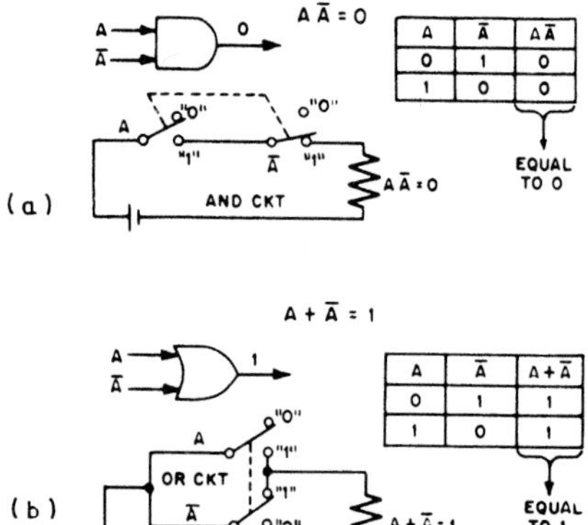

FIG. 9-10 Complementary law. *(Bureau of Naval Personnel)*

Idempotent Law $AA = A$

$$A + A = A$$

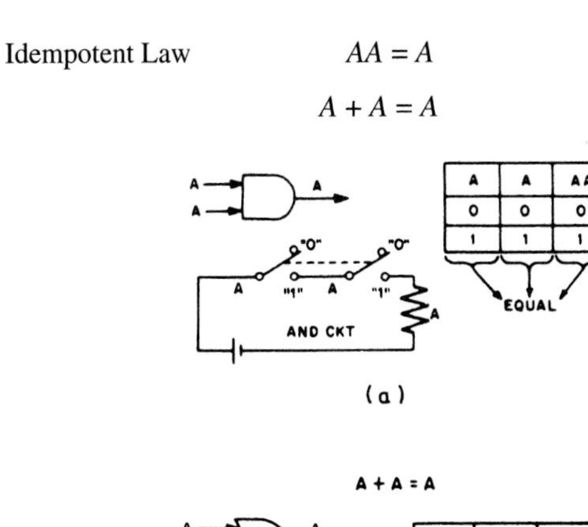

A	A	AA
0	0	0
1	1	1

A	A	A+A
0	0	0
1	1	1

FIG. 9-11 Idempotent law. *(Bureau of Naval Personnel)*

Commutative Law $$AB = BA$$

$$A + B = B + A$$

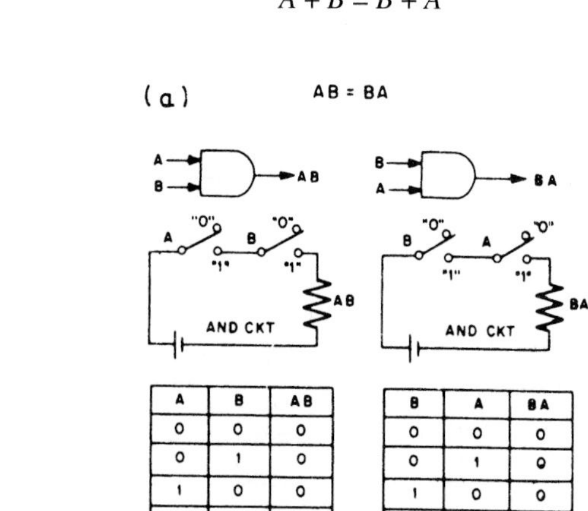

A	B	AB
0	0	0
0	1	0
1	0	0
1	1	1

B	A	BA
0	0	0
0	1	0
1	0	0
1	1	1

FIG. 9-12 Commutative law. *(Bureau of Naval Personnel)*

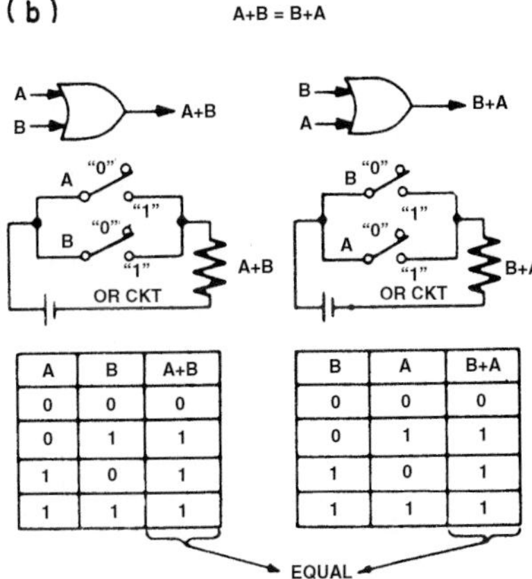

A	B	A+B
0	0	0
0	1	1
1	0	1
1	1	1

B	A	B+A
0	0	0
0	1	1
1	0	1
1	1	1

FIG. 9-12 Commutative law. *(Bureau of Naval Personnel) (Continued)*

Associative Law

$$(AB)C = A(BC) = ABC$$

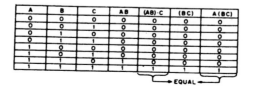

$$(A + B) + C = A + (B + C) = A + B + C$$

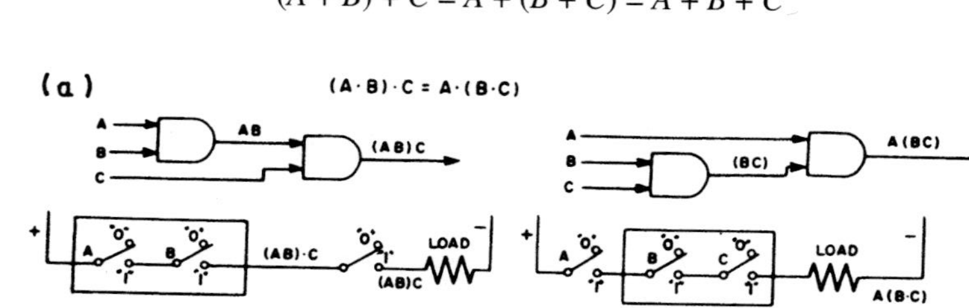

A	B	C	AB	(AB)·C	(BC)	A(BC)
0	0	0	0	0	0	0
0	0	1	0	0	0	0
0	1	0	0	0	0	0
0	1	1	0	0	0	0
1	0	0	0	0	0	0
1	0	1	0	0	0	0
1	1	0	1	0	0	0
1	1	1	1	1	1	1

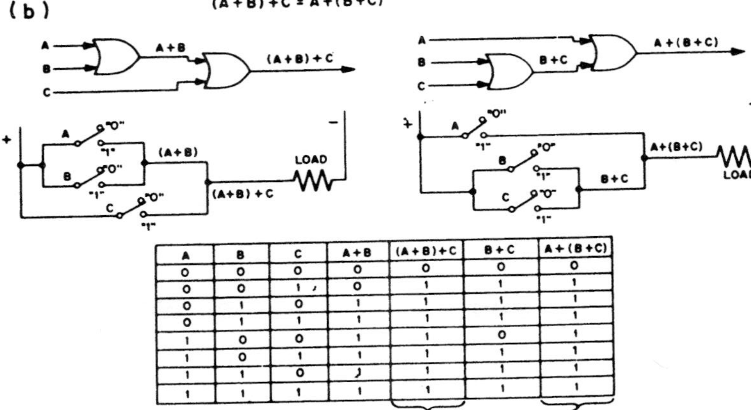

A	B	C	A + B	(A + B) + C	B + C	A + (B + C)
0	0	0	0	0	0	0
0	0	1	0	1	1	1
0	1	0	1	1	1	1
0	1	1	1	1	1	1
1	0	0	1	1	0	1
1	0	1	1	1	1	1
1	1	0	1	1	1	1
1	1	1	1	1	1	1

FIG. 9-13 Associative law. *(Bureau of Naval Personnel)*

Distributive Law

$$A(B + C) = AB + AC$$

$$A + BC = (A + B)(A + C)$$

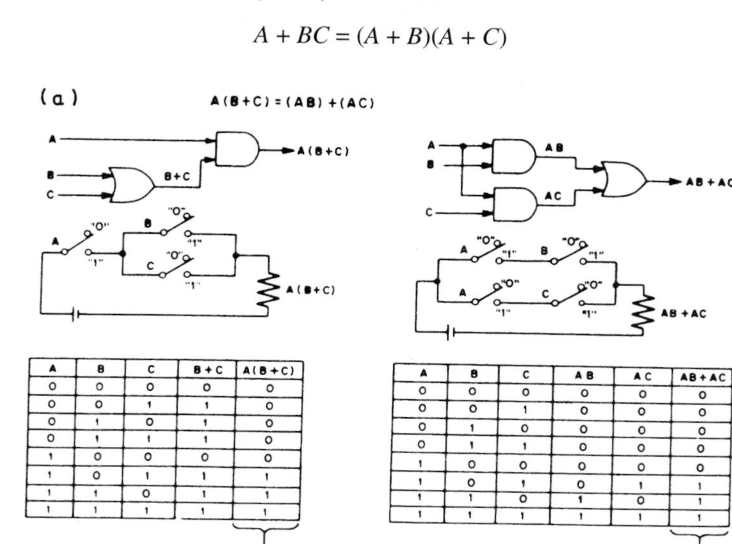

A	B	C	B + C	A(B + C)
0	0	0	0	0
0	0	1	1	0
0	1	0	1	0
0	1	1	1	0
1	0	0	0	0
1	0	1	1	1
1	1	0	1	1
1	1	1	1	1

A	B	C	AB	AC	AB + AC
0	0	0	0	0	0
0	0	1	0	0	0
0	1	0	0	0	0
0	1	1	0	0	0
1	0	0	0	0	0
1	0	1	0	1	1
1	1	0	1	0	1
1	1	1	1	1	1

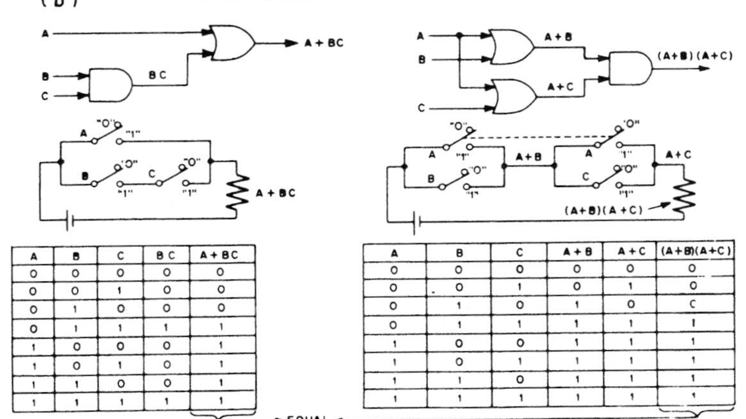

A	B	C	BC	A + BC
0	0	0	0	0
0	0	1	0	0
0	1	0	0	0
0	1	1	1	1
1	0	0	0	1
1	0	1	0	1
1	1	0	0	1
1	1	1	1	1

A	B	C	A + B	A + C	(A+B)(A+C)
0	0	0	0	0	0
0	0	1	0	1	0
0	1	0	1	0	C
0	1	1	1	1	1
1	0	0	1	1	1
1	0	1	1	1	1
1	1	0	1	1	1
1	1	1	1	1	1

FIG. 9-14 Distributive law. *(Bureau of Naval Personnel)*

De Morgan's Theorem

$$\overline{A+B} = \overline{A}\,\overline{B}$$

$$\overline{A+B+C+D\ldots} = \overline{A}\,\overline{B}\,\overline{C}\,\overline{D}\,\overline{E}\ldots$$

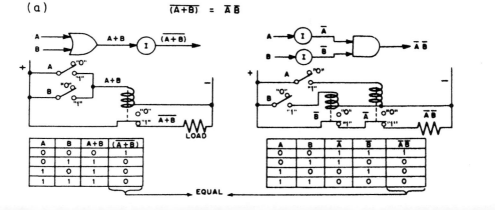

$$\overline{AB} = \overline{A}+\overline{B}$$

$$\overline{A\,B\,C\,D\,E\ldots} = \overline{A}+\overline{B}+\overline{C}+\overline{D}+\overline{E}+\ldots$$

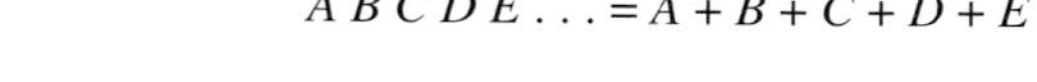

(a) $\overline{(A+B)} = \overline{A}\,\overline{B}$

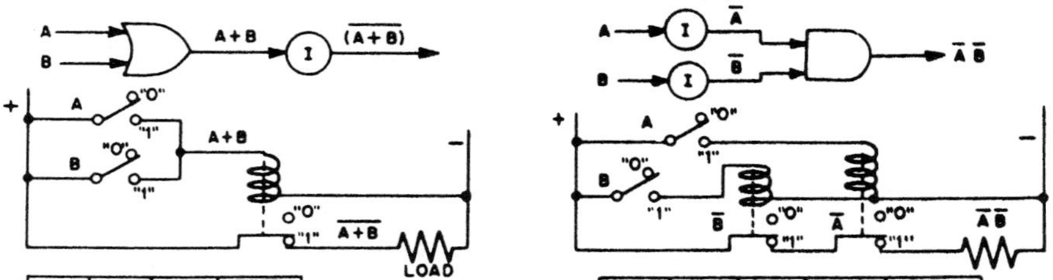

A	B	A+B	$\overline{(A+B)}$
0	0	0	1
0	1	1	0
1	0	1	0
1	1	1	0

A	B	$\overline{A}$	$\overline{B}$	$\overline{A}\,\overline{B}$
0	0	1	1	1
0	1	1	0	0
1	0	0	1	0
1	1	0	0	0

EQUAL

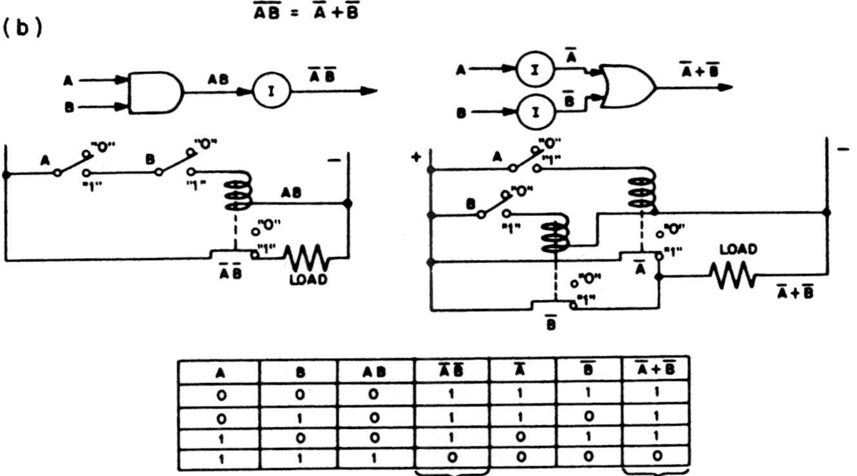

A	B	AB	$\overline{AB}$	$\overline{A}$	$\overline{B}$	$\overline{A} + \overline{B}$
0	0	0	1	1	1	1
0	1	0	1	1	0	1
1	0	0	1	0	1	1
1	1	1	0	0	0	0

FIG. 9-15 De Morgan's theorem. *(Bureau of Naval Personnel)*

Double-Negation Law

$$\overline{\overline{A}} = A$$

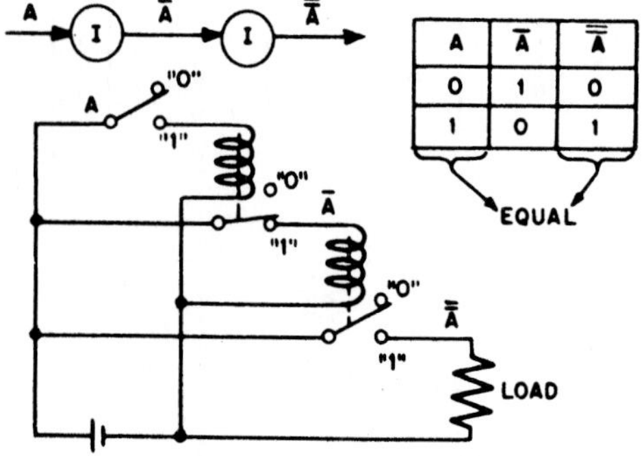

FIG. 9-16 Double-negation law. *(Bureau of Naval Personnel)*

Absorption Law

$$A(A + B) = A$$

$$A + AB = A$$

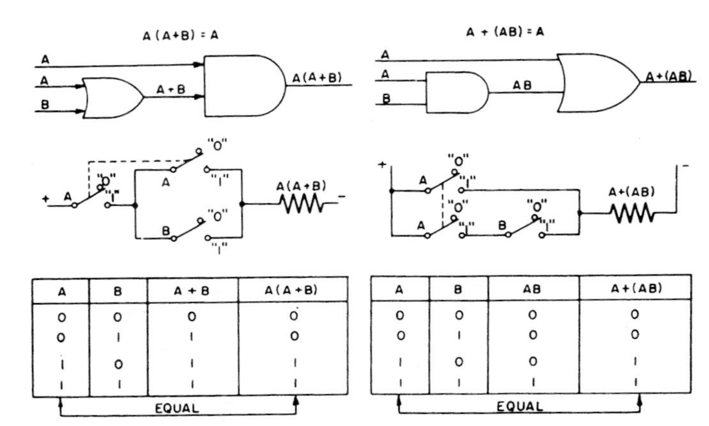

A	B	A + B	A (A + B)
0	0	0	0
0	1	1	0
1	0	1	1
1	1	1	1

A	B	AB	A + (AB)
0	0	0	0
0	1	0	0
1	0	0	1
1	1	1	1

FIG. 9-17 Absorption law. *(Bureau of Naval Personnel)*

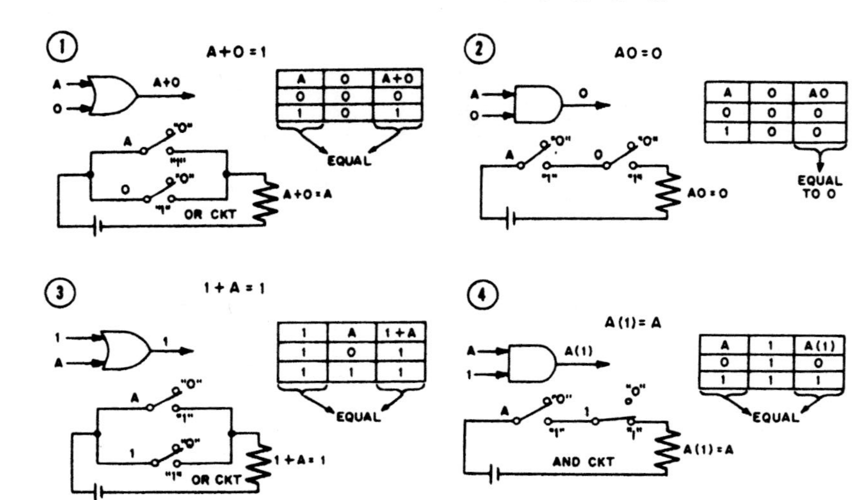

FIG. 9-18 Axioms. *(Bureau of Naval Personnel)*

9-2 NUMBER SYSTEMS

The distinguishing factor of all *positionally weighted number systems* (such as binary, decimal, and hexadecimal) is the *base* or *radix*. Each number system has a symbol set called *digits*, and the number of digits in the system is equal to the base. The special term *bits* is often used to refer to binary digits. In each case, the largest digit is equal to the base minus 1.

The *hexadecimal system* (referred to simply as *hex*) requires symbols for the digits above 9. The first six capital letters serve to represent the hexadecimal digits for 10 through 15.

The base of a number should be indicated by a subscript. Otherwise there is no way to differentiate numbers in different bases, because they all use common digits. The *fractional portion* of a number is separated from the integer portion by a *radix point* (the decimal point in the decimal system).

Polynomial Expansion

A number in any base can be converted to a number in the decimal system by multiplying each digit by its weight and summing the resulting products. This process is called *polynomial expansion*.

EXAMPLE: Express $1\,0111.1_2$ and $32B.9_{16}$ in the decimal system.

$$\begin{aligned} 1\,0111.1_2 &= 1 \times 2^4 + 0 \times 2^3 + 1 \times 2^2 + 1 \times 2^1 + 1 \times 2^0 + 1 \times 2^{-1} \\ &= 1 \times 16 + 0 \times 8 + 1 \times 4 + 1 \times 2 + 1 \times 1 + 1 \times ½ \\ &= 23.5_{10} \end{aligned}$$

$$32B.9_{16} = 3 \times 16^2 + 2 \times 16^1 + 11 \times 16^0 + 9 \times 16^{-1}$$

(*Note:* B_{16} has been changed to 11 to carry out the arithmetic.)

$$= 811.5625_{10}$$

Base Conversion by Grouping

To convert a binary number to hexadecimal just group the binary number in sets of four digits on each side of the radix point. Then change each group into its equivalent hexadecimal digit.

EXAMPLE: Convert $11\ 1111.111_2$ to hexadecimal.

$$11\ 1111.111_2 = 0011|1111.|1110$$

(Note the insertion of leading and trailing zeros to form groups of four digits.)

$$= 3F.E_{16}$$

Changing from hexadecimal to binary is just as easy. Write the binary numbers in groups of four on either side of the radix point for each hexadecimal digit.

EXAMPLE: Convert $7C2.4_{16}$ to binary.

$$7_{16} = 0111_2$$

$$C_{16} = 1100_2$$

$$2_{16} = 0010_2$$

$$4_{16} = 0100_2$$

$$\text{So } 7C2.4_{16} = 0111\ 1100\ 0010.0100_2$$

$$= 111\ 1100\ 0010.01_2$$

Converting Other Bases to Decimal

A more-efficient method than polynomial expansion for converting from other bases to decimal uses separate algorithms for converting the integer and the fractional portions of the number. After the portions are split, the old base is expressed as a decimal number. Then the old base is multiplied by the most-significant digit (MSD) of the number to be converted. That product is added to the next digit to the right. The multiplication and addition are repeated as many times as there are digits. The final sum is the answer.

EXAMPLE: Convert ABC_{16} to decimal.

```
MSD
 ↓
 A = 10          B = 11            C = 12

   × 16       ┌→  + 160        ┌→  + 2736
   ────       │   ─────        │   ──────
    160 ──────┘     171        │     2748
                   × 16        │
                   ────        │
                   2736 ───────┘
```

$$ABC_{16} = 2748_{10}$$

The fractional conversion requires that the least-significant digit (LSD) of the fraction be divided by the old base. The quotient is added to the next digit to the left, and the process repeats as many times as there are digits. The final quotient is the answer.

EXAMPLE: Convert $0.A72_{16}$ to decimal. Starting with the LSD of 2, 2/16 = 0.125
The next digit is 7,

$$\frac{7 + 0.125}{16} = 0.445$$

Changing A_{16} to 10_{10}.

$$\frac{10 + 0.445}{16} = 0.653$$

$$0.A72_{16} = 0.653_{10}$$

Converting Decimal to Other Bases

The algorithm for converting decimal to other bases also requires that the integer and fractional portions be split. For the integer portion, divide by the new base. Save the remainder and divide again into the residue quotient. Repeat the process until the quotient is zero. The MSD of the number in the new base is the last remainder generated.

EXAMPLE: Convert 62_{10} to a hexadecimal number.

	Remainders
$16 \overline{)62}$ = 3	$14 = E_{16}$ (LSD)
$16 \overline{)3}$ = 0	3 (MSD)

$62_{10} = 3E_{16}$

The fractional conversion requires that the fractional number to be converted be multiplied by the new base. The integer that is generated is removed and saved. The process is repeated with the residual product until the desired number of significant digits is generated. The first digit generated is the most significant.

EXAMPLE: Convert 0.29_{10} to a hexadecimal number rounded to two places of accuracy.

$$
\begin{array}{rr}
 & 0.29 \\
 & \times 16 \\
\hline
4 \leftarrow & \boxed{4}.\boxed{64} \\
 & 0.64 \\
 & \times 16 \\
\hline
10 = A \leftarrow & \boxed{10}.\boxed{24} \\
 & 0.24 \\
 & \times 16 \\
\hline
3 \leftarrow & \boxed{3}.84
\end{array}
$$

Rounding to two places:

$$0.29_{10} = 0.4A_{16}$$

TABLE 9-1 $2^{\pm n}$ in Decimal

2^n	n	2^{-n}
1	0	1.0
2	1	0.5
4	2	0.25
8	3	0.125
16	4	0.0625
32	5	0.03125
64	6	0.01562 5
128	7	0.00781 25
256	8	0.00390 625
512	9	0.00195 3125
1024	10	0.00097 65625
2048	11	0.00048 82812 5
4096	12	0.00024 41406 25
8192	13	0.00012 20703 125
16384	14	0.00006 10351 5625
32768	15	0.00003 05175 78125
65536	16	0.00001 52587 89062 5
1 31072	17	0.00000 76293 94531 25
2 62144	18	0.00000 38146 97265 625
5 24288	19	0.00000 19073 48632 8125
10 48576	20	0.00000 09536 74316 40625
20 97152	21	0.00000 04768 37158 20312 5
41 94304	22	0.00000 02384 18579 10156 25
83 88608	23	0.00000 01192 09289 55078 125

2^n	n	2^{-n}
167 77216	24	0.00000 00596 04644 77539 0625
335 54432	25	0.00000 00298 02322 38769 53125
671 08864	26	0.00000 00149 01161 19384 76562 5
1342 17728	27	0.00000 00074 50580 59692 38281 25
2684 35456	28	0.00000 00037 25290 29846 19140 625
5368 70912	29	0.00000 00018 62645 14923 09570 3125
10737 41824	30	0.00000 00009 31322 57461 54785 15625
21474 83648	31	0.00000 00004 65661 28730 77392 57812 5
42949 67296	32	0.00000 00002 32830 64365 38696 28906 25
85899 34592	33	0.00000 00001 16415 32182 69348 14453 125
1 71798 69184	34	0.00000 00000 58207 66091 34674 07226 5625
3 43597 38368	35	0.00000 00000 29103 83045 67337 03613 28125
6 87194 76736	36	0.00000 00000 14551 91522 83668 51806 64062 5
13 74389 53472	37	0.00000 00000 07275 95761 41834 25903 32031 25
27 48779 06944	38	0.00000 00000 03637 97880 70917 12951 66015 625
54 97558 13888	39	0.00000 00000 01818 98940 35458 56475 83007 8125
109 95116 27776	40	0.00000 00000 00909 49470 17729 28237 91503 90625
219 90232 55552	41	0.00000 00000 00454 74735 08864 64118 95751 95312 5
439 80465 11104	42	0.00000 00000 00227 37367 54432 32059 47875 97656 25
879 60930 22208	43	0.00000 00000 00113 68683 77216 16029 73937 98828 125
1759 21860 44416	44	0.00000 00000 00056 84341 88608 08014 86968 99414 0625
3518 43720 88832	45	0.00000 00000 00028 42170 94304 04007 43484 49707 03125
7036 87441 77664	46	0.00000 00000 00014 21085 47152 02003 71742 24853 51562 5
14073 74883 55328	47	0.00000 00000 00007 10542 73576 01001 85871 12426 75781 25
28147 49767 10656	48	0.00000 00000 00003 55271 36788 00500 92935 56213 37890 625
56294 99534 21312	49	0.00000 00000 00001 77635 68394 00250 46467 78106 68945 3125
112589 99068 42624	50	0.00000 00000 00000 88817 84197 00125 23233 89053 34472 65625

Source: *Handbook of Mathematics*, Abramowitz and Stegun, National Bureau of Standards.

TABLE 9-2 2^x in Decimal

x	2^x	x	2^x
0.001	1.00069 33874 62581	0.06	1.04246 57608 41121
0.002	1.00138 72557 11335	0.07	1.04971 66836 23067
0.003	1.00208 16050 79633	0.08	1.05701 80405 61380
0.004	1.00277 64359 01078	0.09	1.06437 01824 53360
0.005	1.00347 17485 09503	0.1	1.07177 34625 36293
0.006	1.00416 75432 38973	0.2	1.14869 83549 97035
0.007	1.00486 38204 23785	0.3	1.23114 44133 44916
0.008	1.00556 05803 98468	0.4	1.31950 79107 72894
0.009	1.00625 78234 97782	0.5	1.41421 35623 73095
0.01	1.00695 55500 56719	0.6	1.51571 65665 10398
0.02	1.01395 94797 90029	0.7	1.62450 47927 12471
0.03	1.02101 21257 07193	0.8	1.74110 11265 92248
0.04	1.02811 38266 56067	0.9	1.86606 59830 73615
0.05	1.03526 49238 41377		

Source: *Handbook of Mathematics,* Abramowitz and Stegun, National Bureau of Standards.

TABLE 9-3 $10^{\pm n}$ in Hexadecimal

10^{+n}	n	10^{-n}
1	0	1.000 000 000 000 000 000 00
12	1	0.063 146 314 631 463 146 31
144	2	0.005 075 341 217 270 243 66
1 750	3	0.000 406 111 564 570 651 77
23 420	4	0.000 032 155 613 530 704 15
303 240	5	0.000 002 476 132 610 706 64
3 641 100	6	0.000 000 206 157 364 055 37
46 113 200	7	0.000 000 015 327 745 152 75
575 360 400	8	0.000 000 001 257 143 561 06
7 346 545 000	9	0.000 000 000 104 560 276 41
112 402 762 000	10	0.000 000 000 006 676 337 66
1 351 035 564 000	11	0.000 000 000 000 537 657 77
16 432 451 210 000	12	0.000 000 000 000 043 136 32
221 411 634 520 000	13	0.000 000 000 000 003 411 35
2 657 142 036 440 000	14	0.000 000 000 000 000 264 11
34 327 724 461 500 000	15	0.000 000 000 000 000 022 01
434 157 115 760 200 000	16	0.000 000 000 000 000 001 63
5 432 127 413 542 400 000	17	0.000 000 000 000 000 000 14
67 405 553 164 731 000 000	18	0.000 000 000 000 000 000 01

Source: *Handbook of Mathematics,* Abramowitz and Stegun, National Bureau of Standards.

TABLE 9-4 $n \log_{10} 2$, $n \log_2 10$ in Decimal

n	$n \log_{10} 2$	$n \log_2 10$
1	0.30102 99957	3.32192 80949
2	0.60205 99913	6.64385 61898
3	0.90308 99870	9.96578 42847
4	1.20411 99827	13.28771 23795
5	1.50514 99783	16.60964 04744
6	1.80617 99740	19.93156 85693
7	2.10720 99696	23.25349 66642
8	2.40823 99653	26.57542 47591
9	2.70926 99610	29.89735 28540
10	3.01029 99566	33.21928 09489

Source: *Handbook of Mathematics,* Abramowitz and Stegun, National Bureau of Standards.

TABLE 9-5 Binary Addition

		Augend	
		0	1
Addend	0	0	1
	1	1	10

TABLE 9-6 Binary Multiplication

		Multiplicand	
		0	1
Multiplier	0	0	0
	1	0	1

TABLE 9-7 Hexadecimal Addition

Addend \ Augend	0	1	2	3	4	5	6	7	8	9	A	B	C	D	E	F
0	0	1	2	3	4	5	6	7	8	9	A	B	C	D	E	F
1	1	2	3	4	5	6	7	8	9	A	B	C	D	E	F	10
2	2	3	4	5	6	7	8	9	A	B	C	D	E	F	10	11
3	3	4	5	6	7	8	9	A	B	C	D	E	F	10	11	12
4	4	5	6	7	8	9	A	B	C	D	E	F	10	11	12	13
5	5	6	7	8	9	A	B	C	D	E	F	10	11	12	13	14
6	6	7	8	9	A	B	C	D	E	F	10	11	12	13	14	15
7	7	8	9	A	B	C	D	E	F	10	11	12	13	14	15	16
8	8	9	A	B	C	D	E	F	10	11	12	13	14	15	16	17
9	9	A	B	C	D	E	F	10	11	12	13	14	15	16	17	18
A	A	B	C	D	E	F	10	11	12	13	14	15	16	17	18	19
B	B	C	D	E	F	10	11	12	13	14	15	16	17	18	19	1A
C	C	D	E	F	10	11	12	13	14	15	16	17	18	19	1A	1B
D	D	E	F	10	11	12	13	14	15	16	17	18	19	1A	1B	1C
E	E	F	10	11	12	13	14	15	16	17	18	19	1A	1B	1C	1D
F	F	10	11	12	13	14	15	16	17	18	19	1A	1B	1C	1D	1E

TABLE 9-8 Hexadecimal Multiplication

Multiplier \ Multiplicand	0	1	2	3	4	5	6	7	8	9	A	B	C	D	E	F
0	0	0	0	0	0	0	0	0	0	0	0	0	0	0	0	0
1	0	1	2	3	4	5	6	7	8	9	A	B	C	D	E	F
2	0	2	4	6	8	A	C	E	10	12	14	16	18	1A	1C	1E
3	0	3	6	9	C	F	12	15	18	1B	1E	21	24	27	2A	2D
4	0	4	8	C	10	14	18	1C	20	24	28	2C	30	34	38	3C
5	0	5	A	F	14	19	1E	23	28	2D	32	37	3C	41	46	4B
6	0	6	C	12	18	1E	24	2A	30	36	3C	42	48	4E	54	5A
7	0	7	E	15	1C	23	2A	31	38	3F	46	4D	54	5B	62	69
8	0	8	10	18	20	28	30	38	40	48	50	58	60	68	70	78
9	0	9	12	1B	24	2D	36	3F	48	51	5A	63	6C	75	7E	87
A	0	A	14	1E	28	32	3C	46	50	5A	64	6E	78	82	8C	96
B	0	B	16	21	2C	37	42	4D	58	63	6E	79	84	8F	9A	A5
C	0	C	18	24	30	3C	48	54	60	6C	78	84	90	9C	A8	B4
D	0	D	1A	27	34	41	4E	5B	68	75	82	8F	9C	A9	B6	C3
E	0	E	1C	2A	38	46	54	62	70	7E	8C	9A	A8	B6	C4	D2
F	0	F	1E	2D	3C	4B	5A	69	78	87	96	A5	B4	C3	D2	E1

TABLE 9-9 Mathematical Constants in Hexadecimal

Constant	Value in hexadecimal
π	3.243 F6A 88
π^{-1}	0.517 CC1 B7
$\sqrt{\pi}$	1.C5B F89 1B
$\ln \pi$	1.250 D04 8E
$\log_2 \pi$	1.A6C 873 49
$\sqrt{10}$	3.298 B07 5B
e	2.B7E 151 62
e^{-1}	0.5E2 D58 D8
$\sqrt{e}$	1.A61 298 E2
$\log_{10} e$	0.6F2 DEC 54
$\log_2 e$	1.715 476 52
$\log_2 10$	3.526 9E1 2F
γ	0.93C 467 E3
$\ln \gamma$	−0.8CA E9B C1
$\log_2 \gamma$	−0.CAF 618 D2
$\sqrt{2}$	1.6A0 9E6 68
$\ln 2$	0.B17 217 F8
$\ln 10$	2.4D7 637 76

9-3 FLIP-FLOPS

The four basic flip-flop circuits are shown in Fig. 9-19. Tables 9-10 through 9-13 are state tables for each flip-flop. In these tables, the notations Q_0 and Q_0 refer to the output states prior to the application of the inputs.

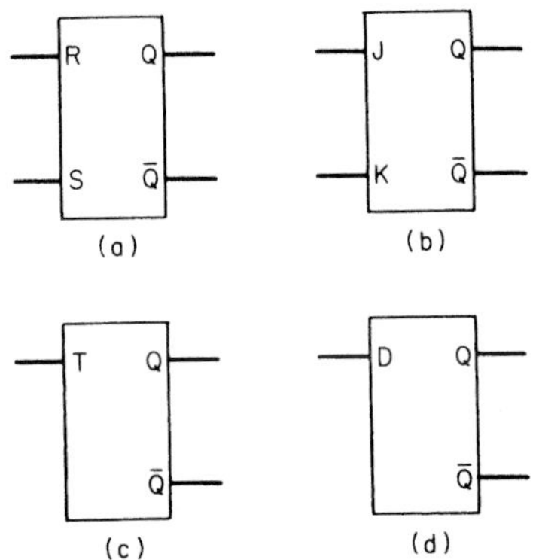

FIG. 9-19 Flip-flop symbology: *(a)* RS, *(b)* JK, *(c)* T, and *(d)* D.

TABLE 9-10 ***RS* Flip-Flop**

Inputs		Outputs		
R	S	Q	$\bar{Q}$	State
1	0	0	1	Reset
0	1	1	0	Set
0	0	Q_o	$\bar{Q}_o$	No change
1	1	$\bar{Q}_o$	Q_o	Forbidden (not stable)

TABLE 9-11 ***JK* Flip-Flop**

Inputs		Outputs		
J	K	Q	$\overline{Q}$	State
1	0	1	0	Set
0	1	0	1	Reset
0	0	Q_o	$\overline{Q}_o$	No change
1	1	$\overline{Q}_o$	Q_o	Toggle

TABLE 9-12 ***T* Flip-Flop**

Input	Outputs		
T	Q	$\overline{Q}$	State
0	Q_o	$\overline{Q}_o$	No change
1	$\overline{Q}_o$	Q_o	Toggle

TABLE 9-13 ***D* Flip-Flop**

Input	Outputs	
D	Q	$\overline{\text{Q}}$
0	0	1
1	1	0

9-4 INFORMATION THEORY CONCEPTS

Nyquist Criterion

A signal must be sampled at twice the rate of its highest frequency component to permit the signal to be accurately reconstructed.

Shannon-Hartley Theorem for Channel Capacity

$$C = \text{BW} \log_2 (1 + \text{SNR})$$

where BW = bandwidth of channel, Hz
SNR = signal-to-noise ratio

EXAMPLE: Find the channel capacity for a 3-kHz telephone channel with an SNR of 15.

$$\begin{aligned} C &= 3000 \log_2 (1 + 15) \\ &= 3000 \log_2 16 \\ &= 3000 \times 4 \\ &= 12{,}000 \text{ bits per second (bps)} \end{aligned}$$

10

Computers

10-1 PC KEYBOARD CODING

The standard PC keyboard sends encoded information to the computer each time a key is pressed. When the key is pressed, a Make Code is sent, then when the key returns to its original position, a Break Code is sent. These codes are tabulated in Table 10-1.

TABLE 10-1 Standard Keyboard Codes

Key	Make Code	Break Code
Main Keyboard		
A	1E	9E
B	30	B0
C	2E	AE
D	20	A0
E	12	92
F	21	A1
G	22	A2
H	23	A3

TABLE 10-1 Standard Keyboard Codes (Continued)

Key	Make Code	Break Code
I	17	97
J	24	A4
K	25	A5
L	26	A6
M	32	B2
N	31	B1
O	18	98
P	19	99
Q	10	90
R	13	93
S	1F	9F
T	14	94
U	16	96
V	2F	AF
W	11	91
X	2D	AD
Y	15	95
Z	2C	AC
0	0B	8B
1	02	82
2	03	83
3	04	84
4	05	85
5	06	86
6	07	87
7	08	88
8	09	89
9	0A	8A
)	0B	8B
!	02	82
@	03	83
#	04	84
$	05	85
%	06	86

TABLE 10-1 Standard Keyboard Codes (Continued)

Key	Make Code	Break Code
^ or ©	07	87
&	08	88
*	09	89
(	0A	8A
• or ™	29	A9
- or _	0C	8C
= or +	0D	8D
[, {, or §	1A	9A
], }, or †	1B	9B
¶ or ®	2B	AB
; or :	27	A7
‘ or “	28	A8
, or <	33	B3
. or >	34	B3
/ or ?	35	B5
\ or |	1C	9C
` or ~	01	81
Control Keys		
Alt (left)	38	B8
Ctrl (left)	1D	9D
Shift (left)	2A	AA
Alt (right)	E0 38	E0 B8
Ctrl (right)	E0 1D	E0 9D
Shift (right)	36	B6
Caps Lock	3A	BA
Tab	0F	8F
Space Bar	39	B9
Enter	E0 1C	E0 9C
Numeric and Cursor Keypad		
Num Lock	45	C5
+	4E	CE

TABLE 10-1 Standard Keyboard Codes (Continued)

Key	Make Code	Break Code
*	37	B7
-	4A	CA
0 or Ins	52	D2
1 or End	4F	CF
2	50	D0
3 or PgDn	51	D1
4	4B	CB
5	4C	CC
6	4D	CD
7 or Home	47	C7
8	48	C8
9 or PgUp	49	C9
Function Keyboard		
Esc	01	81
F1	3B	BB
F2	3C	BC
F3	3D	BD
F4	3E	BE
F5	3F	BF
F6	40	C0
F7	41	C1
F8	42	C2
F9	43	C3
F10	44	C4
F11	57	D7
F12	58	D8
Setup and Arrow Keyboard		
Print Screen	E0 2A E0 37	E0 B7 E0 AA
Scroll Lock	46	C6
Pause	E1 1D E1 9D C5	none
Insert	E0 52	E0 D2

TABLE 10-1 Standard Keyboard Codes (Continued)

Key	Make Code	Break Code
Home	E0 47	E0 C7
Page Up	E0 49	E0 C9
Delete	E0 53	E0 D3
End	E0 4F	E0 CF
Page Down	E0 51	E0 D1
↑	E0 48	E0 C8
←	E0 4B	E0 CB
↓	E0 50	E0 D0
→	E0 4D	E0 CD

10-2 FILE EXTENSIONS

Windows and DOS file names are frequently assigned with a three-character extension, which identifies the type of file. These extensions are often useful for determining the application that created the file—both on local disks and over networks. Extensions are tabulated in Table 10-2.

TABLE 10-2 File Extensions

Extension	Application
AD	After Dark image
ADM	After Dark MultiModule
ADR	After Dark Randomizer
AFM	Adobe Type 1 font metrics
AG4	Access G4 document image
AI	Adobe Illustrator graphic and Encapsulated PostScript header
AIF	Macintosh digital audio
ANN	Windows help annotations

TABLE 10-2 File Extensions (Continued)

Extension	Application
ANS	ANSI text
ARC	ARC compressed archive
ARJ	Jung compressed archive
ASC	ASCII file
ASD	Word document (temporary)
ASM	Assembler source
ASP	Active Server page
ATT	AT&T group IV fax
AU	Sun digital audio
AVI	Microsoft movie format
BAK	Backup file
BAS	BASIC
BAT	DOS and OS/2 batch file
BFC	Windows briefcase document
BIN	Binary file
BMK	Windows help bookmark
BMP	Windows and OS/2 bitmap
C	C source
CAB	Microsoft compressed format
CAL	Windows calendar, SuperCalc worksheet, CALS image format (raster or vector)
CAP	Ventura Publisher captions
CCH	Corel chart
CDR	CorelDraw vector graphic
CDT	CorelDraw template
CDX	CorelDraw compressed drawing
CFG	Configuration
CGM	CGM vector graphic
CH3	Harvard Graphics chart
CHK	DOS chkdsk file
CHP	Ventura Publisher chapter
CIF	Ventura Publisher chapter information
CIT	Intergraph scanned image

TABLE 10-2 File Extensions (Continued)

Extension	Application
CLP	Windows clipboard
CMP	Lead Technologies raster graphic
CMX	Corel clip art
CNT	Windows help content
COB	COBOL source and Truespace 3D file
COM	Executable file
CPI	DOS code page
CPL	Windows control panel applet
CPP	C++ source
CPR	Knowledge Access raster graphic
CSV	Comma delimited file
CUT	Dr. Halo raster graphic
CV5	Canvas 5 raster or vector graphic
DAT	Data
DB	Paradox table
DBF	dBase database
DBT	dBase text
DBX	Databeam raster graphic
DCA	IBM text
DCS	Color separated EPS format
DCT	Dictionary
DG	Autotrol vector graphic
DGN	Intergraph vector graphic
DIB	Windows raster graphic
DIC	Dictionary
DIF	Spreadsheet
DLL	Dynamic link library
DOC	Word and other document
DOT	Word template
DOX	Multimate 4.0 document
DPI	Pointline raster graphic
DRV	Driver
DRW	Designer vector graphic (versions 2.x and 3.x)

TABLE 10-2 File Extensions (Continued)

Extension	Application
DS4	Designer vector graphic (version 4.x)
DSF	Designer vector graphic (versions 6 and 7)
DWG	AutoCAD vector graphic
DX	Autotrol document image
DXF	AutoCAD vector graphic
ED5	EDMICS raster graphic
EMF	Enhanced Windows metafile
EPS	Encapsulated PostScript
ESI	ESRI vector image
EXE	Executable file
FAX	Facsimile file
FDX	Force index
FLC	Autodesk animation
FLI	Autodesk animation
FLT	Graphics conversion filter
FM3	Lotus 1-2-3 format information (version 3)
FMT	dBase screen format
FMV	FrameMaker raster and vector graphic
FNT	Windows font
FON	Windows bitmapped font and phone file
FOR	FORTRAN source
FOT	Windows TrueType font information
FOX	FoxBase compiled program
FRM	dBase report format
FTG	Windows help file link
FTS	Windows help file search index
G4	GTX RasterCAD
GCA	IBM vector graphic
GED	Arts & Letters graphic
GEM	GEM vector graphic
GID	Windows help global index
GIF	CompuServe raster graphic
GP4	CALS Group IV

TABLE 10-2 File Extensions (Continued)

Extension	Application
GRF	Micrografx Charisma vector graphic
GRP	Windows ProgMan group
GX1	Show Partner raster graphic
GX2	Show Partner raster graphic
HLP	Help text
HPL	HP Graphics language
HQX	BinHex format
HTM	HTML document
HYC	WordPerfect hyphen list
ICA	IBM raster graphic
ICO	Windows icon
IDE	Development environment configuration
IDX	FoxBase index
IFF	Amiga
IGF	Insert Systems raster and vector graphic
IL	Icon library
IMG	GEM Paint raster graphic
INF	Setup information
INI	Initialization
JPG	JPEG raster graphic
JT	JT fax
KFX	Kofax Group IV fax
LBL	dBase label
LBM	Deluxe Paint graphic
LIB	Function library
LZH	LHARC compressed
MAC	McPaint raster graphic
MAP	Linkage editor map
MCS	MathCAD format
MDB	Access database
MDX	dBase IV multiple index file
MET	OS/2 metafile
MEU	Menu item

TABLE 10-2 File Extensions (Continued)

Extension	Application
MID	MIDI sound file, Eudora script
MIL	CALS Group IV
MME	MIME encoded file
MMM	Macromind animation
MOD	Eudora script
MOV	QuickTime movie
MPG	MPEG file
MPP	Microsoft project
MRK	Informative Graphics markup file
MSG	Message
MSP	Microsoft Paint raster graphic
NAP	NAPLPS file
NAV	Eudora script
NDX	dBase index
NG	Norton Guides text
NLM	NetWare NLM program
NTZ	InVircible antivirus blueprint
OAZ	OAZ fax
OBJ	Object file and Wavefront 3D file
OVL	Overlay
OVR	Overlay
OZM	Sharp Organizer memo bank
OZP	Sharp Organizer telephone bank
PAS	Pascal source
PCD	Photo CD raster graphic
PCL	HP LaserJet file
PCM	HP LaserJet cartridge information
PCT	PC Paint raster graphic and Macintosh PICT raster or vector graphic
PCW	PC Write document
PCX	PC Paintbrush raster graphic
PDF	Acrobat Portable Document format, printer driver, and QuarkXpress printer description

TABLE 10-2 File Extensions (Continued)

Extension	Application
PDV	PC Paintbrush printer driver
PDW	HiJaak vector graphic
PFA	ASCII Type 1 font
PFB	Encrypted Type 1 font
PGL	HPGL 7475A plotter vector graphic
PIC	Lotus 1-2-3, Micrografx Draw, Macintosh PICT, and IBM Storyboard raster graphic
PIF	DOS information and IBM Picture Interchange
PIX	Insert Systems raster and vector graphic
PM	PageMaker graphics and text
PMx	PageMaker document (x = version number)
PNG	PNG raster graphic
POV	POV-Ray tracing
PPD	PostScript printer description
PPT	PowerPoint
PRD	Microsoft Word printer driver
PRG	dBase source
PRN	Temporary print file and XyWrite printer driver
PRS	WordPerfect printer driver
PRT	Formatted text
PS	PostScript page description
PSD	Photoshop native format
P10	Tektronix Plot10 plotter
PTx	PageMaker template (x = version number)
QLC	ATM font information
R8L	HP LaserJet landscape font
R8P	HP LaserJet portrait font
RA	Real Audio file
RAS	Sun raster graphic
RAW	3D file
RGB	SGI raster graphic
RIA	Alpharel Group IV raster graphic
RIB	Renderman graphic

TABLE 10-2 File Extensions (Continued)

Extension	Application
RIC	Roch FaxNet
RIX	RIX virtual screen
RLC	Image Systems CAD overlay ESP
RLE	Compressed run length encoded file
RND	AutoShade rendering format
RNL	GTX Runlength raster graphic
RTF	Microsoft text and graphic
RV	Real Video
SAM	Ami Pro document
SAT	ACIS 3D model
SBP	IBM Storyboard graphic and Superbase text
SC	Paradox source
SCM	ScreenCam movie
SCR	dBase screen layout, Windows screen saver, and script
SCT	Lotus Manuscript screen caption text
SCx	ColoRIX raster graphic (x = resolution)
SET	Setup parameters
SFL	HP LaserJet landscape font
SFP	HP LaserJet portrait font
SFS	PCL 5 scaleable font
SGI	SGI raster graphic
SLD	AutoCAD slide
SND	Digital audio
SPD	Spedo scaleable font
STY	Ventura Publisher style sheet
SUN	Sun raster graphic
SY3	Harvard Graphics symbol
SYL	SYLK spreadsheet
SYS	DOS and OS/2 driver
TAL	Adobe Type Align shaped text
TDF	Speedo typeface definition
TFM	Intellifont font metric
TGA	Targa raster graphic
TIF	TIFF raster graphic

TABLE 10-2 File Extensions (Continued)

Extension	Application
TMP	Temporary file
TTC	TrueType font compressed
TTF	TrueType font
TXT	ASCII text
USL	HP LaserJet landscape font
USP	HP LaserJet portrait font
VGR	Ventura Publisher chapter information
VOC	Sound Blaster audio
VUE	dBase relational view
WAV	Windows digital audio
WBT	WinBatch file
WK1	Lotus 1-2-3 spreadsheet (version 2.x)
WK3	Lotus 1-2-3 spreadsheet (version 3.x)
WK4	Lotus 1-2-3 spreadsheet (version 4.x)
WKQ	Quattro spreadsheet
WKS	Lotus 1-2-3 spreadsheet (version 1a)
WMF	Windows metafile
WPD	Corel, Windows, and WordPerfect printer description
WPG	WordPerfect raster and vector graphic
WPM	WordPerfect macro
WPS	Microsoft Works document
WRI	Windows Write document
WRK	Symphony spreadsheet
WRL	VRML page
XBM	X Window bitmap
XFX	JetFax
XLC	Excel chart
XLS	Excel spreadsheet
XPM	X Window pixel map
XWD	X Window dump
ZIP	PKZIP compressed file
ZOO	Zoo compressed file
3DS	3D Studio
906	Calcomp plotter
$$$	Temporary file

10-3 TRUSTED COMPUTER SYSTEMS DESIGNATORS

The National Computer Security Center (NCSC) has established levels for secure computer systems. These specifications are contained in DOD Standard 5200.28 (also known as the Orange and Red books). Table 10-3 provides an overview of these designators.

TABLE 10-3 Trusted Computer Systems

Level	Requirements
A1 (highest level)	System can be described by a proven mathematical model
B3	System can be described by a viable mathematical model
B2	Guaranteed security; provides assurance that the system passed test, and security levels cannot be bypassed
B1	Military security level
C2	Requires passwords, logging on, and auditing
C1	Requires user logon with group identifier
D (lowest level)	Non-secure

10-4 PC DISPLAY RESOLUTION

Table 10-4 provides definitions and parameters for various types of personal computer displays.

TABLE 10-4 PC Display Requirements

Designator	*Number of colors	Color depth (bits)
Standard VGA	16	4
Super VGA	256	8

TABLE 10-4 PC Display Requirements (Continued)

Designator	*Number of colors	Color depth (bits)
High color	32K	15
High color	64K	16
True color	16M	24

*K = 1024 bits
M = 1 megabit

10-5 DIGITAL VERSATILE DISK (DIGITAL VIDEO DISK)

A variety of DVD formats have appeared on the market. Table 10-5 describes the capabilities of each type.

TABLE 10-5 DVD Features

Type	Read/Write	Number of sides	Number of layers per side	Capacity (GB)
DVD-ROM	Read only	1	1	4.7
		1	2	8.5
		2	1	9.4
		2	2	17.0
DVD-R	Read/Write once	1	1	3.9
		2	1	7.8
DVD-RAM	Read/Rewritable	1	1	2.6
		2	1	5.2
DVD+RW	Read/Rewritable	1	1	3.0
		2	1	6.0

10-6 TELEPHONE CONNECTORS

Telephone connectors and jacks are used to tie modems into a communications system. The connectors listed in Table 10-6 are the most common.

TABLE 10-6 Telephone Connectors

Designator	Number of pins
RJ-11	4 or 6
RJ-21 (Ethernet)	50 pin telephone on one end and 12 RJ-45 connectors on the other
RJ-45 (10Base-T or Token Ring Type 1)	8

10-7 MICROCOMPUTER BUSES

A comparison of various types of microcomputer bus architectures is provided in Table 10-7. Tables 10-8 through 10-14 list the specifics of each type of bus.

TABLE 10-7 Bus Architecture Comparison

Bus	Width (bits)	Speed (MHz)	Data Rate (Mbps)	Number of Pins
Extended Industry Standard Architecture (EISA)	32	8.3	33.3	188
IEEE-1394 (FireWire)	Serial	100–400	100–400	4

TABLE 10-7 Bus Architecture Comparison (Continued)

Bus	Width (bits)	Speed (MHz)	Data Rate (Mbps)	Number of Pins
Industry Standard Architecture (ISA)	8 16	8 8	4 8	62 98
Microchannel Architecture (MCA)	32	8	33	174
Peripheral Component Interconnect (PCI)	64	33	264	188
Small Computer System Interconnect (SCSI)	8 16 Wide SCSI-II	5 SCSI-I 10 Fast SCSI-II 20 Ultra SCSI-II	—	25–50
Universal Serial Bus (USB)	Serial	12	12	86
Video Electronic Standards Association (VESA) Local Bus (VLB)	3	33 40 50	128–132	116

TABLE 10-8 Extended Industry Standard Architecture (EISA) Bus Description

*Pin Number	Signal Name
A1	Channel Check
A2	System Data 7
A3	System Data 6
A4	System Data 5
A5	System Data 4
A6	System Data 3
A7	System Data 2
A8	System Data 1
A9	System Data 0
A10	Channel Ready
A11	Address Enable
A12	System Address 19
A13	System Address 18
A14	System Address 17
A15	System Address 16
A16	System Address 15
A17	System Address 14
A18	System Address 13
A19	System Address 12
A20	System Address 11
A21	System Address 10
A22	System Address 9
A23	System Address 8
A24	System Address 7
A25	System Address 6
A26	System Address 5
A27	System Address 4
A28	System Address 3
A29	System Address 2
A30	System Address 1
A31	System Address 0

TABLE 10-8 Extended Industry Standard Architecture (EISA) Bus Description (Continued)

*Pin Number	Signal Name
B1	Ground
B2	Reset
B3	+5V
B4	Interrupt Request 2
B5	–5V
B6	DMA Request 2
B7	–12V
B8	No Wait State
B9	+12V
B10	Ground
B11	Standard Memory Write Command
B12	Standard Memory Read Command
B13	I/O Write Command
B14	I/O Read Command
B15	DMA Acknowledge 3
B16	DMA Request 3
B17	DMA Acknowledge 1
B18	DMA Request 1
B19	Refresh
B20	Bus Clock
B21	Interrupt Request 7
B22	Interrupt Request 6
B23	Interrupt Request 5
B24	Interrupt Request 4
B25	Interrupt Request 3
B26	DMA Acknowledge 2
B27	Terminal Count
B28	Bus Address Latch Enable
B29	+5V
B30	Oscillator
B31	Ground

TABLE 10-8 Extended Industry Standard Architecture (EISA) Bus Description (Continued)

*Pin Number	Signal Name
C1	System Bus High Enable
C2	Latchable Address 23
C3	Latchable Address 22
C4	Latchable Address 21
C5	Latchable Address 20
C6	Latchable Address 19
C7	Latchable Address 18
C8	Latchable Address 17
C9	Memory Read Command
C10	Memory Write Command
C11	System Data 8
C12	System Data 9
C13	System Data 10
C14	System Data 11
C15	System Data 12
C16	System Data 13
C17	System Data 14
C18	System Data 15
D1	Memory Access
D2	I/O Size
D3	Interrupt Request 10
D4	Interrupt Request 11
D5	Interrupt Request 12
D6	Interrupt Request 13
D7	Interrupt Request 14
D8	DMA Acknowledge 0
D9	DMA Request 0
D10	DMA Acknowledge 5
D11	DMA Request 5
D12	DMA Acknowledge 6
D13	DMA Request 6

TABLE 10-8 Extended Industry Standard Architecture (EISA) Bus Description (Continued)

*Pin Number	Signal Name
D14	DMA Acknowledge 7
D15	DMA Request 7
D16	+5V
D17	16-bit Bus Master
D18	Ground
E1	Command Phase
E2	Start Phase
E3	EISA Ready
E4	EISA Slave Size 32
E5	Ground
E6	——
E7	EISA Slave Size 16
E8	Slave Burst
E9	Master Burst
E10	Write or Read
E11	Ground
E12	Reserved
E13	Reserved
E14	Reserved
E15	Ground
E16	——
E17	Byte Enable 1
E18	Latchable Address 31
E19	Ground
E20	Latchable Address 30
E21	Latchable Address 28
E22	Latchable Address 27
E23	Latchable Address 25
E24	Ground
E25	——
E26	Latchable Address 15

TABLE 10-8 Extended Industry Standard Architecture (EISA) Bus Description (Continued)

*Pin Number	Signal Name
E27	Latchable Address 13
E28	Latchable Address 12
E29	Latchable Address 11
E30	Ground
E31	Latchable Address 9
F1	Ground
F2	+5V
F3	+5V
F4	Reserved
F5	Reserved
F6	——
F7	Reserved
F8	Reserved
F9	+12V
F10	Memory/Input-Output
F11	Lock
F12	Reserved
F13	Ground
F14	Reserved
F15	Byte Enable 3
F16	——
F17	Byte Enable 2
F18	Byte Enable 0
F19	Ground
F20	+5V
F21	Latchable Address 29
F22	Ground
F23	Latchable Address 26
F24	Latchable Address 24
F25	——
F26	Latchable Address 16

TABLE 10-8 Extended Industry Standard Architecture (EISA) Bus Description (Continued)

*Pin Number	Signal Name
F27	Latchable Address 14
F28	+5V
F29	+5V
F30	Ground
F31	Latchable Address 10
G1	Latchable Address 7
G2	Ground
G3	Latchable Address 4
G4	Latchable Address 3
G5	Ground
G6	——
G7	System Data 17
G8	System Data 19
G9	System Data 20
G10	System Data 22
G11	Ground
G12	System Data 25
G13	System Data 26
G14	System Data 28
G15	——
G16	Ground
G17	System Data 30
G18	System Data 31
G19	Memory Request for Slot
H1	Latchable Address 8
H2	Latchable Address 6
H3	Latchable Address 5
H4	+5V
H5	Latchable Address 2
H6	——
H7	System Data 16

TABLE 10-8 Extended Industry Standard Architecture (EISA) Bus Description (Continued)

*Pin Number	Signal Name
H8	System Data 18
H9	Ground
H10	System Data 21
H11	System Data 23
H12	System Data 24
H13	Ground
H14	System Data 27
H15	——
H16	System Data 29
H17	+5V
H18	+5V
H19	Master Acknowledge for Slot

*The upper row of pins on the EISA Bus (A1 through B31) are identical to those of the ISA bus. The lower row of pins provides the extension signals. PC cards designed for the IBM XT bus, IBM AT bus, and the ISA bus will usually work correctly when inserted in an EISA slot.

TABLE 10-9 IEEE-1394 (FireWire) Bus Description

Pin Number	Signal
1	Cable Power (VP)
2	Cable Ground (VG)
3	Twisted Pair B Differential Data (TPB*)
4	Twisted Pair B Differential Data (TPB)
5	Twisted Pair A Differential Data (TPA*)
6	Twisted Pair A Differential Data (TPA)

TABLE 10-10 Industry Standard Architecture (ISA) Bus Description

*Pin Number	Signal Name
A1	Channel Check
A2	System Data 7
A3	System Data 6
A4	System Data 5
A5	System Data 4
A6	System Data 3
A7	System Data 2
A8	System Data 1
A9	System Data 0
A10	Channel Ready
A11	Address Enable
A12	System Address 19
A13	System Address 18
A14	System Address 17
A15	System Address 16
A16	System Address 15
A17	System Address 14
A18	System Address 13
A19	System Address 12
A20	System Address 11
A21	System Address 10
A22	System Address 9
A23	System Address 8
A24	System Address 7
A25	System Address 6
A26	System Address 5
A27	System Address 4
A28	System Address 3
A29	System Address 2
A30	System Address 1
A31	System Address 0

TABLE 10-10 Industry Standard Architecture (ISA) Bus Description (Continued)

*Pin Number	Signal Name
B1	Ground
B2	Reset
B3	+5V
B4	Interrupt Request 2
B5	–5V
B6	DMA Request 2
B7	–12V
B8	No Wait State
B9	+12V
B10	Ground
B11	Standard Memory Write Command
B12	Standard Memory Read Command
B13	I/O Write Command
B14	I/O Read Command
B15	DMA Acknowledge 3
B16	DMA Request 3
B17	DMA Acknowledge 1
B18	DMA Request 1
B19	Refresh
B20	Bus Clock
B21	Interrupt Request 7
B22	Interrupt Request 6
B23	Interrupt Request 5
B24	Interrupt Request 4
B25	Interrupt Request 3
B26	DMA Acknowledge 2
B27	Terminal Count
B28	Bus Address Latch Enable
B29	+5V
B30	Oscillator
B31	Ground

TABLE 10-10 Industry Standard Architecture (ISA) Bus Description (Continued)

*Pin Number	Signal Name
C1	System Bus High Enable
C2	Latchable Address 23
C3	Latchable Address 22
C4	Latchable Address 21
C5	Latchable Address 20
C6	Latchable Address 19
C7	Latchable Address 18
C8	Latchable Address 17
C9	Memory Read Command
C10	Memory Write Command
C11	System Data 8
C12	System Data 9
C13	System Data 10
C14	System Data 11
C15	System Data 12
C16	System Data 13
C17	System Data 14
C18	System Data 15
D1	Memory Access
D2	I/O Size
D3	Interrupt Request 10
D4	Interrupt Request 11
D5	Interrupt Request 12
D6	Interrupt Request 13
D7	Interrupt Request 14
D8	DMA Acknowledge 0
D9	DMA Request 0
D10	DMA Acknowledge 5
D11	DMA Request 5
D12	DMA Acknowledge 6
D13	DMA Request 6

TABLE 10-10 Industry Standard Architecture (ISA) Bus Description (Continued)

*Pin Number	Signal Name
D14	DMA Acknowledge 7
D15	DMA Request 7
D16	+5V
D17	16-bit Bus Master
D18	Ground

*C and D pins used only on the IBM AT bus.

TABLE 10-11 Peripheral Component Interconnect (PCI) Bus Description

Pin Number	Signal Name
A1	Test Logic Reset
A2	+12V
A3	Test Mode Select
A4	Test Data Input
A5	+5V
A6	Interrupt A
A7	Interrupt C
A8	+5V
A9	Reserved
A10	+5V
A11	Reserved
A12	Ground 3 or Open
A13	Ground 5 or Open
A14	Reserved
A15	Reset
A16	+5V
A17	Grant PCI Use
A18	Ground 8

TABLE 10-11 Peripheral Component Interconnect (PCI) Bus Description (Continued)

Pin Number	Signal Name
A19	Reserved
A20	Address/Data 30
A21	+3.3V
A22	Address/Data 28
A23	Address/Data 26
A24	Ground 10
A25	Address/Data 24
A26	Initialization Device Select
A27	+3.3V
A28	Address/Data 22
A29	Address/Data 20
A30	Ground 12
A31	Address/Data 18
A32	Address/Data 16
A33	+3.3V
A34	Frame
A35	Ground 14
A36	Target Ready
A37	Ground 15
A38	Stop Transfer Cycle
A39	+3.3V
A40	Snoop Done
A41	Snoop Backoff
A42	Ground 17
A43	Parity
A44	Address/Data 15
A45	+3.3V
A46	Address/Data 13
A47	Address/Data 11
A48	Ground 19

TABLE 10-11 Peripheral Component Interconnect (PCI) Bus Description (Continued)

Pin Number	Signal Name
A49	Address/Data 9
A50	——
A51	——
A52	Command Byte Enable 0
A53	+3.3V
A54	Address/Data 6
A55	Address/Data 4
A56	Ground 21
A57	Address/Data 2
A58	Address/Data 0
A59	+5V
A60	Request 64 bit
A61	+5V
A62	+5V
A63	Ground
A64	Command Byte Enable 7
A65	Command Byte Enable 5
A66	+5V
A67	Parity 64
A68	Address/Data 62
A69	Ground
A70	Address/Data 60
A71	Address/Data 58
A72	Ground
A73	Address/Data 56
A74	Address/Data 54
A75	+5V
A76	Address/Data 52
A77	Address/Data 50

TABLE 10-11 Peripheral Component Interconnect (PCI) Bus Description (Continued)

Pin Number	Signal Name
A78	Ground
A79	Address/Data 48
A80	Address/Data 46
A81	Ground
A82	Address/Data 44
A83	Address/Data 42
A84	+5V
A85	Address/Data 40
A86	Address/Data 38
A87	Ground
A88	Address/Data 36
A89	Address/Data 34
A90	Ground
A91	Address/Data 32
A92	Reserved
A93	Ground
A94	Reserved
B1	–12V
B2	Test Clock
B3	Ground
B4	Test Data Output
B5	+5V
B6	+5V
B7	Interrupt B
B8	Interrupt D
B9	Reserved
B10	Reserved
B11	Reserved
B12	Ground

TABLE 10-11 Peripheral Component Interconnect (PCI) Bus Description (Continued)

Pin Number	Signal Name
B13	Ground
B14	Reserved
B15	Ground
B16	Clock
B17	Ground
B18	Request
B19	+5V
B20	Address/Data 31
B21	Address/Data 29
B22	Ground
B23	Address/Data 27
B24	Address/Data 25
B25	+3.3V
B26	Command Byte Enable 3
B27	Address/Data 23
B28	Ground
B29	Address/Data 21
B30	Address/Data 19
B31	+3.3V
B32	Address/Data 17
B33	Command Byte Enable 2
B34	Ground 13
B35	Initiator Ready
B36	+3.3V
B37	Device Select
B38	Ground 16
B39	Lock Bus
B40	Parity Error
B41	+3.3V

TABLE 10-11 Peripheral Component Interconnect (PCI) Bus Description (Continued)

Pin Number	Signal Name
B42	System Error
B43	+3.3V
B44	Command Byte Enable 1
B45	Address/Data 14
B46	Ground 18
B47	Address/Data 12
B48	Address/Data 10
B49	Ground 20
B50	Ground or Open
B51	Ground or Open
B52	Address/Data 8
B53	Address/Data 7
B54	+3.3V
B55	Address/Data 5
B56	Address/Data 3
B57	Ground 22
B58	Address/Data 1
B59	+5V
B60	Acknowledge 64
B61	+5V
B62	+5V
B63	Reserved
B64	Ground
B65	Command Byte Enable 6
B66	Command Byte Enable 4
B67	Ground
B68	Address/Data 63
B69	Address/Data 61
B70	+5V

TABLE 10-11 Peripheral Component Interconnect (PCI) Bus Description (Continued)

Pin Number	Signal Name
B71	Address/Data 59
B72	Address/Data 57
B73	Ground
B74	Address/Data 55
B75	Address/Data 53
B76	Ground
B77	Address/Data 51
B78	Address/Data 49
B79	+5V
B80	Address/Data 47
B81	Address/Data 45
B82	Ground
B83	Address/Data 43
B84	Address/Data 41
B85	Ground
B86	Address/Data 39
B87	Address/Data 37
B88	+5V
B89	Address/Data 35
B90	Address/Data 33
B91	Ground
B92	Reserved
B93	Reserved
B94	Ground

*For the 3.3V card, some +5V power lines are changed to 3.3V.

TABLE 10-12 Small Computer System Interface (SCSI) Description

Pin number	Signal name	Direction
1	D0 RETURN	I/O
2	D0/	I/O
3	D1 RETURN	I/O
4	D1/	I/O
5	D2 RETURN	I/O
6	D2/	I/O
7	D3 RETURN	I/O
8	D3/	I/O
9	D4 RETURN	I/O
10	D4/	I/O
11	D5 RETURN	I/O
12	D5/	I/O
13	D6 RETURN	I/O
14	D6/	I/O
15	D7 RETURN	I/O
16	D7/	I/O
17	PARITY/RETURN	I
18	PARITY	I
19	RETURN	
20	SPARE	
21	RETURN	
22	SPARE	
23	RETURN	
24	SPARE	
25	RETURN	
26	SPARE	
27	RETURN	I
28	SPARE	
29	RETURN	I
30	SPARE	
31	RETURN	I
32	ATTENTION/	O
33	ATTENTION RETURN	O
34	SPARE	

TABLE 10-12 Small Computer System Interface (SCSI) Description (Continued)

Pin number	Signal name	Direction
35	BUSY RETURN	O
36	BSY/(BUSY)	O
37	ACK RETURN	I
38	ACK/ (ACKNOWLEDGE)	I
39	RST RETURN	I
40	RST/ (RESET)	I
41	MSG RETURN	O
42	MSG/ (MESSAGE)	O
43	SEL RETURN	I
44	SEL/ (SELECT)	I
45	C/D RETURN	O
46	C/D/ (COMMAND/DATA)	O
47	REQ RETURN	O
48	REQ/ (REQUEST)	O
49	I/O RETURN	I
50	I/O (IN/OUT)	I/O

TABLE 10-13 Universal Serial Bus (USB) Description

Pin Number	Signal Name
1	+5V
2	Data+
3	Data–
4	–5V (Ground)

TABLE 10-14 Video Electronic Standards Association (VESA) Local Bus (VLB) Description

Pin Number	Signal Name
A1	Data 1
A2	Data 3
A3	Ground
A4	Data 5
A5	Data 7
A6	Data 9
A7	Data 11
A8	Data 13
A9	Data 15
A10	Ground
A11	Data 17
A12	+5V
A13	Data 19
A14	Data 21
A15	Data 23
A16	Data 25
A17	Ground
A18	Data 27
A19	Data 29
A20	Data 31
A21	Address 30
A22	Address 28
A23	Address 26
A24	Ground
A25	Address 24
A26	Address 22
A27	+5V
A28	Address 20
A29	Address 18

TABLE 10-14 Video Electronic Standards Association (VESA) Local Bus (VLB) Description (Continued)

Pin Number	Signal Name
A30	Address 16
A31	Address 14
A32	Address 12
A33	Address 10
A34	Address 8
A35	Ground
A36	Address 6
A37	Address 4
A38	Write Back
A39	Byte Enable 0
A40	+5V
A41	Byte Enable 1
A42	Byte Enable 2
A43	Ground
A44	Byte Enable 3
A45	Address Strobe
A46	——
A47	——
A48	Local Ready
A49	Local Device
A50	Local Request
A51	Ground
A52	Local Grant
A53	+5V
A54	Identification 2
A55	Identification 3
A56	Identification 4
A57	Local Enable
A58	Local Enable Address Strobe

TABLE 10-14 Video Electronic Standards Association (VESA) Local Bus (VLB) Description (Continued)

Pin Number	Signal Name
B1	Data 0
B2	Data 2
B3	Data 4
B4	Data 6
B5	Data 8
B6	Ground
B7	Data 10
B8	Data 12
B9	+5V
B10	Data 14
B11	Data 16
B12	Data 18
B13	Data 20
B14	Ground
B16	Data 22
B16	Data 24
B17	Data 26
B18	Data 28
B19	Data 30
B20	+5V
B21	Address 31
B22	Ground
B23	Address 29
B24	Address 27
B25	Address 25
B26	Address 23
B27	Address 21
B28	Address 19
B29	Ground

TABLE 10-14 Video Electronic Standards Association (VESA) Local Bus (VLB) Description (Continued)

Pin Number	Signal Name
B30	Address 17
B31	Address 15
B32	+5V
B33	Address 13
B34	Address 11
B35	Address 9
B36	Address 7
B37	Address 5
B38	Ground
B39	Address 3
B40	Address 2
B41	Not Connected
B42	Reset
B43	Data/Command
B44	Memory/Input-Output
B45	Write/Read
B46	——
B47	——
B48	Ready Return
B49	Ground
B50	Interrupt 9
B51	Burst Ready
B52	Burst Last
B53	Identification 0
B54	Identification 1
B55	Ground
B56	Local Clock
B57	+5V
B58	Local Bus Size 16 Bits

10-8 COMMON COMPUTER CONNECTORS

Tables 10-15 through 10-22 provide the pin configurations for a number of connectors frequently encountered on computers.

TABLE 10-15 Color Graphics Adapter (CGA) Connector

Pin Number	Signal Name
1	Ground
2	Ground
3	Red
4	Green
5	Blue
6	Intensity
7	Reserved
8	Horizontal Sync
9	Vertical Sync

TABLE 10-16 Enhanced Graphics Adapter (EGA) Connector

Pin Number	Signal Name
1	Ground
2	Secondary Red
3	Primary Red
4	Primary Green
5	Primary Blue
6	Secondary Green/Intensity
7	Secondary Blue
8	Horizontal Sync
9	Vertical Sync

TABLE 10-17 Video Graphics Adapter (VGA) Connector

Pin Number	Signal Name
1	Red
2	Green
3	Blue
4	Monitor Identification Bit 2
5	Ground
6	Red Ground
7	Green Ground
8	Blue Ground
9	Key (No Pin)
10	Sync Ground
11	Monitor Identification Bit 0
12	Monitor Identification Bit 1
13	Horizontal or Composite Sync
14	Vertical Sync
15	Monitor Identification Bit 3

TABLE 10-18 Keyboard Connectors

Pin Number	Signal Name
6 Pin Connector	
1	Data
2	Not Connected
3	Ground
4	+5V
5	Clock
6	Not Connected

TABLE 10-18 Keyboard Connectors (Continued)

Pin Number	Signal Name
5 Pin (XT) Connector	
1	Clock
2	Data
3	Reset
4	Ground
5	+5V

TABLE 10-19 Mouse Connectors

Pin Number	Signal Name
1	Data
2	Not Connected
3	Ground
4	+5V
5	Clock
6	Not Connected

TABLE 10-20 Music Instrument Digital Interface (MIDI) Connectors

Pin Number	Signal Name
Input	
1	Not Connected
2	Not Connected
3	Not Connected
4	Current Source
5	Current Sink
Output	
1	Not Connected
2	Ground
3	Not Connected
4	Current Sink
5	Current Source

TABLE 10-21 Winchester Disk Interface Control (SI 506/412) Connector

Pin number	Signal name	Direction
1	RETURN	I
2	REDUCE WRITE CURRENT/	O
3	RETURN	I
4	HEAD SELECT 2/	O
5	RETURN	I
6	WRITE GATE/	O
7	RETURN	I
8	SEEK COMPLETE/	I
9	RETURN	I
10	TRACK 000/	I
11	RETURN	I
12	WRITE FAULT/	I
13	RETURN	I
14	HEAD SELECT 0/	O
15	RETURN	I
16	RESERVED	
17	RETURN	
18	HEAD SELECT 1/	O
19	RETURN	I
20	INDEX/	I
21	RETURN	I
22	DRIVE READY/	I
23	RETURN	I
24	STEP/	O
25	RETURN	I
26	DRIVE SELECT 1/	O
27	RETURN	I
28	DRIVE SELECT 2/	O
29	RETURN	I
30	DRIVE SELECT 3/	O
31	RETURN	I
32	DRIVE SELECT 4/	O
33	RETURN	I
34	DIRECTION IN/	O

TABLE 10-22 Winchester Disk Interface Data Connector

Pin number	Signal name	Direction
1	DRIVE SELECTED	I
2	RETURN	I
3	RESERVED	
4	RETURN	I
5	RESERVED	
6	RETURN	I
7	RESERVED	
8	RETURN	I
9	RESERVED	
10	RESERVED	
11	RETURN	I
12	RETURN	I
13	RETURN	I
14	+ MFM WRITE DATA	O
15	− MFM WRITE DATA	O
16	RETURN	I
17	RETURN	I
18	+ MFM READ DATA	I
19	− MFM READ DATA	I
20	RETURN	I

10-9 UNIX SUMMARY

Table 10-23 provides a summary of UNIX commands and their associated arguments or options.

TABLE 10-23 UNIX Commands

Command	Argument/Option
alias	[name=['command']]
assign	[-d][-u][-v][device]
at	[-f filename][m][time [date]
	-l
	-r job
awk	[-f program] file
banner	string
bash	[script]
bc	[file]
bg	[job]
cal	[month][year]
calendar	——
cancel	requestID
cat	filename
cd	[directory]
chmod	[-R] permissions filenames
clear	——
cmp	file1 file2
compress	[-v] filenames
cp	[-i] file1 file2
	[-i][-R] file1 directory[/file2]
cpio	-i [-c][-d][-E listname][-u][-v][-V][filenames]
	-o[-c][-v][-V]
	-p[-d][-l][-u][-v][-V] directory
csh	[script]
date	——
df	[-k][directory]

TABLE 10-23 UNIX **Commands (Continued)**

Command	Argument/Option
diff	[-b][-i][-w] file1 file2
	[-b][-i][-w] file1 directory1
	[-b][-i][-r][-w] directory1 directory2
diff3	file1 file2 file3
dircmp	[-d][-s] directory1 directory2
du	[-a][-s] directories
echo	[-n] text
exit	——
fg	[%job]
file	filename
find	[-name filenames][-user username][-atime+days] [-mtimes +days][-print][-exec command { } \;][-ok command { } \;]
finger	[username]
	[@hostname]
grep	[-i][-l][-n][-v] text filenames
head	[-lines] filename
history	——
jobs	——
kill	%job
	[-9] pid
ln	[-n][-s] file newname
	[-n][-s] files directory
lp	[-c][-d printer][-m][-n copies][-o options] [-p pagenumbers][-w] filename
ls	[-a][-l][-p][-r][-R][-t][-x][pathname]
man	[-][-k keywords] topic
mesg	[n|y]
mkdir	directory
more	[-s][-u][+linenum][+/text][filename]
mv	[-i] oldname newname
	[-i] filename directory[/newname]

TABLE 10-23 UNIX **Commands (Continued)**

Command	Argument/Option
nice	command [arguments] &
pack	filenames
passwd	——
pr	[-a][-d][-f][-F][-h text][-l lines][-m][-n][-o offset][-t] [-w width][+pagenumber][-columns] filenames
ps	[-a][-e][-f][-t ttys][-u usernames]
pwd	——
rm	[-i][-r] filenames
rmdir	directory
script	[-a][filenames]
sdiff	[-s][-w width] file1 file2
sed	[-f commandfile][commands] filenames
sleep	time
sort	[-b][-d][-f][-i][-m][-n][-r][-u][+fields][-o outputfile] filenames
spell	[-b][+wordlist] filenames
stty	[charnaem char][sane][[-]tostop][-a]
tail	[-r][-lines] filename
talk	user[@computer]
tar	c|r|t|u|x [v][w][0-9][f tarfile] filenames
tee	[-a] filenames
time	command [arguments]
touch	[-a][-c][-m][-t date] filenames
tty	——
umask	[permissions]
unalias	names
uname	[-s][-a]
uncompress	[-c] filenames
uniq	[-c][-d][-u][-f fields][-s chars][oldfile[newfile]]
unpack	filenames
uucp	[-m] ownfile computer!newfile

TABLE 10-23 UNIX Commands (Continued)

Command	Argument/Option
uudecode	filenames
uuencode	file1 file2
wc	[-c][-l][-w][filename]
who	[-q][am i]
write	username [terminal]

11

Networks

11-1 MODEMS

A series of somewhat standardized acronyms for modem status signals has been developed. These acronyms are listed in Table 11-1.

TABLE 11-1 Modem Status Acronyms

Acronym	Meaning
AA	Auto answer
CA	Compression active
CD	Carrier detect
CTS	Clear to send
DC	Data compression in use
DSR	Data set ready
EC	Error control
FAX	Fax connection
FC	Flow control
HS	High speed
LB	Low battery

TABLE 11-1 Modem Status Acronyms (Continued)

Acronym	Meaning
MR	Modem ready (same as DSR)
OH	Off hook
PWR	Power
RI or RNG	Ringing
RTS	Request to send
RXD or RD	Receiving data
SD	Sending data
TD	Transmitting data (same as SD)
TM	Test mode
TR	Terminal ready (same as DTR)
TXD	Transmitting data (same as SD and TD)
VFC	V.fast mode

11-2 ETHERNET CONFIGURATIONS

Ethernets can be configured to support a variety of architectures characterized by different data rates and maximum separations between segments. Table 11-2 provides a summary of Ethernet configurations.

TABLE 11-2 Ethernet Configurations

Nomenclature	Data Rate (Mbps)	Architecture	Maximum Segment Length (m)	Number of Devices per Segment
10Base-5 thick	10	Bus	500	100
10Base-2 thin	10	Bus	185	30
10Base-T	10	Star	100	1
FOIRL fiber	10	Star	1,000	1

TABLE 11-2 Ethernet Configurations (Continued)

Nomenclature	Data Rate (Mbps)	Architecture	Maximum Segment Length (m)	Number of Devices per Segment
10Base-F fiber	10	Star	2,000	1
100Base-T	100	Star	100	1
100Base-FX	100	Star		
Multimode fiber			2,000	1
Single mode fiber			10,000	1
Gigabit Ethernet	1,000	—	—	—

11-3 DIGITAL SIGNAL CIRCUITS

Table 11-3 specifies digital-signal circuits used in North America, Asia, and Europe. Table 11-4 identifies Synchronous Optical Network (SONET) circuits in terms of the types of service and data rates supported.

TABLE 11-3 Digital Signal Circuits

Service	Number of Voice Channels	*Data Rate (bps)
North America and Asia		
DS0 (T0)	1	64K
DS1 (T1)	24	1.544M
DS1C (T1C)	48	3.152M
DS2 (T2)	96	6.312M
DS3 (T3)	672	44.736M
DS4 (T4)	4,032	274.176M

TABLE 11-3 Digital Signal Circuits (Continued)

Service	Number of Voice Channels	*Data Rate (bps)
Europe		
E1	30	2.048M
E2	120	8.448M
E3	480	34.368M
E4	1,920	139.264M
E5	7,680	565.148M

*K = 1,000, M = 1,000,000

TABLE 11-4 Synchronous Optical Network (SONET) Circuits

Synchronous Transport Signal Service	Optical Carrier Service	Data Rate (Mbps)
STS-1	OC-1	51.84
STS-3	OC-3	155.52
STS-12	OC-12	622.08
STS-48	OC-48	2,488.32
STS-192	OC-192	9,953.28

11-4 DIGITAL SUBSCRIBER LINES (DSL)

DSL circuits are rapidly becoming available either as asymmetric digital subscriber lines (ADSL) or symmetric digital subscriber lines (SDSL). Table 11-5 provides an overview of the capabilities of each type.

TABLE 11-5 Digital Subscriber Lines

Type	*Upstream Rate (bps)	**Downstream Rate (bps)	Cable Pairs Required	Maximum Distance (feet)
Asymmetric				
Carrierless	64K	1.544M	1	8,000
Amplitude Phase	640K	6.312M	1	12,000
(CAP ADSL)				
Discrete	176K	1.544M	1	18,000
MultiTone	224–260K	6.312M	1	12,000
(DMT ADSL)				
Rate Adaptive	128K–1M	600K–7M	1	18,000-25,000
(RADSL)				
Very High Bit	1.6-2.3M	12.96M	1	4,500
Rate (VDSL)		25.82M	1	3,000
		51.84M	1	1,000
Symmetric				
High Bit Rate	1.544M	1.544M	2	12,000
(HDSL)	2.048M	2.048M	3	12,000
Single Line				
(SDSL)	1.544–2.048M	1.544–2.048M	1	10,000
ISDN (IDSL)	128K	128K	1	18,000

*Upstream refers to traffic sent from the subscriber terminal (K = 1,000, M = 1,000,000)

**Downstream refers to traffic received by the subscriber terminal (K = 1,000, M = 1,000,000)

11-5 OPEN SYSTEMS INTERCONNECTION DIAGRAMS

Open System Interconnection (OSI) architecture is a seven-layered model that finds wide application in networking applications. Tables 11-6 through 11-10 show a variety of uses for this approach.

TABLE 11-6 Transmission Control Protocol (TCP)/Internet Protocol (IP)

OSI Layer	Function
7 Application	Network File System (NFS)
6 Presentation	Remote File Service (RFS)
	Server Message Block (SMB)
	Network File System (NFS)
5 Session	Simple Mail Transfer Protocol (SMTP)
	File Transfer Protocol (FTP)
	TELNET Virtual Terminal Protocol
	Simple Network Management Protocol (SNMP)
4 Transport	Transmission Control Protocol (TCP)
	User Datagram Protocol (UDP)
3 Network	Internet Protocol (IP)
	X.25 Packet Protocol
2 Data Link	Various
1 Physical	Various

TABLE 11-7 Windows NT

OSI Layer	Function
7 Application	Redirector
6 Presentation	Server Message Block (SMB)
5 Session 4 Transport 3 Network	Internet Packet Exchange (IPX) Transmission Control Protocol (TCP)
2 Data Link	Network Driver Interface Specification (NDIS)
1 Physical	Various

TABLE 11-8 Novell NetWare

OSI Layer	Function
7 Application	NetWare Shell
6 Presentation	NetWare Core Protocol (NCP)
5 Session	NetBIOS Emulator
	Named Pipes
4 Transport	Sequenced Packet Exchange (SPX)
3 Network	Internet Packet Exchange (IPX)
2 Data Link	Open Data Link Interface (ODI)
	Network Driver Interface Specification (NDIS)
1 Physical	Various

TABLE 11-9 UNIX

OSI Layer	Function
7 Application 6 Presentation	Network Filing System (NFS)
5 Session	Simple Mail Transfer Protocol (SMTP)
	File Transfer Protocol (FTP)
	TELNET Virtual Terminal Protocol
	Simple Network Management Protocol (SNMP)
4 Transport	Transmission Control Protocol (TCP)
3 Network	Internet Protocol (IP)
2 Data Link	Media Access Control
1 Physical	Various

TABLE 11-10 Fiber Distributed Data Interface (FDDI)

OSI Layer	Function
7 Application	NA
6 Presentation	NA
5 Session	NA
4 Transport	NA
3 Network	NA
*2 Data Link	Media Access Control (MAC)
	Logical Link Control (LLC)
	Station Management Function (SMT)
*1 Physical	Station Management Function (SMT)
	Media Interface Connector (MIC)
	Physical Layer Medium Dependent (PMD)
	Physical Layer Protocol (PHY)

*The Station Management Function spans both layers 1 and 2

11-6 VIDEOCONFERENCING

Table 11-11 provides a brief description of International Telecommunications Union (ITU) standards for videoconferencing.

TABLE 11-11 ITU Videoconferencing Standards

Standard	Content
H.245	Protocol for audio and videoconferencing (including flow control, encryption, and jitter management)
H.261	Compression for H.320 videoconferencing transmission
H.262	MPEG-2 compression
H.263	Enhanced H.261 compression
H.310	ATM and broadcast ISDN network videoconferencing with MPEG compression

TABLE 11-11 ITU Videoconferencing Standards (Continued)

Standard	Content
H.320	Digital line videoconferencing with H.261 compression
H.321	ATM and broadband ISDN network videoconferencing
H.322	LAN videoconferencing
H.323	Packet-switched network videoconferencing
H.324	Analog telephone line videoconferencing

11-7 INTERNET ADDRESSES

The Internet applies a two-character code to addresses to identify the country. Table 11-12 lists those codes.

TABLE 11-12 Internet Country Codes

Code	Country/Region
AD	Andorra
AE	United Arab Emirates
AF	Afghanistan
AG	Antiqua and Barbuda
AI	Anguilla
AL	Albania
AM	Armenia
AN	Netherlands Antilles
AO	Angola
AQ	Antarctica
AR	Argentina
AS	American Samoa
AT	Austria
AU	Australia
AW	Aruba
AZ	Azerbaijan
BA	Bosnia-Herzegovina

TABLE 11-12 Internet Country Codes (Continued)

Code	Country/Region
BB	Barbados
BD	Bangladesh
BE	Belgium
BF	Burkina Faso
BG	Bulgaria
BH	Bahrain
BI	Burundi
BJ	Benin
BM	Bermuda
BN	Brunei Darussalam
BO	Bolivia
BR	Brazil
BS	Bahamas
BT	Bhutan
BV	Bouvet Island
BW	Botswana
BY	Belarus
BZ	Belize
CA	Canada
CC	Cocos (Keeling) Islands
CF	Central African Republic
CG	Congo
CH	Switzerland
CI	Ivory Coast
CK	Cook Islands
CL	Chile
CM	Cameroon
CN	China
CO	Colombia
CR	Costa Rica
CU	Cuba
CV	Cape Verde
CX	Christmas Island

TABLE 11-12 Internet Country Codes (Continued)

Code	Country/Region
CY	Cyprus
CZ	Czech Republic
DE	Germany
DJ	Djibouti
DK	Denmark
DM	Dominica
DO	Dominican Republic
DZ	Algeria
EC	Ecuador
EE	Estonia
EG	Egypt
EH	Western Sahara
ER	Eritrea
ES	Spain
ET	Ethiopia
FI	Finland
FJ	Fiji
FK	Falkland Islands (Malvinas)
FM	Micronesia
FO	Faroe Islands
FR	France
FX	France (European Territory)
GA	Gabon
GB	Great Britain
GD	Grenada
GE	Georgia
GF	Guyana (French)
GG	Guernsey (Channel Islands)
GH	Ghana
GI	Gibraltar
GL	Greenland
GM	Gambia
GN	Guinea

TABLE 11-12 Internet Country Codes (Continued)

Code	Country/Region
GP	Guadeloupe (French)
GQ	Equatorial Guinea
GR	Greece
GS	South Georgia and South Sandwich Islands
GT	Guatemala
GU	Guam (US)
GW	Guinea Bissau
GY	Guyana
HK	Hong Kong
HM	Heard and McDonald Islands
HN	Honduras
HR	Croatia
HT	Haiti
HU	Hungary
ID	Indonesia
IE	Ireland
IL	Israel
IM	Isle of Man
IN	India
IO	British Indian Ocean Territory
IQ	Iraq
IR	Iran
IS	Iceland
IT	Italy
JE	Jersey (Channel Islands)
JM	Jamaica
JO	Jordan
JP	Japan
KE	Kenya
KG	Kyrgyz Republic
KH	Cambodia
KI	Kiribati
KM	Comoros

TABLE 11-12 Internet Country Codes (Continued)

Code	Country/Region
KN	St. Kitts Nevis Anguilla
KP	Korea (North)
KR	Korea (South)
KW	Kuwait
KY	Cayman Islands
KZ	Kazakstan
LA	Laos
LB	Lebanon
LC	Saint Lucia
LI	Liechtenstein
LK	Sri Lanka
LR	Liberia
LS	Lesotho
LT	Lithuania
LU	Luxembourg
LV	Latvia
LY	Libya
MA	Morocco
MC	Monaco
MD	Moldova
MG	Madagascar
MH	Marshall Islands
MK	Macedonia
ML	Mali
MM	Myanmar
MN	Mongolia
MO	Macau
MP	Northern Mariana Islands
MQ	Martinique (French)
MR	Mauritania
MS	Montserrat
MT	Malta
MU	Mauritius

TABLE 11-12 Internet Country Codes (Continued)

Code	Country/Region
MV	Maldives
MW	Malawi
MX	Mexico
MY	Malaysia
MZ	Mozambique
NA	Namibia
NC	New Caledonia (French)
NE	Niger
NF	Norfolk Island
NG	Nigeria
NI	Nicaragua
NL	Netherlands
NO	Norway
NP	Nepal
NR	Nauru
NU	Niue
NZ	New Zealand
OM	Oman
PA	Panama
PE	Peru
PF	Polynesia (French)
PG	Papua New Guinea
PH	Philippines
PK	Pakistan
PL	Poland
PM	St. Pierre and Miquelon
PN	Pitcairn
PR	Puerto Rico (US)
PT	Portugal
PW	Palau
PY	Paraguay
QA	Qatar
RE	Reunion (French)

TABLE 11-12 Internet Country Codes (Continued)

Code	Country/Region
RO	Romania
RU	Russian Federation
RW	Rwanda
SA	Saudi Arabia
SB	Solomon Islands
SC	Seychelles
SD	Sudan
SE	Sweden
SG	Singapore
SH	St. Helena
SI	Slovenia
SJ	Svalbard and Jan Mayen Islands
SK	Slovakia (Slovak Republic)
SL	Sierra Leone
SM	San Marino
SN	Senegal
SO	Somalia
SR	Surinam
ST	St. Tome and Principe
SU	Soviet Union
SV	El Salvador
SY	Syria
SZ	Swaziland
TC	Turks and Caicos Islands
TD	Chad
TF	French Southern Territories
TG	Togo
TH	Thailand
TJ	Tadjikistan
TK	Tokelau
TM	Turkmenistan
TN	Tunisia
TO	Tonga

TABLE 11-12 Internet Country Codes (Continued)

Code	Country/Region
TP	East Timor
TR	Turkey
TT	Trinidad and Tobago
TV	Tuvalu
TW	Taiwan
TZ	Tanzania
UA	Ukraine
UG	Uganda
UK	United Kingdom
UM	United States Minor Outlying Islands
US	United States
UY	Uruguay
UZ	Uzbekistan
VA	Vatican City State
VC	St. Vincent and Grenadines
VE	Venezuela
VG	Virgin Islands (British)
VI	Virgin Islands (US)
VN	Vietnam
VU	Vanuatu
WF	Wallis and Futuna Islands
WS	Western Samoa
YE	Yemen
YT	Mayotte
YU	Yugoslavia
ZA	South Africa
ZM	Zambia
ZR	Congo (formerly Zaire)
ZW	Zimbabwe

11-8 DATA COMMUNICATIONS STANDARDS

TABLE 11-13 Data Communications Standards

Organization	Standard	Brief Description
American National Standards Institute (ANSI)	X3.4	ASCII coded character set used extensively in communications
	X3.15	Serial bit and character transmission sequences
	X3.16	Character structure and parity for digital communications
	X3.24	Signal quality and timing for data terminal equipment (DTE) and data communication equipment (DCE) RS-232C interface (identical to EIA RS-334)
	X3.28	Link control for character-oriented protocols
	X3.44	System elements for information channel performance criteria
	X3.66	Advanced data communications control procedure (ADCCP)
Consultative Committee for International Telegraph and Telephone (CCITT)	CCITT V.10	Electrical characteristics for unbalanced serial interface (equivalent to EIA RS-423)
	CCITT V.11	Electrical characteristics for balanced serial interface (equivalent to EIA RS-422)
	CCITT V.24	Functions of DTE/DCE interface lines
	CCITT V.28	Electrical characteristics compatible with RS-232C

TABLE 11-13 Data Communications Standards (Continued)

Organization	Standard	Brief Description
	CCITT X.20	Asynchronous character-oriented equipment for public data network
	CCITT X.21	Synchronous character-oriented equipment for public data network
	CCITT X.25	Public data network interface, link control, and packet exchange
Electronic Industries Association (EIA)	RS-232C, D	Serial interface between DTE and DCE
	RS-334	Signal quality and timing for DTE/DCE RS-232C interface
	RS-422	Electrical characteristics of balanced serial interface
	RS-423	Electrical characteristics of unbalanced serial interface
	RS-449	Mechanical and functional requirements of RS-422/423 interfaces
	RS-485	Multipoint Communications
	RS-530	Connector for RS-422 and RS-423
International Standards Organization (ISO)	ISO DP 7498	Reference model for open systems interconnection (OSI) used as a framework for network architectures
Institute of Electrical and Electronic Engineers (IEEE)	IEEE-802	Local and metropolitan area networks FireWire

11-9 NETWORKING CONNECTORS

Tables 11-14 through 11-20 describe the configurations of connectors frequently used in network applications.

TABLE 11-14 Comparison of EIA Serial Communication Standards

Standard	Signal levels*	Format†	Maximum data rate, bps	Maximum distance, m/ft
RS-232	M: −3 to −25 V S: +3 to +25 V	U	19.2K	15/50
RS-422/RS-449	M: −4 to −6 V S: +4 to +6 V	B	10M	1200/4000
RS-423/RS-449	M: −4 to −6 V S: +4 to +6 V	U	100K	60/200

*M—mark; S—space.
†U—unbalanced; B—balanced.

TABLE 11-15 RS-232 Interface

Pin number	EIA circuit	CCITT equivalent	Description	Signal source
1	AA		Protective ground	
2	BA	103	Transmitted data (TxD)	DTE
3	BB	104	Received data (RxD)	DCE
4	CA	105	Request to send (RTS)	DTE
5	CB	106	Clear to send (CTS)	DCE
6	CC	107	Data set ready (DSR)	DCE
7	AB	102	Signal ground	
8	CF	109	Receive line signal detector	DCE
9			Reserved for testing	
10			Reserved for testing	

TABLE 11-15 RS-232 Interface (Continued)

Pin number	EIA circuit	CCITT equivalent	Description	Signal source
11			Unassigned	
12	SCF		Secondary receive signal detect	DCE
13	SCB	121	Secondary clear to send	DCE
14	SBA	118	Secondary transmit data	DTE
15	DB	114	Transmit signal element timing (DCE)	DCE
16	SBB	119	Secondary receive data	DCE
17	DD	115	Receive signal element timing	DCE
18			Unassigned	
19	SCA	120	Secondary request to send	DTE
20	CD	108.2	Data terminal ready (DTR)	DTE
21	CG	110	Signal quality detector	DCE
22	CE	125	Ring indicator	DCE
23	CH/CI	111/112	Data signal rate select (DTE/DCE)	DTE/DCE
24	DA	113	Transmit signal element timing (DTE)	DTE
25			Unassigned	

TABLE 11-16 RS-449 Interface (Primary Channel)

Pin number	EIA circuit	Description	Signal source	Circuit type*/ category†
1	Shield			
2	SI	Signaling rate indicator	DCE	PC/II
3	Spare			
4	SD	Send data	DTE	PD/I
5	ST	Send timing	DCE	PT/I
6	RD	Receive data	DCE	PD/I
7	RS	Request to send	DTE	PC/I
8	RT	Receive timing	DCE	PT/I

TABLE 11-16 RS-449 Interface (Primary Channel) (Continued)

Pin number	EIA circuit	Description	Signal source	Circuit type*/ category†
9	CS	Clear to send	DCE	PC/I
10	LL	Local loopback	DTE	C/II
11	DM	Data mode	DCE	C/I
12	TR	Terminal ready	DTE	C/I
13	RR	Receiver ready	DCE	PC/I
14	RL	Remote loopback	DTE	C/II
15	IC	Incoming call	DCE	C/II
16	SF/SR	Select frequency/signaling rate	DTE	PC/II
17	TT	Terminal timing	DTE	PT/I
18	TM	Test mode	DCE	C/II
19	SG	Signal ground		CM/I
20	RC	Receive common	DCE	CM/II
21	Spare			
22	SD	Send data common	DTE	PD/I
23	ST	Send timing common	DCE	PT/I
24	RD	Receive data common	DCE	PD/I
25	RS	Request to send common	DTE	PC/I
26	RT	Receive timing common	DCE	PT/I
27	CS	Clear to send common	DCE	PC/I
28	IS	Terminal in service	DTE	C/II
29	DM	Data mode common	DCE	C/I
30	TR	Terminal ready common	DTE	C/I
31	RR	Receiver ready common	DCE	PC/I
32	SS	Select standby	DTE	C/II
33	SQ	Signal quality	DCE	PC/II
34	NS	New signal	DTE	PC/II
35	TT	Terminal timing common	DTE	PT/I
36	SB	Standby indicator	DCE	C/II
37	SC	Send common	DTE	CM/II

*Circuit type: PC—primary channel control; PD—primary channel data; PT—primary channel timing; C—control; CM—common.

†Category I circuits are compatible with RS-422 and RS-423; category II circuits are compatible with RS-423 only.

TABLE 11-17 RS-449 Interface (Secondary Channel)

Pin number	EIA circuit	Description	Signal source	Circuit type*/ category†
1	Shield			
2	SRR	Secondary receiver ready	DCE	SC/II
3	SSD	Secondary send data	DTE	SD/II
4	SRD	Secondary receive data	DCE	SD/II
5	SG	Signal ground		
6	RC	Receive common	DCE	CM/II
7	SRS	Secondary request to send	DTE	SC/II
8	SCS	Secondary clear to send	DCE	SC/II
9	SC	Send common	DTE	CB/II

*Circuit type: SC—secondary channel control; SD—secondary channel data; CM—common.

†Category I circuits are compatible with RS-422 and RS-423; category II circuits are compatible with RS-423 only.

TABLE 11-18 Ethernet 10Base-T and 100Base-T Connector Description

Pin Number	Signal Name
1	Transceive Data+
2	Transceive Data–
3	Receive Data+
4	Not Connected
5	Not Connected
6	Receive Data–
7	Not Connected
8	Not Connected

TABLE 11-19 Ethernet 10Base-T4 Connector Description

Pin Number	Signal Name
1	Transceive Data+
2	Transceive Data–
3	Receive Data+
4	Bidirectional Data+
5	Bidirectional Data–
6	Receive Data–
7	Bidirectional Data+
8	Bidirectional Data–

TABLE 11-20 Infrared Data Association (IrDA) Connector Description

Pin Number	Signal Name
1	+5V
2	Not Connected
3	Receive Data
4	Ground
5	Transmit Data

11-10 FIBRE CHANNEL

Table 11-21 provides a description of the Fibre Channel organizational channel.

TABLE 11-21 Fibre Channel Overview

*Layer	Function	
FC-4	Channels • High Performance Parallel Interface (HIPPI) Framing Protocol • Intelligent Peripheral Interface (IPI) • Single Byte Command Code Set Mappings (SBCCS) • Small Computer System Interface (SCSI)	Networks • ATM Adaptation Layer for computer data (AAL5) • Internet Protocol • IEEE 802.2
FC-3	Common Services	
FC-2	Framing Protocol/Flow Control	
FC-1	Encode/Decode	
FC-0	Physical Link • 133 Mbps • 266 Mbps • 531 Mbps • 1.062 Gbps	

*Note: Although the Fibre Channel specifications comprises a layered hierarchy similar to that of the ISO Open System Interconnection, the two structures differ considerably.

12

Electronics Mathematics

12-1 ALGEBRA

Laws of Exponents

$$a^x a^y = a^{x+y}$$

$$\frac{a^x}{a^y} = a^{x-y}$$

$$(a^x)^y = a^{xy}$$

$$\sqrt[y]{a^x} = a^{x/y}$$

Ratios

$$\text{If } \frac{a}{b} = \frac{c}{d} \text{ then } ad = bc \text{ and } \frac{a}{c} = \frac{b}{d}$$

Solutions to Quadratic Equations

$$\text{If } ax^2 + bx + c = 0 \text{ then } x = \frac{-b \pm \sqrt{b^2 - 4ac}}{2a}$$

Expansions and Factors

$$(a \pm b)^2 = a^2 \pm 2ab + b^2$$

$$(a \pm b)^3 = a^3 \pm 3a^2b + 3ab^2 \pm b^3$$

$$(a \pm b)^4 = a^4 \pm 4a^3b + 6a^2b^2 \pm 4ab^3 + b^4$$

$$(a + b + c)^2 = a^2 + b^2 + c^2 + 2ab + 2ac + 2bc$$

$$a^2 - b^2 = (a + b)(a - b)$$

$$a^2 + b^2 = (a + jb)(a - jb) \text{ where } j = \sqrt{-1}$$

$$a^3 - b^3 = (a - b)(a^2 + ab + b^2)$$

$$a^3 + b^3 = (a + b)(a^2 - ab + b^2)$$

Matrices

A *matrix* is a rectangular array of the form:

$$\begin{pmatrix} a_{11}\ a_{12} \dots a_{1n} \\ a_{21}\ a_{22} \dots a_{2n} \\ \text{-----------------} \\ a_{m1}\ a_{m2} \dots a_{mn} \end{pmatrix}$$

where elements a_{ij} are called *scalars*. Each horizontal line of scalars is a *row*, and a vertical line of scalars is a *column*. In each of the following operations, the matrices A and B, defined below, are used as examples.

$$A = \begin{pmatrix} a_{11}\ a_{12} \ldots a_{1n} \\ a_{21}\ a_{22} \ldots a_{2n} \\ \text{-------------------} \\ a_{m1}\ a_{m2} \ldots a_{mn} \end{pmatrix}$$

$$B = \begin{pmatrix} b_{11}\ b_{12} \ldots b_{1n} \\ b_{21}\ b_{22} \ldots b_{2n} \\ \text{-------------------} \\ b_{m1}\ b_{m2} \ldots b_{mn} \end{pmatrix}$$

Addition of Matrices

$$A + B = \begin{pmatrix} a_{11} + b_{11} & a_{12} + b_{12} \ldots a_{1n} + b_{1n} \\ a_{21} + b_{21} & a_{22} + b_{22} \ldots a_{2n} + b_{2n} \\ \text{----------------} & \text{------------------------} \\ a_{m1} + b_{m1} & a_{m2} + b_{m2} \ldots a_{mn} + b_{mn} \end{pmatrix}$$

Scalar Multiplication

$$kA = \begin{pmatrix} ka_{11} & ka_{12} \ldots ka_{1n} \\ ka_{21} & ka_{22} \ldots ka_{2n} \\ \text{--------} & \text{-------------} \\ ka_{m1} & ka_{m2} \ldots ka_{mn} \end{pmatrix}$$

where k = scalar

Matrix Multiplication
If A is an $m \times p$ matrix (m rows and p columns) and B is a $p \times n$ matrix, then the product of AB is an $m \times n$ matrix. The ij element

in the product matrix is obtained by multiplying the ith row of $A(A_i)$ by the jth column of $B(B_j)$.

$$AB = \begin{pmatrix} a_{11} \dots a_{1p} \\ \text{-------------} \\ a_{i1} \dots a_{ip} \\ \text{-------------} \\ a_{m1} \dots b_{mp} \end{pmatrix} \begin{pmatrix} b_{11} \dots b_{1j} \dots b_{1n} \\ \text{-------------------} \\ \text{-------------------} \\ \text{-------------------} \\ b_{p1} \dots b_{pj} \dots b_{pn} \end{pmatrix}$$

$$= \begin{pmatrix} c_{11} \dots c_{1n} \\ \text{-------------} \\ \dots c_{ij} \dots \\ \text{-------------} \\ c_{m1} \dots c_{mn} \end{pmatrix}$$

where $c_{ij} = \sum_{k=1}^{p} a_{ik} b_{kj}$

$$= a_{i1} b_{1j} + a_{i2} b_{2j} + \dots + a_{ip} b_{pj}$$

Laws of Matrix Addition and Multiplication

$(A + B) + C = A + (B + C)$	$k(A + B) = kA + kB$
$A + 0 = A$	$(k + m)A = kA + mA$
$A + (-A) = 0$	$(km)A = k(mA)$
$A + B = B + A$	$(1)(A) = A$
	$(0)(A) = 0$
$(AB)C = A(BC)$	$(B + C)A = BA + CA$
$A(B + C) = AB + AC$	$k(AB) = (kA)B = A(kB)$

where A, B, C = matrices
k, m = scalars

Determinants

A determinant is a square matrix ($n \times n$) that is summed over all permutations of element products.

1 × 1 Determinant

$$A = |a_{11}| = a_{11}$$

2 × 2 Determinant

$$A = \begin{vmatrix} a_{11} & a_{12} \\ a_{21} & a_{22} \end{vmatrix} = a_{11}a_{22} - a_{12}a_{21}$$

3 × 3 Determinant

$$A = \begin{vmatrix} a_{11} & a_{12} & a_{13} \\ a_{21} & a_{22} & a_{23} \\ a_{31} & a_{32} & a_{33} \end{vmatrix}$$

$$= a_{11}a_{22}a_{33} + a_{12}a_{23}a_{31} + a_{13}a_{21}a_{32} - a_{13}a_{22}a_{31} - a_{12}a_{21}a_{33} - a_{11}a_{23}a_{32}$$

Minors and Cofactors

If we delete the ith row and jth column of a determinant, we obtain, a new *matrix* called the *minor* $|M_{ij}|$. The *cofactor* A_{ij} of the determinant is formed by placing the correct sign on the minor. The cofactor is *scalar*.

$$A_{ij} = (-1)^{i+j} |M_{ij}|$$

EXAMPLE:

$$A = \begin{vmatrix} 2 & 3 & 1 \\ 1 & 2 & 1 \\ 3 & 7 & 8 \end{vmatrix}$$

$$M_{21} = \begin{pmatrix} 3 & 1 \\ 7 & 8 \end{pmatrix}$$

$$A_{21} = (-1)^{2+1} \begin{pmatrix} 3 & 1 \\ 7 & 8 \end{pmatrix} = -(24 - 7) = -17$$

Note that the signs of the minor form a checkerboard pattern:

$$\begin{vmatrix} + & - & + & - & \ldots \\ - & + & - & + & \ldots \\ + & - & + & - & \ldots \\ \text{-----} & & & & \end{vmatrix}$$

Expansion by Cofactors

We use cofactors to compute (expand) the values of a general $n \times n$ determinant. The value of a determinant is equal to the sum of products formed by multiplying the elements of any row (or column) by their respective cofactors. That is:

$$|A| = a_{i1}A_{i1} + a_{i2}A_{i2} + \ldots + a_{in}A_{in}$$

$$= \sum_{j=1}^{n} a_{ij}A_{ij}$$

EXAMPLE:

$$|A| = \begin{vmatrix} 2 & 3 & 1 \\ 1 & 2 & 1 \\ 3 & 7 & 8 \end{vmatrix}$$

Expansion by the Second Row

$$\begin{aligned}
&= (1)(-1)^{2+1}\begin{pmatrix} 3 & 1 \\ 7 & 8 \end{pmatrix} + 2(-1)^{2+2}\begin{pmatrix} 2 & 1 \\ 3 & 8 \end{pmatrix} + 1(-1)^{2+3}\begin{pmatrix} 2 & 3 \\ 3 & 7 \end{pmatrix} \\
&= -(24 - 7) + 2(16 - 3) - (14 - 9) \\
&= -17 + 26 - 5 \\
&= 4
\end{aligned}$$

Rules of Determinants

1. If $|A|$ has a row (or column) of all zeros, then $|A| = 0$.
2. If $|A|$ has two identical rows (or columns), then $|A| = 0$.
3. Interchanging two rows (or columns) of $|A|$ changes its value to $-|A|$.
4. Adding a multiple of a row (or column) of $|A|$ to another row (column) does not change the value of the determinant.

Solving Simultaneous Linear Equations with Determinants

A system of linear equations, such as:

$$\begin{aligned}
a_{11}x_1 + a_{12}x_2 + \ldots a_{1n}x_n &= b_1 \\
a_{21}x_1 + a_{22}x_2 + \ldots a_{2n}x_n &= b_2 \\
&\cdots\cdots \\
a_{n1}x_1 + a_{n2}x_2 + \ldots a_{nn}x_n &= b_n
\end{aligned}$$

can be solved, if $|A| \neq 0$, in this fashion:

$$x_1 = \frac{|\Delta_1|}{|A|}$$

$$x_2 = \frac{|\Delta_2|}{|A|}$$

$$\vdots$$

$$x_n = \frac{|\Delta_n|}{|A|}$$

where $|\Delta_i|$ = determinant obtained by replacing the ith column of $|A|$ by the column of constant terms.

EXAMPLE: Solve:

$$-3x + 2y = 7$$

$$5x + 3y = 1$$

$$|A| = \begin{vmatrix} -3 & 2 \\ 5 & 3 \end{vmatrix} = -9 - (10) = -19$$

Because $|A| \neq 0$ we can proceed:

$$x = \frac{\begin{vmatrix} 7 & 2 \\ 1 & 3 \end{vmatrix}}{-19} = \frac{21 - (2)}{-19} = \frac{19}{-19} = -1$$

$$y = \frac{\begin{vmatrix} -3 & 7 \\ 5 & 1 \end{vmatrix}}{-19} = \frac{-3-35}{-19} = \frac{-38}{-19} = 2$$

12-2 GEOMETRY

Formulas of common plane and solid objects are given in Table 12-1.

TABLE 12-1 Formulas of Plane and Solid Objects

Figure	Properties
Triangle (base b, height h)	$A = bh/2$
Square (side x)	$A = x^2$ $P = 4x$
Rectangle (base b, height h)	$A = bh$ $P = 2(b+h)$

TABLE 12-1 Formulas of Plane and Solid Objects (Continued)

Figure	Properties
Parallelogram 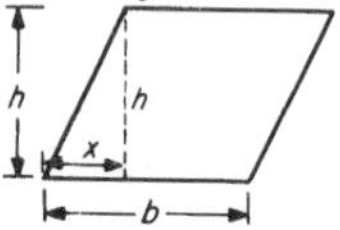	$A = bh$ $P = 2(b\sqrt{h^2+x^2})$
Regular n-sided polygon 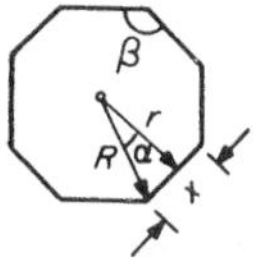	$A = \frac{xrn}{2}$ $P = nx$ $x = 2R \sin \alpha = 2r \tan \alpha$ $\alpha = \frac{180°}{n}$ $\beta = \frac{(n-2)\,180°}{n}$
Circle 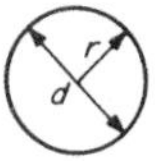	$A = \pi r^2 = \frac{\pi d^2}{4}$ $C = 2\pi r = \pi d$
Trapezoid 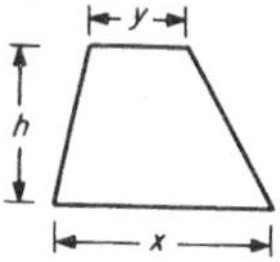	$A = \frac{h(x+y)}{2}$
Sector 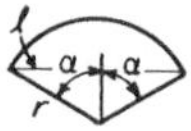	$A = \frac{\alpha r^2}{2}$ α is in radians $= \frac{\alpha \pi r^2}{180}$ α is in degrees Chord length $\ell = 2r \sin \alpha$

TABLE 12-1 Formulas of Plane and Solid Objects (Continued)

Figure	Properties
Ellipse 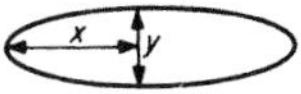	$A = \frac{\pi xy}{2}$ $P = \pi(x + y)\left(1 + \frac{k^2}{4} + \frac{k^4}{64} + \frac{k^6}{256} + \cdots\right)$ where $k = \frac{x - y}{x + y}$
Parabola 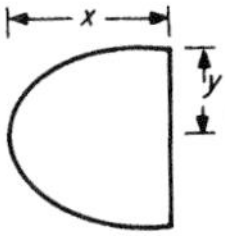	$A = \frac{4xy}{3}$
Rod 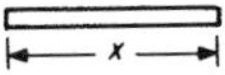	$V = A_c x$
Cube 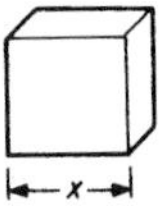	$V = x^3$ $A = 6x^2$
Rectangular parallelopiped 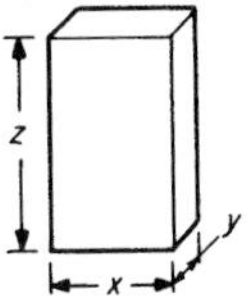	$V = xyz$ $A = 2(xy + xz + yz)$

TABLE 12-1 Formulas of Plane and Solid Objects (Continued)

Figure	Properties
Sphere	$V = \frac{4}{3}\pi r^3$ $A = 4\pi r^2$
Cone	$V = \frac{\pi r^2 h}{3}$ $A = \pi r\sqrt{r^2 + h^2}$
Cylinder	$V = \pi r^2 h$ $A = 2\pi r(r + h)$ (including ends)

12-3 TRIGONOMETRY*

Formulas for Solution of Plane Right Triangles

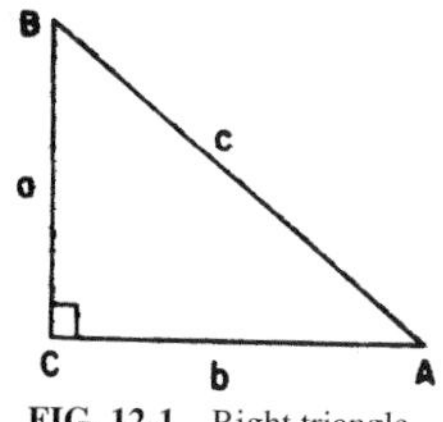

FIG. 12-1 Right triangle.

*This section of Chapter 12 was reprinted from *Handbook of Mathematical Functions*, Abramowitz and Stegun, National Bureau of Standards.

If A, B, and C are the vertices (C the right angle), and a, b and c the sides opposite respectively,

$$\sin A = \frac{a}{c} = \frac{1}{\csc A}$$

$$\cos A = \frac{b}{c} = \frac{1}{\sec A}$$

$$\tan A = \frac{a}{b} = \frac{1}{\cot A}$$

$$\text{versine } A = \text{vers } A = 1 - \cos A$$

$$\text{coversine } A = \text{covers } A = 1 - \sin A$$

$$\text{haversine } A = \text{hav } A = \tfrac{1}{2} \text{ vers } A$$

$$\text{exsecant } A = \text{exsec } A = \sec A - 1$$

Formula for Solution of Plane Triangles

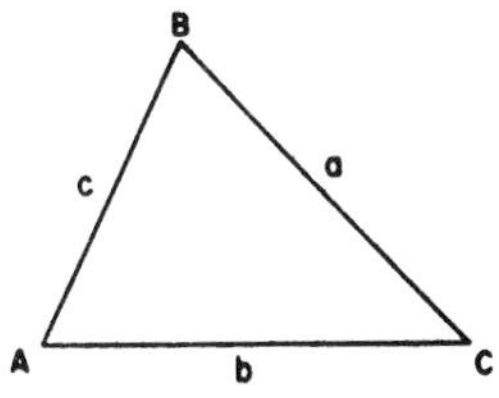

FIG. 12-2 Plane triangle.

In a triangle with angles A, B, and C and sides opposite a, b, and c, respectively,

$$\frac{a}{\sin A} = \frac{b}{\sin B} = \frac{c}{\sin C}$$

$$\cos A = \frac{c^2 + b^2 - a^2}{2bc}$$

$$a = b \cos C + c \cos B$$

$$\frac{a+b}{a-b} = \frac{\tan \frac{1}{2}(A+B)}{\tan \frac{1}{2}(A-B)}$$

$$\text{area} = \frac{bc \sin A}{2} = [s(s-a)(s-b)(s-c)]^{\frac{1}{2}}$$

$$s = \tfrac{1}{2}(a+b+c)$$

Formulas for Solution of Spherical Triangles

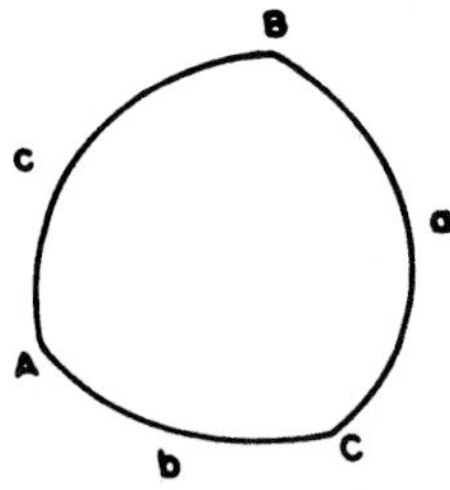

FIG. 12-3 Spherical triangle.

If A, B, and C are the three angles and a, b, and c the opposite sides,

$$\frac{\sin A}{\sin a} = \frac{\sin B}{\sin b} = \frac{\sin C}{\sin c}$$

$$\cos a = \cos b \cos c + \sin b \sin c \cos A$$

$$= \frac{\cos b \cos (c \pm \theta)}{\cos \theta}$$

where $\tan \theta = \tan b \cos A$

$$\cos A = -\cos B \cos C + \sin B \sin C \cos a$$

Circular Functions

Definitions

$$\sin z = \frac{e^{jz} - e^{-jz}}{2j} \quad (z = x + jy) \quad \text{where } j = (-1)^{1/2}$$

$$\cos z = \frac{e^{jz} + e^{-jz}}{2}$$

$$\tan z = \frac{\sin z}{\cos z}$$

$$\csc z = \frac{1}{\sin z}$$

$$\sec z = \frac{1}{\cos z}$$

$$\cot z = \frac{1}{\tan z}$$

Periodic Properties

$$\sin (z + 2k\pi) = \sin z \quad (k \text{ any integer})$$

$$\cos (z + 2k\pi) = \cos z$$

$$\tan (z + k\pi) = \tan z$$

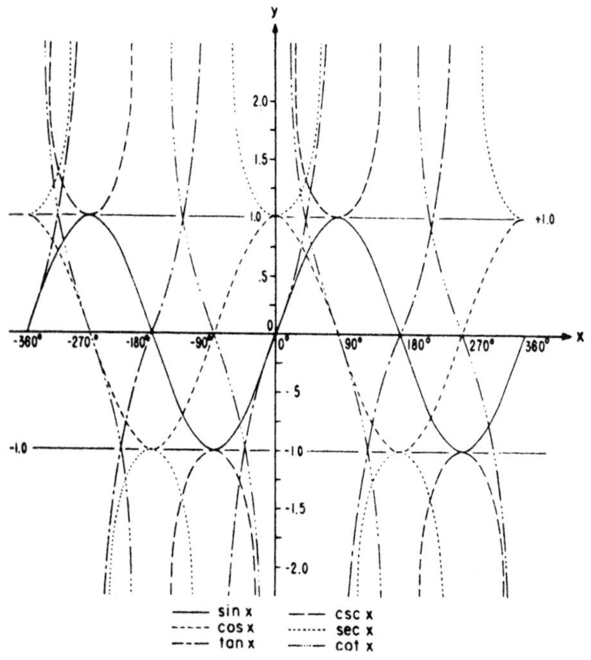

FIG. 12-4 Circular functions.

Relations Between Circular Functions

$$\sin^2 z + \cos^2 z = 1$$

$$\sec^2 z - \tan^2 z = 1$$

$$\csc^2 z - \cot^2 z = 1$$

Negative Angle Formulas

$$\sin(-z) = -\sin z$$

$$\cos(-z) = \cos z$$

$$\tan(-z) = -\tan z$$

Addition Formulas

$$\sin(z_1 + z_2) = \sin z_1 \cos z_2 + \cos z_1 \sin z_2$$

$$\cos(z_1 + z_2) = \cos z_1 \cos z_2 - \sin z_1 \sin z_2$$

$$\tan(z_1 + z_2) = \frac{\tan z_1 + \tan z_2}{1 - \tan z_1 \tan z_2}$$

$$\cot(z_1 + z_2) = \frac{\cot z_1 \cot z_2 - 1}{\cot z_2 + \cot z_1}$$

Half-Angle Formulas

$$\sin \frac{z}{2} = \pm\left(\frac{1 - \cos z}{2}\right)^{1/2}$$

$$\cos\frac{z}{2} = \pm\left(\frac{1+\cos z}{2}\right)^{1/2}$$

$$\tan\frac{z}{2} = \pm\left(\frac{1-\cos z}{1+\cos z}\right)^{1/2} = \frac{1-\cos z}{\sin z} = \frac{\sin z}{1+\cos z}$$

The ambiguity in sign can be resolved with the aid of a diagram.

Transformation of Trigonometric Integrals

If $\tan\frac{u}{2} = z$ then:

$$\sin u = \frac{2z}{1+z^2},\ \cos u = \frac{1-z^2}{1+z^2},\ du = \frac{2}{1+z^2}dz$$

Multiple-Angle Formulas

$$\sin 2z = 2\sin z\cos z = \frac{2\tan z}{1+\tan^2 z}$$

$$\cos 2z = 2\cos^2 z - 1 = 1 - 2\sin^2 z$$

$$= \cos^2 z - \sin^2 z = \frac{1-\tan^2 z}{1+\tan^2 z}$$

$$\tan 2z = \frac{2\tan z}{1-\tan^2 z} = \frac{2\cot z}{\cot^2 z - 1} = \frac{2}{\cot z - \tan z}$$

$$\sin 3z = 3\sin z - 4\sin^3 z$$

$$\cos 3z = -3\cos z + 4\cos^3 z$$

$$\sin 4z = 8 \cos^3 z \sin z - 4 \cos z \sin z$$

$$\cos 4z = 8 \cos^4 z - 8 \cos^2 z + 1$$

Products of Sines and Cosines

$$2 \sin z_1 \sin z_2 = \cos(z_1 - z_2) - \cos (z_1 + z_2)$$

$$2 \cos z_1 \cos z_2 = \cos(z_1 - z_2) + \cos(z_1 + z_2)$$

$$2 \sin z_1 \cos z_2 = \sin(z_1 - z_2) + \sin(z_1 + z_2)$$

Addition and Subtraction of Two Circular Functions

$$\sin z_1 + \sin z_2 = 2 \sin \left(\frac{z_1 + z_2}{2}\right) \cos \left(\frac{z_1 - z_2}{2}\right)$$

$$\sin z_1 - \sin z_2 = 2 \cos \left(\frac{z_1 + z_2}{2}\right) \sin \left(\frac{z_1 - z_2}{2}\right)$$

$$\cos z_1 + \cos z_2 = 2 \cos \left(\frac{z_1 + z_2}{2}\right) \cos \left(\frac{z_1 - z_2}{2}\right)$$

$$\cos z_1 - \cos z_2 = -2 \sin \left(\frac{z_1 + z_2}{2}\right) \sin \left(\frac{z_1 - z_2}{2}\right)$$

$$\tan z_1 \pm \tan z_2 = \frac{\sin (z_1 \pm z_2)}{\cos z_1 \cos z_2}$$

$$\cot z_1 \pm \cot z_2 = \frac{\sin (z_2 \pm z_1)}{\sin z_1 \sin z_2}$$

Relations Between Squares of Sines and Cosines

$$\sin^2 z_1 - \sin^2 z_2 = \sin(z_1 + z_2)\sin(z_1 - z_2)$$

$$\cos^2 z_1 - \cos^2 z_2 = -\sin(z_1 + z_2)\sin(z_1 - z_2)$$

$$\cos^2 z_1 - \sin^2 z_2 = \cos(z_1 + z_2)\cos(z_1 - z_2)$$

TABLE 12-2 Signs of the Circular Functions in the Four Quadrants

Quadrant	sin csc	cos sec	tan cot
I	+	+	+
II	+	−	−
III	−	−	+
IV	−	+	−

TABLE 12-3 Functions of Angles in Any quadrant in Terms of Angles in the First Quadrant. ($0 \le \theta \le \frac{\pi}{2}$, k any integer)

	$-\theta$	$\frac{\pi}{2} \pm \theta$	$\pi \pm \theta$	$\frac{3\pi}{2} \pm \theta$	$2k\pi \pm \theta$
sin	$-\sin\theta$	$\cos\theta$	$\mp\sin\theta$	$-\cos\theta$	$\pm\sin\theta$
cos	$\cos\theta$	$\mp\sin\theta$	$-\cos\theta$	$\pm\sin\theta$	$+\cos\theta$
tan	$-\tan\theta$	$\mp\cot\theta$	$\pm\tan\theta$	$\mp\cot\theta$	$\pm\tan\theta$
csc	$-\csc\theta$	$+\sec\theta$	$\mp\csc\theta$	$-\sec\theta$	$\pm\csc\theta$
sec	$\sec\theta$	$\mp\csc\theta$	$-\sec\theta$	$\pm\csc\theta$	$+\sec\theta$
cot	$-\cot\theta$	$\mp\tan\theta$	$\pm\cot\theta$	$\mp\tan\theta$	$\pm\cot\theta$

TABLE 12-4 Relations Between Circular (or Inverse Circular) Functions

	$\sin x=a$	$\cos x=a$	$\tan x=a$	$\csc x=a$	$\sec x=a$	$\cot x=a$
$\sin x$	a	$(1-a^2)^{\frac{1}{2}}$	$a(1+a^2)^{-\frac{1}{2}}$	a^{-1}	$a^{-1}(a^2-1)^{\frac{1}{2}}$	$(1+a^2)^{-\frac{1}{2}}$
$\cos x$	$(1-a^2)^{\frac{1}{2}}$	a	$(1+a^2)^{-\frac{1}{2}}$	$a^{-1}(a^2-1)^{\frac{1}{2}}$	a^{-1}	$a(1+a^2)^{-\frac{1}{2}}$
$\tan x$	$a(1-a^2)^{-\frac{1}{2}}$	$a^{-1}(1-a^2)^{\frac{1}{2}}$	a	$(a^2-1)^{-\frac{1}{2}}$	$(a^2-1)^{\frac{1}{2}}$	a^{-1}
$\csc x$	a^{-1}	$(1-a^2)^{-\frac{1}{2}}$	$a^{-1}(1+a^2)^{\frac{1}{2}}$	a	$a(a^2-1)^{-\frac{1}{2}}$	$(1+a^2)^{\frac{1}{2}}$
$\sec x$	$(1-a^2)^{-\frac{1}{2}}$	a^{-1}	$(1+a^2)^{\frac{1}{2}}$	$a(a^2-1)^{-\frac{1}{2}}$	a	$a^{-1}(1+a^2)^{\frac{1}{2}}$
$\cot x$	$a^{-1}(1-a^2)^{\frac{1}{2}}$	$a(1-a^2)^{-\frac{1}{2}}$	a^{-1}	$(a^2-1)^{\frac{1}{2}}$	$(a^2-1)^{-\frac{1}{2}}$	a

$\left(0\leq x\leq\frac{\pi}{2}\right)$ Illustration: If $\sin x=a$, $\cot x=a^{-1}(1-a^2)^{\frac{1}{2}}$

$\operatorname{arcsec} a=\operatorname{arccot}(a^2-1)^{-\frac{1}{2}}$

TABLE 12-5 Circular Functions for Certain Angles

	0 0°	$\pi/12$ 15°	$\pi/6$ 30°	$\pi/4$ 45°	$\pi/3$ 60°
sin	0	$\frac{\sqrt{2}}{4}(\sqrt{3}-1)$	1/2	$\sqrt{2}/2$	$\sqrt{3}/2$
cos	1	$\frac{\sqrt{2}}{4}(\sqrt{3}+1)$	$\sqrt{3}/2$	$\sqrt{2}/2$	1/2
tan	0	$2-\sqrt{3}$	$\sqrt{3}/3$	1	$\sqrt{3}$
csc	∞	$\sqrt{2}(\sqrt{3}+1)$	2	$\sqrt{2}$	$2\sqrt{3}/3$
sec	1	$\sqrt{2}(\sqrt{3}-1)$	$2\sqrt{3}/3$	$\sqrt{2}$	2
cot	∞	$2+\sqrt{3}$	$\sqrt{3}$	1	$\sqrt{3}/3$

TABLE 12-5 Circular Functions for Certain Angles (Continued)

	$5\pi/12$ 75°	$\pi/2$ 90°	$7\pi/12$ 105°	$2\pi/3$ 120°
sin	$\frac{\sqrt{2}}{4}(\sqrt{3}+1)$	1	$\frac{\sqrt{2}}{4}(\sqrt{3}+1)$	$\sqrt{3}/2$
cos	$\frac{\sqrt{2}}{4}(\sqrt{3}-1)$	0	$\frac{-\sqrt{2}}{4}(\sqrt{3}-1)$	$-1/2$
tan	$2+\sqrt{3}$	∞	$-(2+\sqrt{3})$	$-\sqrt{3}$
csc	$\sqrt{2}(\sqrt{3}-1)$	1	$\sqrt{2}(\sqrt{3}-1)$	$2\sqrt{3}/3$
sec	$\sqrt{2}(\sqrt{3}+1)$	∞	$-\sqrt{2}(\sqrt{3}+1)$	-2
cot	$2-\sqrt{3}$	0	$-(2-\sqrt{3})$	$-\sqrt{3}/3$

TABLE 12-5 Circular Functions for Certain Angles (Continued)

	$3\pi/4$ 135°	$5\pi/6$ 150°	$11\pi/12$ 165°	π 180°
sin	$\sqrt{2/2}$	$1/2$	$\frac{\sqrt{2}}{4}(\sqrt{3}-1)$	0
cos	$-\sqrt{2}/2$	$-\sqrt{3}/2$	$\frac{-\sqrt{2}}{4}(\sqrt{3}+1)$	-1
tan	-1	$-\sqrt{3}/3$	$-(2-\sqrt{3})$	0
csc	$\sqrt{2}$	2	$\sqrt{2}(\sqrt{3}+1)$	∞
sec	$-\sqrt{2}$	$-2\sqrt{3}/3$	$-\sqrt{2}(\sqrt{3}-1)$	-1
cot	-1	$-\sqrt{3}$	$-(2+\sqrt{3})$	∞

12-4 LOGARITHMS*

Logarithmic Identities

$$\ln (z_1 z_2) = \ln z_1 + \ln z_2.$$

*This section of Chap. 12 was reprinted from *Handbook of Mathematical Functions*, Abramowitz and Stegun, National Bureau of Standards.

$$\ln \frac{z_1}{z_2} = \ln z_1 - \ln z_2$$

$$\ln z^n = n \ln z \quad (n \text{ integer})$$

Special Values

$$\ln 1 = 0$$

$$\ln 0 = -\infty$$

$$\ln(-1) = \pi j \qquad \text{where } j = (-1)^{1/2}$$

$$\ln(\pm j) = \pm \tfrac{1}{2}\pi j$$

$\ln e = 1$, e is the real number such that

$$\int_1^e \frac{dt}{t} = 1$$

$$e = \lim_{n \to \infty} \left(1 + \frac{1}{n}\right)^n = 2.71828\ 18284\ldots$$

Logarithms to General Base

$$\log_a z = \ln z / \ln a$$

$$\log_a z = \frac{\log_b z}{\log_b a}$$

$$\log_a b = \frac{1}{\log_b a}$$

$$\log_e z = \ln z$$

$$\log_{10} z = \ln z/\ln 10 = \log_{10} e \ln z$$

$$= (.43429\ 44819\ldots) \ln z$$

$$\ln z = \ln 10 \log_{10} z = (2.30258\ 50929\ldots) \log_{10} z$$

($\log_e x = \ln x$, called *natural*, *Napierian*, or *hyperbolic logarithm*; $\log_{10} x$, called *common* or *Briggs logarithm*.)

Definition of General Powers

$$\text{If } N = a^z, \text{ then } z = \log_a N$$

$$a^z = \exp(z \ln a)$$

$$\text{If } a = |a| \exp(j \arg a) \quad (-\pi < \arg a \leq \pi)$$

$$|a^z| = |a|^x e^{-y \arg a}$$

$$\arg(a^z) = y \ln |a| + x \arg a$$

$$\ln a^z = z \ln a \quad \text{for one of the values of } \ln a^z$$

$$\ln a^x = x \ln a \quad (a \text{ real and positive})$$

$$|e^z| = e^x$$

$$\arg(e^z) = y$$

$$a^{z1} a^{z2} = a^{z1 + z2}$$

$$a^z b^z = (ab)^z \quad (-\pi < \arg a + \arg b \leq \pi)$$

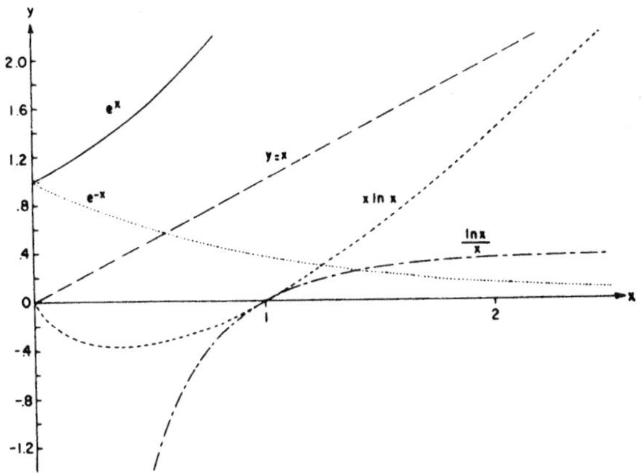

FIG. 12-5 Logarithmic and exponential functions.

Periodic Property

$$e^{z+2\pi kj} = e^{z} \quad (k \text{ any integer})$$

Exponential Identities

$$e^{z1}e^{z2} = e^{z1+z2}$$

$$(e^{z1})^{z2} = e^{z1z2} \quad (-\pi < \mathscr{I} z_1 \leq \pi)$$

The restriction $(-\pi < \mathscr{I} z_1 \leq \pi)$ can be removed if z_2 is an integer.

Special Values

$$e = 2.71828\ 18284\ldots$$

$$e^0 = 1$$

$$e^{\infty} = \infty$$

$$e^{-\infty} = 0$$

$$e^{\pm\pi j} = -1$$

$$e^{\pm\frac{\pi j}{2}} = \pm j$$

$$e^{2\pi kj} = 1 \quad (k \text{ any integer})$$

12-5 DIFFERENTIAL AND INTEGRAL CALCULUS*

Rules for Differentiation and Integration

Derivatives

$$\frac{d}{dx}(cu) = c\,\frac{du}{dx},\ c \text{ constant}$$

$$\frac{d}{dx}(u+v) = \frac{du}{dx} + \frac{dv}{dx}$$

$$\frac{d}{dx}(uv) = u\,\frac{dv}{dx} + v\,\frac{du}{dx}$$

*This section of Chap. 12 was reprinted from *Handbook of Mathematical Functions*, Abramowitz and Stegun, National Bureau of Standards.

$$\frac{d}{dx}(u/v) = \frac{vdu/dx - udv/dx}{v^2}$$

$$\frac{d}{dx}u(v) = \frac{du}{dv}\frac{dv}{dx}$$

$$\frac{d}{dx}(u^v) = u^v\left(\frac{v}{u}\frac{du}{dx} + \ln u\frac{dv}{dx}\right)$$

Leibniz's Theorem for Differentiation of a Product

$$\frac{d^n}{dx^n}(uv) = \frac{d^n u}{dx^n}v + \binom{n}{1}\frac{d^{n-1}u}{dx^{n-1}}\frac{dv}{dx} + \binom{n}{2}\frac{d^{n-2}u}{dx^{n-2}}\frac{d^2v}{dx^2}$$

$$+ \ldots + \binom{n}{r}\frac{d^{n-r}u}{dx^{n-r}}\frac{d^r v}{dx^r} + \ldots + u\frac{d^n v}{dx^n}$$

$$\frac{dx}{dy} = 1/\frac{dy}{dx}$$

$$\frac{d^2x}{dy^2} = \frac{-d^2y}{dx^2}\left(\frac{dy}{dx}\right)^{-3}$$

$$\frac{d^3x}{dy^3} = -\left[\frac{d^3y}{dx^3}\frac{dy}{dx} - 3\left(\frac{d^2y}{dx^2}\right)^2\right]\left(\frac{dy}{dx}\right)^{-5}$$

Integration by Parts

$$\int udv = uv - \int vdu$$

$$\int uvdx = \left(\int udx\right)v - \int\left(\int udx\right)\frac{dv}{dx}\,dx$$

Integrals of Rational Algebraic Functions
(Integration constants are omitted)

$$\int (ax + b)^n dx = \frac{(ax + b)^{n+1}}{a(n + 1)} \quad (n \neq -1)$$

$$\int \frac{dx}{ax + b} = \frac{1}{a} \ln |ax + b|$$

Differentiation Formulas

$$\frac{d}{dz} \sin z = \cos z$$

$$\frac{d}{dz} \cos z = -\sin z$$

$$\frac{d}{dz} \tan z = \sec^2 z$$

$$\frac{d}{dz} \csc z = -\csc z \cot z$$

$$\frac{d}{dz} \sec z = \sec z \tan z$$

$$\frac{d}{dz} \cot z = -\csc^2 z$$

$$\frac{d^n}{dz^n} \sin z = \sin \left(z + \frac{1}{2} n\pi \right)$$

$$\frac{d^n}{dz^n} \cos z = \cos \left(z + \frac{1}{2} n\pi \right)$$

Integration Formulas

$$\int \sin z\, dz = -\cos z$$

$$\int \cos z\, dz = \sin z$$

$$\int \tan z\, dz = -\ln \cos z = \ln \sec z$$

$$\int \csc z\, dz = \ln \tan \frac{z}{2} = \ln (\csc z - \cot z) = \frac{1}{2} \ln \frac{1 - \cos z}{1 + \cos z}$$

$$\int \sec z\, dz = \ln (\sec z + \tan z) = \ln \tan \left(\frac{\pi}{4} + \frac{z}{2}\right) = \text{gd}^{-1} (z)$$

= Inverse Gudermannian Function

$$\text{gd } z = 2 \text{ arc tan } e^{z} - \frac{\pi}{2}$$

$$\int \cot z\, dz = \ln \sin z = -\ln \csc z$$

$$\int z^n \sin z\, dz = -z^n \cos z + n \int z^{n-1} \cos z\, dz$$

$$\int \frac{\sin z}{z^n} dz = \frac{-\sin z}{(n-1)z^{n-1}} + \frac{1}{n-1} \int \frac{\cos z}{z^{n-1}} dz \quad (n > 1)$$

$$\int \frac{z}{\sin^2 z} dz = -z \cot z + \ln \sin z$$

$$\int \tan^n z\, dz = \frac{\tan^{n-1} z}{n-1} - \int \tan^{n-2} z\, dz \quad (n \neq 1)$$

$$\int \cot^n z\, dz = -\frac{\cot^{n-1} z}{n-1} - \int \cot^{n-2} z\, dz \ (n \neq 1)$$

Differentiation Formulas

$$\frac{d}{dz}\ln z = \frac{1}{z}$$

$$\frac{d^n}{dz^n}\ln z = (-1)^{n-1}(n-1)!z^{-n}$$

Integration Formulas

$$\int \frac{dz}{z} = \ln z$$

$$\int \ln z\, dz = z\ln z - z$$

$$\int z^n \ln z\, dz = \frac{z^{n+1}}{n+1}\ln z - \frac{z^{n+1}}{(n+1)^2} \quad (n \neq -1,\ n \text{ integer})$$

$$\int z^n (\ln z)^m\, dz = \frac{z^{n+1}(\ln z)^m}{n+1} - \frac{m}{n+1}\int z^n (\ln z)^{m-1} dz \quad (n \neq -1)$$

TABLE 12-6 Some One-Dimensional Continuous Distribution Functions

Name	Domain	Probability Density Function $f(x)$	Restrictions on Parameters	Mean	Variance
Error function	$-\infty < x < \infty$	$\frac{h}{\sqrt{\pi}} e^{-h^2 x^2}$	$0 < h < \infty$	0	$\frac{1}{2h^2}$
Normal	$-\infty < x < \infty$	$\frac{1}{\sigma\sqrt{2\pi}} e^{-\frac{1}{2}\left(\frac{x-m}{\sigma}\right)^2}$	$-\infty < m < \infty$ $0 < \sigma < \infty$	m	σ^2
Cauchy	$-\infty < x < \infty$	$\frac{1}{\pi\beta} \frac{1}{1+\left(\frac{x-\alpha}{\beta}\right)^2}$	$-\infty < \alpha < \infty$ $0 < \beta < \infty$	not defined	not defined
Exponential	$\alpha \leq x < \infty$	$\frac{1}{\beta} e^{-\left(\frac{x-\alpha}{\beta}\right)}$	$-\infty < \alpha < \infty$ $0 < \beta < \infty$	$\alpha+\beta$	β^2
Laplace, or double exponential	$-\infty < x < \infty$	$\frac{1}{2\beta} e^{-\left\|\frac{x-\alpha}{\beta}\right\|}$	$-\infty < \alpha < \infty$ $0 < \beta < \infty$	α	$2\beta^2$
Extreme-Value,[4] (Fisher-Tippett Type I or doubly exponential)	$-\infty < x < \infty$	$\frac{1}{\beta} \exp(-y-e^{-y})$ with $y = \frac{x-\alpha}{\beta}$	$-\infty < \alpha < \infty$ $0 < \beta < \infty$	$\alpha+\gamma\beta$	$\frac{(\pi\beta)^2}{6}$

[4] γ (Euler's constant) = .57721 56649 . . .

TABLE 12-6 Some One-Dimensional Continuous Distribution Functions (Continued)

Name	Domain	Probability Density Function $f(x)$	Restrictions on Parameters	Mean	Variance
Pearson Type III	$\alpha \le x < \infty$	$\frac{1}{\beta\Gamma(p)} y^{p-1}e^{-y}$ with $y=\frac{x-\alpha}{\beta}$	$-\infty<\alpha<\infty$ $0<\beta<\infty$ $0<p<\infty$	$\alpha+p\beta$	$p\beta^2$
Gamma distribution	$0 \le x < \infty$	$\frac{1}{\Gamma(p)} x^{p-1}e^{-x}$	$0<p<\infty$	p	p
Beta distribution	$0 \le x \le 1$	$\frac{1}{B(a,\ b)} x^{a-1}(1-x)^{b-1}$	$1 \le a<\infty$ $1 \le b<\infty$	$\frac{a}{a+b}$	$\frac{ab}{(a+b)^2(a+b+1)}$
Rectangular, or uniform	$m-\frac{h}{2} \le x \le m+\frac{h}{2}$	$\frac{1}{h}$	$-\infty<m<\infty$ $0<h<\infty$	m	$\frac{h^2}{12}$

Error function	0	0	$e^{\frac{-t^2}{4h^2}}$	$\kappa_1=0,\ \kappa_2=\frac{1}{2h^2}$ $\kappa_n=0$ for $n>2$
Normal	0	0	$e^{imt-\frac{\sigma^2t^2}{2}}$	$\kappa_1=m,\ \kappa_2=\sigma^2,\ \kappa_n=0$ for $n>2$
Cauchy	not defined	not defined	$e^{i\alpha t-\beta\|t\|}$	not defined
Exponential	2	6	$e^{i\alpha t}(1-i\beta t)^{-1}$	$\kappa_1=\alpha+\beta,\ \kappa_n=\beta^n\Gamma(n)$ for $n>1$
Laplace, or double exponential	0	3	$e^{i\alpha t}(1+\beta^2t^2)^{-1}$	$\kappa_1=\alpha,\ \kappa_2=2\beta^2$ $\kappa_{2n+1}=0,\ \kappa_{2n}=\frac{(2n)!}{n}\beta^{2n}$ for $n=1, 2, \ldots$
Extreme-Value,[4] (Fisher-Tippett Type I or doubly exponential)	1.3	2.4	$\Gamma(1-i\beta t)e^{i\alpha t}$	$\kappa_1=\gamma,\ \kappa_2=\frac{(\pi\beta)^2}{6}$ $\kappa_n=\beta^n\Gamma(n)\sum_{r=1}^{\infty}\frac{1}{r^n}$ for $n>2$

TABLE 12-6 Some One-Dimensional Continuous Distribution Functions (Continued)

Name	Skewness γ_1	Excess γ_2	Characteristic function	Cumulants
Pearson Type III	$\frac{2}{\sqrt{p}}$	$6/p$	$e^{i\alpha t}(1-i\beta t)^{-p}$	$\kappa_1=\alpha+\beta p$, $\kappa_n=\beta^n p\Gamma(n)$ for $n>1$
Gamma distribution	$\frac{2}{\sqrt{p}}$	$6/p$	$(1-it)^{-p}$	$\kappa_1=p$, $\kappa_n=p\Gamma(n)$ for $n>1$
Beta distribution	$\frac{2(a-b)}{(a+b+2)}$	See footnote 1.	$M(a, a+b, it)$	
Rectangular, or uniform	0	-1.2	$\frac{2}{ht}\sin\left(\frac{ht}{2}\right)e^{imt}$	$\kappa_1=m$, $\kappa_{2n+1}=0$ $\kappa_{2n}=\frac{h^{2n}B_{2n}}{2n}$ B_{2n} (Bernoulli numbers), $B_2=\frac{1}{6}$, $B_4=-\frac{1}{30}$, . . .

Single point or degenerate	$x=c$ (c a constant)	$p=1$	$-\infty<c<+\infty$	c	0
Binomial	$x_s=s$, for $s=0, 1, 2, \ldots, n$	$\binom{n}{s} p^s(1-p)^{n-s}$	$0<p<1$ $(q=1-p)$	np	npq
Hypergeometric	$x_s=s$, for $s=0, 1, \ldots \min(n, N_1)$	$\dfrac{\binom{N_1}{s}\binom{N_2}{n-s}}{\binom{N_1+N_2}{n}}$	N_1 and N_2 integers, and $n\leq N_1+N_2$ $(N=N_1+N_2$, $p=N_1/N$ and $q=1-p=N_2/N)$	np	$npq\left(\dfrac{N-n}{N-1}\right)$
Poisson	$x_s=s$, for $s=0, 1, 2, \ldots, \infty$	$\dfrac{e^{-m}m^s}{s!}$	$0<m<\infty$	m	m

$$^1\gamma_2=\sqrt{\frac{a+b+1}{ab}}\left\{\frac{3(a+b+1)[2(a+b)^2+ab(a+b-6)]}{ab(a+b+2)(a+b+3)}-3\right\}.$$

TABLE 12-6 Some One-Dimensional Continuous Distribution Functions (Continued)

Name	Domain	Point Probabilities	Restrictions on Parameters	Mean	Variance
Negative binomial	$x_s = s$, for $s = 0, 1, 2, \ldots, \infty$	$\binom{n+s-1}{s} p^n (1-p)^s$	$n \geq 0$ and $0 < p < 1$ ($p = 1/Q$, and $1-p = P/Q$)	nP	nPQ
Geometric	$x_s = s$, for $s = 0, 1, 2, \ldots, \infty$	$p(1-p)^s$	$0 < p < 1$	$\frac{1-p}{p}$	$\frac{1-p}{p^2}$

Single point or degenerate	-------	-------	$e^{i\lambda t}$	$\kappa_1=\lambda,\ \kappa_r=0$ for $r>1$
Binomial	$\dfrac{q-p}{\sqrt{npq}}$	$\dfrac{1-6pq}{npq}$	$(q+pe^{it})^n$	$\kappa_1=np$ $\kappa_{r+1}=pq\,\dfrac{d\kappa_r}{dp}$ for $r\geq 1$
Hypergeometric	$\dfrac{q-p}{\sqrt{npq}}\left(\dfrac{N-1}{N-n}\right)^{\frac{1}{2}}\left(\dfrac{N-2n}{N-2}\right)$	Complicated	$\dfrac{\binom{N_2}{n}}{\binom{N}{n}}\,F(-n,\,-N_1;\,N_2-n+1;\,e^{it})$	Complicated
Poisson	$m^{-\frac{1}{2}}$	m^{-1}	$e^{m(e^{it}-1)}$	$\kappa_r=m$ for $r=1,2,\ldots$

TABLE 12-6 Some One-Dimensional Continuous Distribution Functions (Continued)

Name	Skewness γ_1	Excess γ_2	Characteristic function	Cumulants
Negative binomial	$\dfrac{Q+P}{\sqrt{nPQ}}$	$\dfrac{1+6PQ}{nPQ}$	$(Q-Pe^{it})^{-n}$	$\kappa_1=nP$ $\kappa_{r+1}=PQ\,\dfrac{d\kappa_r}{dQ}$ for $r\geq 1$
Geometric	$\dfrac{2-p}{\sqrt{1-p}}$	$6+\dfrac{p^2}{1-p}$	$p[1-(1-p)e^{it}]^{-1}$	$\kappa_1=\dfrac{1-p}{p}$, $\kappa_{r+1}=-(1-p)\,\dfrac{d\kappa_r}{dp}$, $r\geq 1$

Source: Reprinted from *Handbook of Mathematical Functions*, Abramowitz and Stegun, National Bureau of Standards.

12-6 FOURIER SERIES

$$\tfrac{1}{2}a_0 + \sum_{k=1}^{\infty} (a_k \cos k\omega_0 t + b_k \sin \omega_0 t)$$

where $\omega_0 = \dfrac{2\pi}{T}$

$T = \text{period}$

$$a_k = \frac{2}{T}\int_{-T/2}^{T/2} f(\tau) \sin k\omega_0 \tau d\tau$$

$$b_k = \frac{2}{T}\int_{-T/2}^{T/2} f(\tau) \sin k\omega_0 \tau d\tau$$

See also Table 12-7.

12-7 HYPERBOLIC FUNCTIONS*

Definitions

$$\sinh z = \frac{e^z - e^{-z}}{2} \quad (z = x + iy)$$

$$\cosh z = \frac{e^z + e^{-z}}{2}$$

$$\tanh z = \sinh z/\cosh z$$

$$\operatorname{csch} z = 1/\sinh z$$

$$\operatorname{sech} z = 1/\cosh z$$

$$\coth z = 1/\tanh z$$

*This section of Chap. 12 was reprinted from *Handbook of Mathematical Functions*, Abramowitz and Stegun, National Bureau of Standards.

TABLE 12-7 Fourier Coefficients for Periodic Functions

Function $f(t) = f(t) + T$	Type	Fourier coefficients*
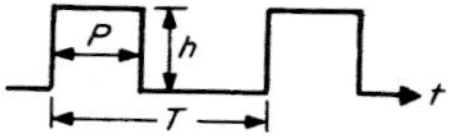	Rectangular pulses	$a_n = \frac{2Ph}{T} \sin c \frac{nP}{T}$ $b_n = 0$
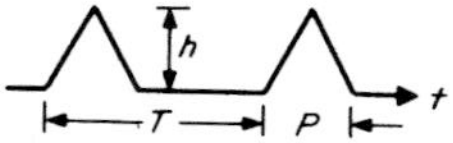	Triangular pulses	$a_n = \frac{Ph}{T} \sin c^2 \frac{nP}{2T}$ $b_n = 0$
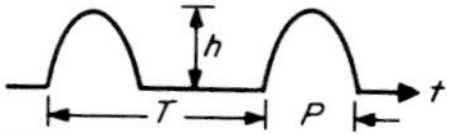	Rectified sine wave	$a_n = \frac{Ph}{T} \left[\sin c \left(\frac{nP}{T} - \frac{1}{2} \right) + \sin c \left(\frac{nP}{T} + \frac{1}{2} \right) \right]$ $b_n = 0$
	Sawtooth wave	$a_n = 0$ $b_n = -\frac{b}{n\pi}$
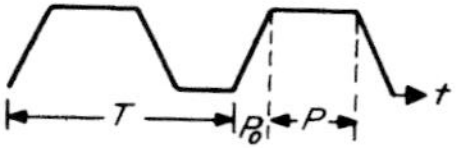	Trapezoid wave	$a_n = \frac{2(P+P_0)h}{T} \sin c \frac{hP_0}{T} \sin c \frac{n(P_0+P)}{T}$ $b_n = 0$

*$\sin x = \frac{\sin \pi x}{\pi x}$

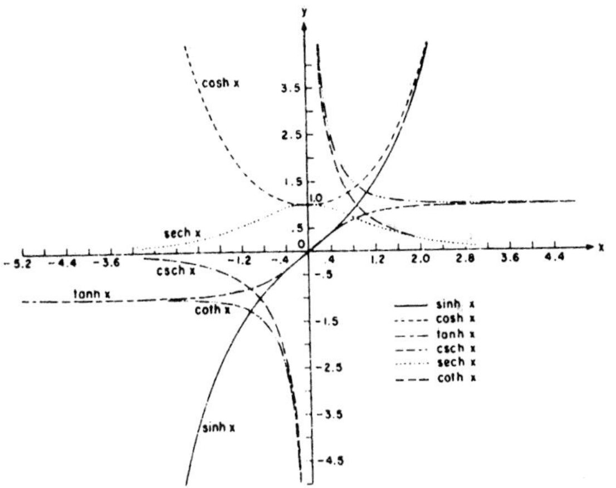

FIG. 12-6 Hyperbolic function.

Relation to Circular Functions

Hyperbolic formulas can be derived from trigonometric identities by replacing z by jz

$$\sinh z = -j \sin jz$$

$$\cosh z = \cos jz$$

$$\tanh z = -j \tan jz$$

$$\operatorname{csch} z = j \csc jz$$

$$\operatorname{sech} z = \sec jz$$

$$\coth z = j \cot jz$$

TABLE 12-8 Special Values of the Hyperbolic Function

z	0	$\frac{\pi}{2}j$	πj	$\frac{3\pi}{2}j$	∞
sinh z	0	j	0	$-j$	∞
cosh z	1	0	-1	0	∞
tanh z	0	∞j	0	$-\infty j$	1
csch z	∞	$-j$	∞	j	0
sech z	1	∞	-1	∞	0
coth z	∞	0	∞	0	1

where $j = (-1)^{1/2}$

12-8 COMPLEX ALGEBRA

In the following equations $j = (-1)^{1/2}$.

Rectangular Form

$$(A + jB) + (C + jD) = (A + C) + j(B + D)$$

$$(A + jB)(C + jD) = (AC - BD) + j(BC + AD)$$

$$\frac{A + jB}{C + jD} = \frac{AC + BD}{C^2 + D^2} + \frac{j(BD - AD)}{C^2 + D^2}$$

$$\frac{1}{A + jB} = \frac{A}{A^2 + B^2} - \frac{jB}{A^2 + B^2}$$

Polar Form

$$A + jB = \rho(\cos\theta + j\sin\theta) = \rho e^{j\theta}$$

Complex Conjugate

$$(A + jB)^* = A - jB$$

12-9 PROBABILITY AND STATISTICS

Definitions

Size of sample population (N)—total number of items in population. Mode (Mo)—score that occurs most frequently in sample population. Median (Mdn)—score below which and above which half the scores in an ordered distribution fall.

Summary Measures of Univariate Distributions

Mean—The Average

$$\overline{X} = \frac{\sum_{i=1}^{N} X_i}{N}$$

Geometric Mean

$$\overline{G} = \sqrt[N]{X_1 X_2 \ldots X_N}$$

Harmonic Mean

$$\overline{H} = \frac{N}{\sum_{i=1}^{N} \frac{1}{X_i}}$$

Variance

$$s^2 = \frac{\sum_{i=1}^{N}(X_i - \overline{X})^2}{N}$$

Standard Deviation

$$s = \sqrt{s^2}$$

Standard Score

$$z = \frac{X_i - \overline{X}}{s}$$

Central Moments

First moment (m_1) = mean

$$\overline{X} = \frac{\sum_{i=1}^{N} X_i}{N}$$

Second moment (m_2) = variance

$$s^2 = \frac{\sum_{i=1}^{N}(X_i - \overline{X})^2}{N}$$

Third moment (m_3) = measure of skewness

$$m_3 = \frac{\sum_{i=1}^{N} X^3_i}{N}$$

Fourth moment (m_4) = measure of degree of kurtosis

$$m_4 = \frac{\sum_{i=1}^{N} X^4_i}{N}$$

Skewness

$$S_k = \frac{m_3}{m_2^{3/2}} = \frac{m_3}{s^3}$$

Kurtosis

$$K_u = \frac{m_4}{m_2^2}$$

Normal Distribution

See Table 12-9.

TABLE 12-9 Normal Probability Function

x	$P(x)$	x	$P(x)$
0.00	0.50000 00000 00000	0.70	0.75803 63477 76927
0.02	0.50797 83137 16902	0.72	0.76423 75022 20749
0.04	0.51595 34368 52831	0.74	0.77035 00028 35210
0.06	0.52392 21826 54107	0.76	0.77637 27075 62401
0.08	0.53188 13720 13988	0.78	0.78230 45624 14267
0.10	0.53982 78372 77029	0.80	0.78814 46014 16604
0.12	0.54775 84260 20584	0.82	0.79389 19464 14187
0.14	0.55567 00048 05907	0.84	0.79954 58067 39551
0.16	0.56355 94628 91433	0.86	0.80510 54787 48192
0.18	0.57142 37159 00901	0.88	0.81057 03452 23288
0.20	0.57925 97094 39103	0.90	0.81593 98746 53241
0.22	0.58706 44226 48215	0.92	0.82121 36203 85629
0.24	0.59483 48716 97796	0.94	0.82639 12196 61376
0.26	0.60256 81132 01761	0.96	0.83147 23925 33162
0.28	0.61026 12475 55797	0.98	0.83645 69406 72308
0.30	0.61791 14221 88953	1.00	0.84134 47460 68543
0.32	0.62551 58347 23320	1.02	0.84613 57696 27265
0.34	0.63307 17360 36028	1.04	0.85083 00496 69019
0.36	0.64057 64332 17991	1.06	0.85542 77003 36091
0.38	0.64802 72924 24163	1.08	0.85992 89099 11231
0.40	0.65542 17416 10324	1.10	0.86433 39390 53618
0.42	0.66275 72731 51751	1.12	0.86864 31189 57270
0.44	0.67003 14463 39407	1.14	0.87285 68494 37202
0.46	0.67724 18897 49653	1.16	0.87697 55969 48657
0.48	0.68438 63034 83778	1.18	0.88099 98925 44800
0.50	0.69146 24612 74013	1.20	0.88493 03297 78292
0.52	0.69846 82124 53034	1.22	0.88876 75625 52166
0.54	0.70540 14837 84302	1.24	0.89251 23029 25413
0.56	0.71226 02811 50973	1.26	0.89616 53188 78700
0.58	0.71904 26911 01436	1.28	0.89972 74320 45558
0.60	0.72574 68822 49927	1.30	0.90319 95154 14390
0.62	0.73237 11065 31017	1.32	0.90658 24910 06528
0.64	0.73891 37003 07139	1.34	0.90987 73275 35548
0.66	0.74537 30853 28664	1.36	0.91308 50380 52915
0.68	0.75174 77695 46430	1.38	0.91620 66775 84986

TABLE 12-9 Normal Probability Function (Continued)

x	$P(x)$	x	$P(x)$
1.40	0.91924 33407 66229	2.10	0.98213 55794 37184
1.42	0.92219 61594 73454	2.12	0.98299 69773 52367
1.44	0.92506 63004 65673	2.14	0.98382 26166 27834
1.46	0.92785 49630 34106	2.16	0.98461 36652 16075
1.48	0.93056 33766 66669	2.18	0.98537 12692 24011
1.50	0.93319 27987 31142	2.20	0.98609 65524 86502
1.52	0.93574 45121 81064	2.22	0.98679 06161 92744
1.54	0.93821 98232 88188	2.24	0.98745 45385 64054
1.56	0.94062 00594 05207	2.26	0.98808 93745 81453
1.58	0.94294 65667 62246	2.28	0.98869 61557 61447
1.60	0.94520 07083 00442	2.30	0.98927 58899 78324
1.62	0.94738 38615 45748	2.32	0.98982 95613 31281
1.64	0.94949 74165 25897	2.34	0.99035 81300 54642
1.66	0.95154 27737 33277	2.36	0.99086 25324 69428
1.68	0.95352 13421 36280	2.38	0.99134 36809 74484
1.70	0.95543 45372 41457	2.40	0.99180 24640 75404
1.72	0.95728 37792 08671	2.42	0.99223 97464 49447
1.74	0.95907 04910 21193	2.44	0.99265 63690 44652
1.76	0.96079 60967 12518	2.46	0.99305 31492 11376
1.78	0.96246 20196 51483	2.48	0.99343 08808 64453
1.80	0.96406 96808 87074	2.50	0.99379 03346 74224
1.82	0.96562 04975 54110	2.52	0.99413 22582 84668
1.84	0.96711 58813 40836	2.54	0.99445 73765 56918
1.86	0.96855 72370 19248	2.56	0.99476 63918 36444
1.88	0.96994 59610 38800	2.58	0.99505 99842 42230
1.90	0.97128 34401 83998	2.60	0.99533 88119 76281
1.92	0.97257 10502 96163	2.62	0.99560 35116 51879
1.94	0.97381 01550 59548	2.64	0.99585 46986 38964
1.96	0.97500 21048 51780	2.66	0.99609 29674 25147
1.98	0.97614 82356 58492	2.68	0.99631 88919 90825
2.00	0.97724 98680 51821	2.70	0.99653 30261 96960
2.02	0.97830 83062 32353	2.72	0.99673 59041 84109
2.04	0.97932 48371 33930	2.74	0.99692 80407 81350
2.06	0.98030 07295 90623	2.76	0.99710 99319 23774
2.08	0.98123 72335 65062	2.78	0.99728 20550 77299

TABLE 12-9 Normal Probability Function (Continued)

x	$P(x)$	x	$P(x)$
2.80	0.99744 48696 69572	3.75	0.99991 15827
2.82	0.99759 88175 25811	3.80	0.99992 76520
2.84	0.99774 43233 08458	3.85	0.99994 09411
2.86	0.99788 17949 59596	3.90	0.99995 19037
2.88	0.99801 16241 45106	3.95	0.99996 09244
2.90	0.99813 41866 99616	4.00	0.99996 83288
2.92	0.99824 98430 71324	4.05	0.99997 43912
2.94	0.99835 89387 65843	4.10	0.99997 93425
2.96	0.99846 18047 88262	4.15	0.99998 33762
2.98	0.99855 87580 82660	4.20	0.99998 66543
3.00	0.99865 01020	4.25	0.99998 93115
3.05	0.99885 57932	4.30	0.99999 14601
3.10	0.99903 23968	4.35	0.99999 31931
3.15	0.99918 36477	4.40	0.99999 45875
3.20	0.99931 28621	4.45	0.99999 57065
3.25	0.99942 29750	4.50	0.99999 66023
3.30	0.99951 65759	4.55	0.99999 73177
3.35	0.99959 59422	4.60	0.99999 78875
3.40	0.99966 30707	4.65	0.99999 83403
3.45	0.99971 97067	4.70	0.99999 86992
3.50	0.99976 73709	4.75	0.99999 89829
3.55	0.99980 73844	4.80	0.99999 92067
3.60	0.99984 08914	4.85	0.99999 93827
3.65	0.99986 88798	4.90	0.99999 95208
3.70	0.99989 22003	4.95	0.99999 96289
		5.00	0.99999 97133

Source: Reprinted from *Handbook of Mathematical Functions*, Abramowitz and Stegun, National Bureau of Standards.

12-10 VECTOR ANALYSIS

Definitions

$$\mathrm{a} = a\boldsymbol{a}$$

where $\mathbf{a}$ = vector
a = magnitude of $\mathbf{a}$
$\boldsymbol{a}$ = unit vector in direction of $\mathbf{a}$

Associative Law

$$\mathbf{a} + (\mathbf{b} + \mathbf{c}) = (\mathbf{a} + \mathbf{b}) + \mathbf{c} = \mathbf{a} + \mathbf{b} + \mathbf{c}$$

Commutative Law

$$\mathbf{a} + \mathbf{b} = \mathbf{b} + \mathbf{a}$$

Dot or Scalar Product

$$\mathbf{a} \cdot \mathbf{b} = \mathbf{b} \cdot \mathbf{a} = ab \cos \theta$$

where θ = angle included by $\mathbf{a}$ and $\mathbf{b}$

Cross or Vector Product

$$\mathbf{a} \times \mathbf{b} = \mathbf{b} \times \mathbf{a} = ab \sin \theta \mathbf{c}$$

where θ = smallest angle formed by rotating $\mathbf{a}$ into $\mathbf{b}$
c = unit vector perpendicular to plane of $\mathbf{a}$ and $\mathbf{b}$ in direction of a right-hand screw rotating from $\mathbf{a}$ to $\mathbf{b}$ through angle θ

Distributive Law

Scalar Multiplication

$$\mathbf{a}\cdot(\mathbf{b} + \mathbf{c}) = \mathbf{a}\cdot\mathbf{b} + \mathbf{a}\cdot\mathbf{c}$$

Vector Multiplication

$$\mathbf{a} \times (\mathbf{b} + \mathbf{c}) = \mathbf{a} \times \mathbf{b} + \mathbf{a} \times \mathbf{c}$$

Triple Product

Scalar

$$\mathbf{a}\cdot(\mathbf{b} \times \mathbf{c}) = (\mathbf{a} \times \mathbf{b})\cdot\mathbf{c} = \mathbf{c}\cdot(\mathbf{a} \times \mathbf{b}) = \mathbf{b}\cdot(\mathbf{c} \times \mathbf{a})$$

Vector

$$\mathbf{a} \times (\mathbf{b} \times \mathbf{c}) = (\mathbf{a}\cdot\mathbf{c})\mathbf{b} - (\mathbf{a}\cdot\mathbf{b})\mathbf{c}$$

$$(\mathbf{a} \times \mathbf{b})\cdot(\mathbf{c} \times \mathbf{d}) = (\mathbf{a}\cdot\mathbf{c})(\mathbf{b}\cdot\mathbf{d}) - (\mathbf{a}\cdot\mathbf{d})(\mathbf{b}\cdot\mathbf{c})$$

$$(\mathbf{a} \times \mathbf{b}) \times (\mathbf{c} \times \mathbf{d}) = (\mathbf{a} \times \mathbf{b}\cdot\mathbf{d})\mathbf{c} - (\mathbf{a} \times \mathbf{b}\cdot\mathbf{c})\mathbf{d}$$

Del Operator

$$\nabla \equiv \mathbf{i}\left(\frac{\partial}{\partial x}\right) + \mathbf{j}\left(\frac{\partial}{\partial y}\right) + \mathbf{k}\left(\frac{\partial}{\partial z}\right)$$

where **i**, **j**, **k** = unit vectors in directions of x, y, z axes, respectively

Gradient

$$\text{grad}\ \phi = \nabla\ \phi$$

$$= \mathbf{i}\left(\frac{\partial \phi}{\partial \mathbf{x}}\right) + \mathbf{j}\left(\frac{\partial \phi}{\partial \mathbf{y}}\right) + \mathbf{k}\left(\frac{\partial \phi}{\partial \mathbf{z}}\right) \text{ in cartesian coordinates}$$

$$\text{grad}\ (\phi + \psi) = \text{grad}\ \phi + \text{grad}\ \psi$$

$$\text{grad}\ (\phi\ \psi) = \psi\ \text{grad}\ \phi + \phi\ \text{grad}\ \psi$$

Divergence

$$\text{div}\ \mathbf{a} = \nabla \bullet \mathbf{a} = \left(\frac{\partial a_x}{\partial \mathbf{x}}\right) + \left(\frac{\partial a_y}{\partial \mathbf{y}}\right) + \left(\frac{\partial a_z}{\partial \mathbf{z}}\right) \text{ in cartesian coordinates}$$

where a_x, a_y, a_z = components of **a** in directions of x, y, z axes, respectively

$$\text{div}\ (\mathbf{a} + \mathbf{b}) = \text{div}\ \mathbf{a} + \text{div}\ \mathbf{b}$$

$$\text{div}\ (\phi\mathbf{a}) = \phi\ \text{div}\ \mathbf{a} + \mathbf{a}\cdot\text{grad}\ \phi$$

Curl

$$\text{curl}\ \mathbf{a} = \nabla \times \mathbf{a}$$

$$= \mathbf{i}\left(\frac{\partial a_z}{\partial \mathbf{y}} - \frac{\partial a_y}{\partial \mathbf{z}}\right) + \mathbf{j}\left(\frac{\partial a_x}{\partial \mathbf{z}} - \frac{\partial a_z}{\partial \mathbf{x}}\right) + \mathbf{k}\left(\frac{\partial a_y}{\partial \mathbf{x}} - \frac{\partial a_x}{\partial \mathbf{y}}\right)$$

$$= \begin{vmatrix} \mathbf{i} & \mathbf{j} & \mathbf{k} \\ \dfrac{\partial}{\partial \mathbf{x}} & \dfrac{\partial}{\partial \mathbf{y}} & \dfrac{\partial}{\partial \mathbf{z}} \\ a_x & a_y & a_z \end{vmatrix} \text{ in cartesian coordinates}$$

$$\text{curl } (\mathbf{a} + \mathbf{b}) = \text{curl } \mathbf{a} + \text{curl } \mathbf{b}$$

$$\text{curl } (\phi\mathbf{a}) = \text{grad } \phi \times \mathbf{a} + \phi \text{ curl } \mathbf{a}$$

$$\text{curl grad } \phi = 0$$

$$\text{div curl } \mathbf{a} = 0$$

$$\text{div } (\mathbf{a} \times \mathbf{b}) = \mathbf{b} \bullet \text{curl } \mathbf{a} - \mathbf{a} \bullet \text{curl } \mathbf{b}$$

Laplacian

$$\nabla^2 \equiv \nabla \bullet \nabla$$

$$\nabla^2\phi = \frac{\partial^2\phi}{\partial \mathbf{x}^2} + \frac{\partial^2\phi}{\partial \mathbf{y}^2} + \frac{\partial^2\phi}{\partial \mathbf{z}^2} \text{ in cartesian coordinates}$$

$$\text{curl curl } \mathbf{a} = \text{grad div } \mathbf{a} - (\mathbf{i}\,\nabla^2 a_x + \mathbf{j}\nabla^2 a_y + \mathbf{k}\nabla^2 a_z)$$

$$= \nabla(\nabla \bullet \mathbf{a}) - \nabla^2\mathbf{a}$$

13

Mathematical Tables and Formulas

TABLE 13-1 Mathematical constants

Constant	Value
π	3.14159 26536
$\log_{10}\pi$	0.49714 98727
$\log_e\pi$	1.14472 98858
$1/\pi$	0.31830 99
e	2.71828 18285
$\log_{10}e$	0.43429 44819
$1/e$	0.36787 94412
$\log_e 2$	0.69314 71806
$\log_{10}2$	0.30102 99957
γ (Euler's constant)	0.57721 56649

13-1 DECIBELS

Definitions

$$\text{dB} = 10 \log \frac{P_2}{P_1}$$

$$= 20 \log \frac{V_2}{V_1}$$

$$= 20 \log \frac{I_2}{I_1}$$

$$= 20 \log \frac{V_2 \sqrt{Z_1}}{V_1 \sqrt{Z_2}} \quad Z_1 \neq Z_2$$

$$= 20 \log \frac{I_2 \sqrt{Z_2}}{I_1 \sqrt{Z_1}} \quad Z_1 \neq Z_2$$

Zero dB Standards

Television

0 dB = 1 mV across a 75-Ω resistor
= 1.333×10^{-8} W

Automobiles

0 dB = 1 mW into a 600-Ω resistor
= 0.7746 V_{rms} across 600 Ω

Reference Levels

dBj = 1 mV
dBk = 1 kW
dBm = 1 mW into 600 Ω
dBs = 1 mW into 600 Ω (Japanese)

dBv = 1 V (not common)

dBw = 1 W

dBvg = voltage gain

VU = 1 mW into 600 Ω

TABLE 13-2 Conversion of Decibels to Voltage and Power Ratios

dB	P_2/P_1		V_2/V_1 or I_2/I_1	
	Gain	Loss	Gain	Loss
0	1.000	1.000	1.000	1.000
0.5	1.122	0.8913	1.059	0.9441
1	1.259	0.7943	1.122	0.8913
1.5	1.413	0.7079	1.189	0.8414
2	1.585	0.6310	1.259	0.7943
2.5	1.778	0.5623	1.344	0.7499
3	1.995	0.5012	1.413	0.7079
3.5	2.239	0.4467	1.496	0.6683
4	2.512	0.3981	1.585	0.6310
4.5	2.818	0.3548	1.679	0.5957
5	3.162	0.3162	1.778	0.5623
6	3.981	0.2512	1.995	0.5012
7	5.012	0.1995	2.239	0.4467
8	6.310	0.1585	2.512	0.3981
9	7.943	0.1259	2.818	0.3548
10	10.000	0.1000	3.162	0.3162
20	10^2	10^{-2}	10.000	0.1000
30	10^3	10^{-3}	31.62	0.03162
40	10^4	10^{-4}	10^2	10^{-2}
50	10^5	10^{-5}	3.162×10^2	3.162×10^{-2}
60	10^6	10^{-6}	10^3	10^{-3}
70	10^7	10^{-7}	3.162×10^3	3.162×10^{-3}
80	10^8	10^{-8}	10^4	10^{-4}
90	10^9	10^{-9}	3.162×10^4	3.162×10^{-4}
100	10^{10}	10^{-10}	10^5	10^{-5}

TABLE 13-3 Operations for the Laplace Transform

Original Function F(t)	*Image Function f(s)*
$F(t)$	$\int_0^\infty e^{-st}F(t)dt$
Inversion Formula	
$\frac{1}{2\pi i}\int_{c-i\infty}^{c+i\infty} e^{ts}f(s)ds$	$f(s)$
Linearity Property	
$AF(t)+BG(t)$	$Af(s)+Bg(s)$
Differentiation	
$F'(t)$	$sf(s)-F(+0)$
$F^{(n)}(t)$	$s^nf(s)-s^{n-1}F(+0)-s^{n-2}F'(+0)-\dots -F^{(n-1)}(+0)$
Integration	
$\int_0^t F(\tau)d\tau$	$\frac{1}{s}f(s)$

$\int_0^t \int_0^\tau F(\lambda) d\lambda d\tau$ $\qquad$ $\frac{1}{s^2} f(s)$

Convolution (Faltung) Theorem

$\int_0^t F_1(t-\tau) F_2(\tau) d\tau = F_1 * F_2$ $\qquad$ $f_1(s) f_2(s)$

Differentiation

$-tF(t)$ $\qquad$ $f'(s)$

$(-1)^n t^n F(t)$ $\qquad$ $f^{(n)}(s)$

Integration

$\frac{1}{t} F(t)$ $\qquad$ $\int_s^\infty f(x) dx$

Linear Transformation

$e^{at} F(t)$ $\qquad$ $f(s-a)$

$\frac{1}{c} F\left(\frac{t}{c}\right) \quad (c>0)$ $\qquad$ $f(cs)$

TABLE 13-3 Operations for the Laplace Transform (Continued)

Original Function $F(t)$	*Image Function $f(s)$*
$\frac{1}{c} e^{(b/c)t} F\left(\frac{t}{c}\right) \quad (c>0)$	$f(cs-b)$
Translation	
$F(t-b)u(t-b) \quad (b>0)$	$e^{-bs}f(s)$
Periodic Functions	
$F(t+a)=F(t)$	$\frac{\int_0^a e^{-st}F(t)dt}{1-e^{-as}}$
$F(t+a)=-F(t)$	$\frac{\int_0^a e^{-st}F(t)dt}{1+e^{-as}}$

Half-Wave Rectification of $F(t)$	
$F(t) \sum_{n=0}^{\infty} (-1)^n u(t-na)$	$\dfrac{f(s)}{1-e^{-as}}$
Full-Wave Rectification of $F(t)$	
$\lvert F(t) \rvert$	$f(s) \coth \dfrac{as}{2}$

Source: Reprinted from *Handbook of Mathematical Functions,* Abramowitz and Stegun, National Bureau of Standards.

TABLE 13-4 Laplace Transforms

$f(s)$	$F(t)$
$\frac{1}{s}$	1
$\frac{1}{s^2}$	t
$\frac{1}{s^n} \quad (n=1, 2, 3, \ldots)$	$\frac{t^{n-1}}{(n-1)!}$
$\frac{1}{\sqrt{s}}$	$\frac{1}{\sqrt{\pi t}}$
$s^{-3/2}$	$2\sqrt{t/\pi}$
$s^{-(n+\frac{1}{2})} \quad (n=1, 2, 3, \ldots)$	$\frac{2^n t^{n-\frac{1}{2}}}{1 \cdot 3 \cdot 5 \ldots (2n-1)\sqrt{\pi}}$
$\frac{\Gamma(k)}{s^k} \quad (k>0)$	t^{k-1}

$\frac{1}{s+a}$	e^{-at}
$\frac{1}{(s+a)^2}$	te^{-at}
$\frac{1}{(s+a)^n}$ $(n=1, 2, 3, \ldots)$	$\frac{t^{n-1}e^{-at}}{(n-1)!}$
$\frac{\Gamma(k)}{(s+a)^k}$ $(k>0)$	$t^{k-1}e^{-at}$
$\frac{1}{(s+a)(s+b)}$ $(a\neq b)$	$\frac{e^{-at}-e^{-bt}}{b-a}$
$\frac{s}{(s+a)(s+b)}$ $(a\neq b)$	$\frac{ae^{-at}-be^{-bt}}{a-b}$
$\frac{1}{(s+a)(s+b)(s+c)}$ (a, b, c distinct constants)	$-\frac{(b-c)e^{-at}+(c-a)e^{-bt}+(a-b)e^{-ct}}{(a-b)(b-c)(c-a)}$

TABLE 13-4 Laplace Transforms (Continued)

$f(s)$	$F(t)$
$\frac{1}{s^2+a^2}$	$\frac{1}{a}\sin at$
$\frac{s}{s^2+a^2}$	$\cos at$
$\frac{1}{s^2-a^2}$	$\frac{1}{a}\sinh at$
$\frac{s}{s^2-a^2}$	$\cosh at$
$\frac{1}{s(s^2+a^2)}$	$\frac{1}{a^2}(1-\cos at)$
$\frac{1}{s^2(s^2+a^2)}$	$\frac{1}{a^3}(at-\sin at)$

$\dfrac{1}{(s^2+a^2)^2}$	$\dfrac{1}{2a^3}(\sin at - at\cos at)$
	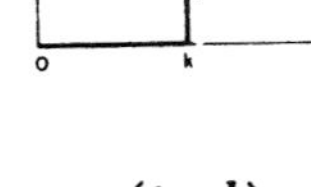
$\dfrac{1}{s}e^{-ks}$	$u(t-k)$
$\dfrac{1}{s^2}e^{-ks}$	$(t-k)u(t-k)$
$\dfrac{1}{s^\mu}e^{-ks} \qquad (\mu>0)$	$\dfrac{(t-k)^{\mu-1}}{\Gamma(\mu)}u(t-k)$

TABLE 13-4 Laplace Transforms (Continued)

$f(s)$	$F(t)$
$\dfrac{1-e^{-ks}}{s}$	$u(t)-u(t-k)$
$\dfrac{1}{s(1-e^{-ks})}=\dfrac{1+\coth \frac{1}{2}ks}{2s}$	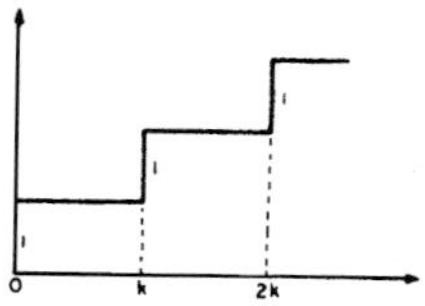 $\sum_{n=0}^{\infty} u(t-nk)$
$\dfrac{1}{s(e^{ks}-a)}$	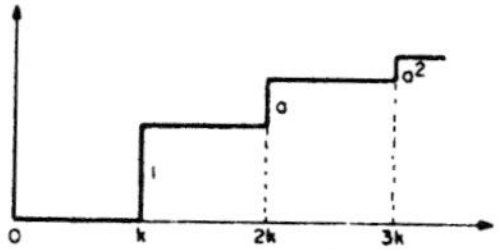 $\sum_{n=1}^{\infty} a^{n-1}u(t-nk)$

	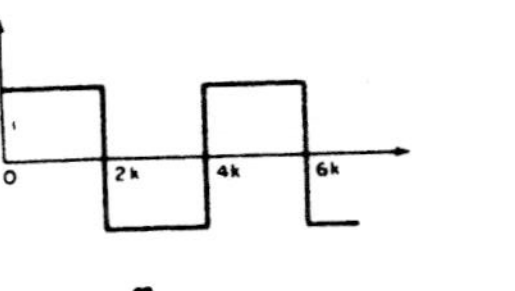
$\frac{1}{s}\tanh ks$	$u(t)+2\sum_{n=1}^{\infty}(-1)^n u(t-2nk)$
	0 k 2k 3k
$\frac{1}{s(1+e^{-ks})}$	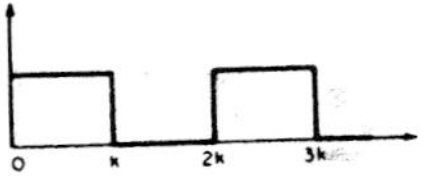 $\sum_{n=0}^{\infty}(-1)^n u(t-nk)$
	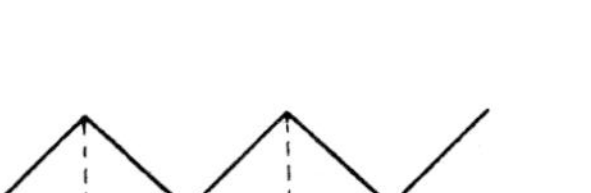
$\frac{1}{s^2}\tanh ks$	$tu(t)+2\sum_{n=1}^{\infty}(-1)^n(t-2nk)u(t-2nk)$

TABLE 13-4 Laplace Transforms (Continued)

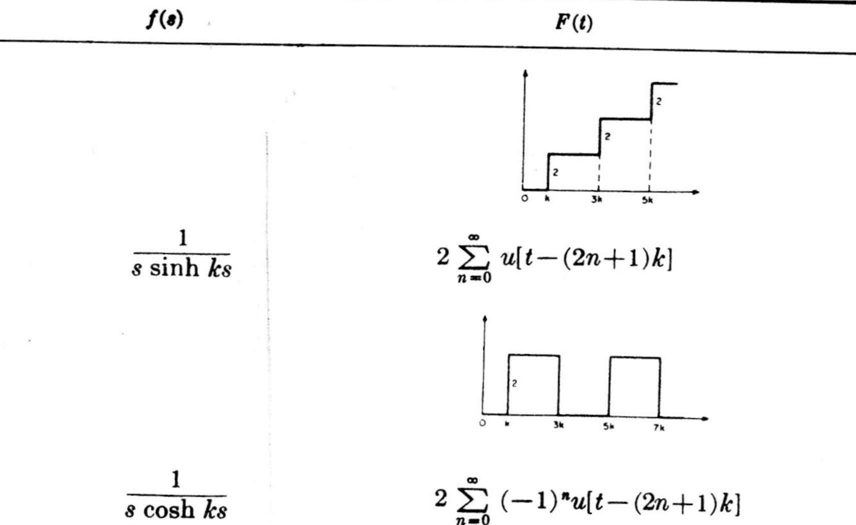

$f(s)$	$F(t)$
$\dfrac{1}{s \sinh ks}$	$2\sum_{n=0}^{\infty} u[t-(2n+1)k]$
$\dfrac{1}{s \cosh ks}$	$2\sum_{n=0}^{\infty} (-1)^n u[t-(2n+1)k]$

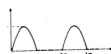

$F(s)$	$f(t)$
$\dfrac{1}{s}\coth ks$	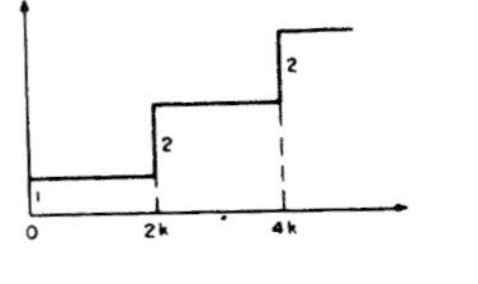 $u(t)+2\sum_{n=1}^{\infty} u(t-2nk)$
$\dfrac{k}{s^2+k^2}\coth\dfrac{\pi s}{2k}$	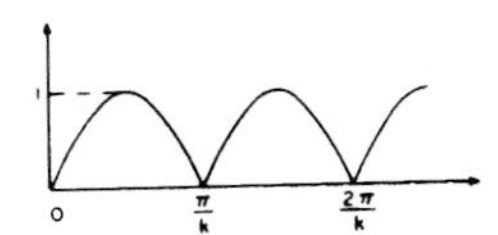 $\lvert\sin kt\rvert$
$\dfrac{1}{(s^2+1)(1-e^{-\pi s})}$	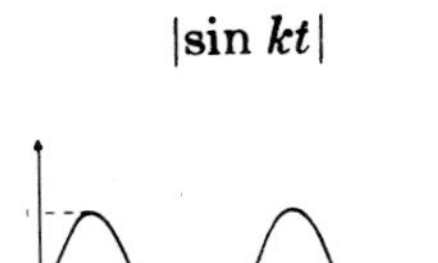 $\sum_{n=0}^{\infty}(-1)^n u(t-n\pi)\sin t$

TABLE 13-4 Laplace Transforms (Continued)

$f(s)$	$F(t)$
$\frac{1}{s} e^{-\frac{k}{s}}$	$J_0(2\sqrt{kt})$
$\frac{1}{\sqrt{s}} e^{-\frac{k}{s}}$	$\frac{1}{\sqrt{\pi t}} \cos 2\sqrt{kt}$
$\frac{1}{s} \ln s$	$-\gamma - \ln t$ ($\gamma = .57721\ 56649 \ldots$ Euler's constant)

Source: Reprinted from *Handbook of Mathematical Functions,* Abramowitz and Stegun, National Bureau of Standards.

TABLE 13-5 2500 Five Digit Random Numbers

53479	81115	98036	12217	59526	40238	40577	39351	43211	69255
97344	70328	58116	91964	26240	44643	83287	97391	92823	77578
66023	38277	74523	71118	84892	13956	98899	92315	65783	59640
99776	75723	03172	43112	83086	81982	14538	26162	24899	20551
30176	48979	92153	38416	42436	26636	83903	44722	69210	69117
81874	83339	14988	99937	13213	30177	47967	93793	86693	98854
19839	90630	71863	95053	55532	60908	84108	55342	48479	63799
09337	33435	53869	52769	18801	25820	96198	66518	78314	97013
31151	58295	40823	41330	21093	93882	49192	44876	47185	81425
67619	52515	03037	81699	17106	64982	60834	85319	47814	08075
61946	48790	11602	83043	22257	11832	04344	95541	20366	55937
04811	64892	96346	79065	26999	43967	63485	93572	80753	96582
05763	39601	56140	25513	86151	78657	02184	29715	04334	15678
73260	56877	40794	13948	96289	90185	47111	66807	61849	44686
54909	09976	76580	02645	35795	44537	64428	35441	28318	99001
42583	36335	60068	04044	29678	16342	48592	25547	63177	75225
27266	27403	97520	23334	36453	33699	23672	45884	41515	04756
49843	11442	66682	36055	32002	78600	36924	59962	68191	62580
29316	40460	27076	69232	51423	58515	49920	03901	26597	33068
30463	27856	67798	16837	74273	05793	02900	63498	00782	35097
28708	84088	65535	44258	33869	82530	98399	26387	02836	36838
13183	50652	94872	28257	78547	55286	33591	61965	51723	14211

TABLE 13-5 2500 Five Digit Random Numbers (Continued)

60796	76639	30157	40295	99476	28334	15368	42481	60312	42770
13486	46918	64683	07411	77842	01908	47796	65796	44230	77230
34914	94502	39374	34185	57500	22514	04060	94511	44612	10485
28105	04814	85170	86490	35695	03483	57315	63174	71902	71182
59231	45028	01173	08848	81925	71494	95401	34049	04851	65914
87437	82758	71093	36833	53582	25986	46005	42840	81683	21459
29046	01301	55343	65732	78714	43644	46248	53205	94868	48711
62035	71886	94506	15263	61435	10369	42054	68257	14385	79436
38856	80048	59973	73368	52876	47673	41020	82295	26430	87377
40666	43328	87379	86418	95841	25590	54137	94182	42308	07361
40588	90087	37729	08667	37256	20317	53316	50982	32900	32097
78237	86556	50276	20431	00243	02303	71029	49932	23245	00862
98247	67474	71455	69540	01169	03320	67017	92543	97977	52728
69977	78558	65430	32627	28312	61815	14598	79728	55699	91348
39843	23074	40814	03713	21891	96353	96806	24595	26203	26009
62880	87277	99895	99965	34374	42556	11679	99605	98011	48867
56138	64927	29454	52967	86624	62422	30163	76181	95317	39264
90804	56026	48994	64569	67465	60180	12972	03848	62582	93855
09665	44672	74762	33357	67301	80546	97659	11348	78771	45011
34756	50403	76634	12767	32220	34545	18100	53513	14521	72120
12157	73327	74196	26668	78087	53636	52304	00007	05708	63538
69384	07734	94451	76428	16121	09300	67417	68587	87932	38840
93358	64565	43766	45041	44930	69970	16964	08277	67752	60292

38879	35544	99563	85404	04913	62547	78406	01017	86187	22072
58314	60298	72394	69668	12474	93059	02053	29807	63645	12792
83568	10227	99471	74729	22075	10233	21575	20325	21317	57124
28067	91152	40568	33705	64510	07067	64374	26336	79652	31140
05730	75557	93161	80921	55873	54103	34801	83157	04534	81368
26687	74223	43546	45699	94469	82125	37370	23966	68926	37664
60675	75169	24510	15100	02011	14375	65187	10630	64421	66745
45418	98635	83123	98558	09953	60255	42071	40930	97992	93085
69872	48026	89755	28470	44130	59979	91063	28766	85962	77173
03765	86366	99539	44183	23886	89977	11964	51581	18033	56239
84686	57636	32326	19867	71345	42002	96997	84379	27991	21459
91512	49670	32556	85189	28023	88151	62896	95498	29423	38138
10737	49307	18307	22246	22461	10003	93157	66984	44919	30467
54870	19676	58367	20905	38324	00026	98440	37427	22896	37637
48967	49579	65369	74305	62085	39297	10309	23173	74212	32272
91430	79112	03685	05411	23027	54735	91550	06250	18705	18909
92564	29567	47476	62804	73428	04535	86395	12162	59647	97726
41734	12199	77441	92415	63542	42115	84972	12454	33133	48467
25251	78110	54178	78241	09226	87529	35376	90690	54178	08561
91657	11563	66036	28523	83705	09956	76610	88116	78351	50877
00149	84745	63222	50533	50159	60433	04822	49577	89049	16162
53250	73200	84066	59620	61009	38542	05758	06178	80193	26466
25587	17481	56716	49749	70733	32733	60365	14108	52573	39391
01176	12182	06882	27562	75456	54261	38564	89054	96911	88906
83531	15544	40834	20296	88576	47815	96540	79462	78666	25353

TABLE 13-5 2500 Five Digit Random Numbers (Continued)

19902	98866	32805	61091	91587	30340	84909	64047	67750	87638
96516	78705	25556	35181	29064	49005	29843	68949	50506	45862
99417	56171	19848	24352	51844	03791	72127	57958	08366	43190
77699	57853	93213	27342	28906	31052	65815	21637	49385	75406
32245	83794	99528	05150	27246	48263	62156	62469	97048	16511
12874	72753	66469	13782	64330	00056	73324	03920	13193	19466
63899	41910	45484	55461	66518	82486	74694	07865	09724	76490
16255	43271	26540	41298	35095	32170	70625	66407	01050	44225
75553	30207	41814	74985	40223	91223	64238	73012	83100	92041
41772	18441	34685	13892	38843	69007	10362	84125	08814	66785
09270	01245	81765	06809	10561	10080	17482	05471	82273	06902
85058	17815	71551	36356	97519	54144	51132	83169	27373	68609
80222	87572	62758	14858	36350	23304	70453	21065	63812	29860
83901	88028	56743	25598	79349	47880	77912	52020	84305	02897
36303	57833	77622	02238	53285	77316	40106	38456	92214	54278
91543	63886	60539	96334	20804	72692	08944	02870	74892	22598
14415	33816	78231	87674	96473	44451	25098	29296	50679	07798
82465	07781	09938	66874	72128	99685	84329	14530	08410	45953
27306	39843	05634	96368	72022	01278	92830	40094	31776	41822
91960	82766	02331	08797	33858	21847	17391	53755	58079	48498
59284	96108	91610	07483	37943	96832	15444	12091	36690	58317
10428	96003	71223	21352	78685	55964	35510	94805	23422	04492

65527	41039	79574	05105	59588	02115	33446	56780	18402	36279
59688	43078	93275	31978	08768	84805	50661	18523	83235	50602
44452	10188	43565	46531	93023	07618	12910	60934	53403	18401
87275	82013	59804	78595	60553	14038	12096	95472	42736	08573
94155	93110	49964	27753	85090	77677	69303	66323	77811	22791
26488	76394	91282	03419	68758	89575	66469	97835	66681	03171
37073	34547	88296	68638	12976	50896	10023	27220	05785	77538
83835	89575	55956	93957	30361	47679	83001	35056	07103	63072
55034	81217	90564	81943	11241	84512	12288	89862	00760	76159
25521	99536	43233	48786	49221	06960	31564	21458	88199	06312
85421	72744	97242	66383	00132	05661	96442	37388	57671	27916
61219	48390	47344	30413	39392	91365	56203	79204	05330	31196
20230	03147	58854	11650	28415	12821	58931	30508	65989	26675
95776	83206	56144	55953	89787	64426	08448	45707	80364	60262
07603	17344	01148	83300	96955	65027	31713	89013	79557	49755
00645	17459	78742	39005	36027	98807	72666	54484	68262	38827
62950	83162	61504	31557	80590	47893	72360	72720	08396	33674
79350	10276	81933	26347	08068	67816	06659	87917	74166	85519
48339	69834	59047	82175	92010	58446	69591	56205	95700	86211
05842	08439	79836	50957	32059	32910	15842	13918	41365	80115
25855	02209	07307	59942	71389	76159	11263	38787	61541	22606
25272	16152	82323	70718	98081	38631	91956	49909	76253	33970
73003	29058	17605	49298	47675	90445	68919	05676	23823	84892
81310	94430	22663	06584	38142	00146	17496	51115	61458	65790

TABLE 13-5 2500 Five Digit Random Numbers (Continued)

10024	44713	59832	80721	63711	67882	25100	45345	55743	67618
84671	52806	89124	37691	20897	82339	22627	06142	05773	03547
29296	58162	21858	33732	94056	88806	54603	00384	66340	69232
51771	94074	70630	41286	90583	87680	13961	55627	23670	35109
42166	56251	60770	51672	36031	77273	85218	14812	90758	23677
78355	67041	22492	51522	31164	30450	27600	44428	96380	26772
09552	51347	33864	89018	73418	81538	77399	30448	97740	18158
15771	63127	34847	05660	06156	48970	55699	61818	91763	20821
13231	99058	93754	36730	44286	44326	15729	37500	47269	13333
50583	03570	38472	73236	67613	72780	78174	18718	99092	64114
99485	57330	10634	74905	90671	19643	69903	60950	17968	37217
54676	39524	73785	48864	69835	62798	65205	69187	05572	74741
99343	71549	10248	76036	31702	76868	88909	69574	27642	00336
35492	40231	34868	55356	12847	68093	52643	32732	67016	46784
98170	25384	03841	23920	47954	10359	70114	11177	63298	99903
02670	86155	56860	02592	01646	42200	79950	37764	82341	71952
36934	42879	81637	79952	07066	41625	96804	92388	88860	68580
56851	12778	24309	73660	84264	24668	16686	02239	66022	64133
05464	28892	14271	23778	88599	17081	33884	88783	39015	57118
15025	20237	63386	71122	06620	07415	94982	32324	79427	70387
95610	08030	81469	91066	88857	56583	01224	28097	19726	71465
09026	40378	05731	55128	74298	49196	31669	42605	30368	96424

81431	99955	52462	67667	97322	69808	21240	65921	12629	92896
21431	59335	58627	94822	65484	09641	41018	85100	16110	32077
95832	76145	11636	80284	17787	97934	12822	73890	66009	27521
99813	44631	43746	99790	86823	12114	31706	05024	28156	04202
77210	31148	50543	11603	50934	02498	09184	95875	85840	71954
13268	02609	79833	66058	80277	08533	28676	37532	70535	82356
44285	71735	26620	54691	14909	52132	81110	74548	78853	31996
70526	45953	79637	57374	05053	31965	33376	13232	85666	86615
88386	11222	25080	71462	09818	46001	19065	68981	18310	74178
83161	73994	17209	79441	64091	49790	11936	44864	86978	34538
50214	71721	33851	45144	05696	29935	12823	01594	08453	52825
97689	29341	67747	80643	13620	23943	49396	83686	37302	95350
12367	23891	31506	90721	18710	89140	58595	99425	22840	08267
38890	30239	34237	22578	74420	22734	26930	40604	10782	80128
80788	55410	39770	93317	18270	21141	52085	78093	85638	81140
02395	77585	08854	23562	33544	45796	10976	44721	24781	09690
73720	70184	69112	71887	80140	72876	38984	23409	63957	44751
61383	17222	55234	18963	39006	93504	18273	49815	52802	69675
39161	44282	14975	97498	25973	33605	60141	30030	77677	49294
80907	74484	39884	19885	37311	04209	49675	39596	01052	43999
09052	65670	63660	34035	06578	87837	28125	48883	50482	55735
33425	24226	32043	60082	20418	85047	53570	32554	64099	52326
72651	69474	73648	71530	55454	19576	15552	20577	12124	50038
04142	32092	83586	61825	35482	32736	63403	91499	37196	02762

TABLE 13-5 2500 Five Digit Random Numbers (Continued)

85226	14193	52213	60746	24414	57858	31884	51266	82293	73553
54888	03579	91674	59502	08619	33790	29011	85193	62262	28684
33258	51516	82032	45233	39351	33229	59464	65545	76809	16982
75973	15957	32405	82081	02214	57143	33526	47194	94526	73253
90638	75314	35381	34451	49246	11465	25102	71489	89883	99708
65061	15498	93348	33566	19427	66826	03044	97361	08159	47485
64420	07427	82233	97812	39572	07766	65844	29980	15533	90114
27175	17389	76963	75117	45580	99904	47160	55364	25666	25405
32215	30094	87276	56896	15625	32594	80663	08082	19422	80717
54209	58043	72350	89828	02706	16815	89985	37380	44032	59366
59286	66964	84843	71549	67553	33867	83011	66213	69372	23903
83872	58167	01221	95558	22196	65905	38785	01355	47489	28170
83310	57080	03366	80017	39601	40698	56434	64055	02495	50880
64545	29500	13351	78647	92628	19354	60479	57338	52133	07114
39269	00076	55489	01524	76568	22571	20328	84623	30188	43904
29763	05675	28193	65514	11954	78599	63902	21346	19219	90286
06310	02998	01463	27738	90288	17697	64511	39552	34694	03211
97541	47607	57655	59102	21851	44446	07976	54295	84671	78755
82968	85717	11619	97721	53513	53781	98941	38401	70939	11319
76878	34727	12524	90642	16921	13669	17420	84483	68309	85241
87394	78884	87237	92086	95633	66841	22906	64989	86952	54700
74040	12731	59616	33697	12592	44891	67982	72972	89795	10587
47896	41413	66431	70046	50793	45920	96564	67958	56369	44725

87778	71697	64148	54363	92114	34037	59061	62051	62049	33526
96977	63143	72219	80040	11990	47698	95621	72990	29047	85893
43820	13285	77811	81697	29937	70750	02029	32377	00556	86687
57203	83960	40096	39234	65953	59911	91411	55573	88427	45573
49065	72171	80939	06017	90323	63687	07932	99587	49014	26452
94250	84270	95798	13477	80139	26335	55169	73417	40766	45170
68148	81382	82383	18674	40453	92828	30042	37412	43423	45138
12208	97809	33619	28868	41646	16734	88860	32636	41985	84615
88317	89705	26119	12416	19438	65665	60989	59766	11418	18250
56728	80359	29613	63052	15251	44684	64681	42354	51029	77680
07138	12320	01073	19304	87042	58920	28454	81069	93978	66659
21188	64554	55618	36088	24331	84390	16022	12200	77559	75661
02154	12250	88738	43917	03655	21099	60805	63246	26842	35816
90953	85238	32771	07305	36181	47420	19681	33184	41386	03249
80103	91308	12858	41293	00325	15013	19579	91132	12720	92603
92630	78240	19267	95457	53497	23894	37708	79862	76471	66418
79445	78735	71549	44843	26104	67318	00701	34986	66751	99723
59654	71966	27386	50004	05358	94031	29281	18544	52429	06080
31524	49587	76612	39789	13537	48086	59483	60680	84675	53014
06348	76938	90379	51392	55887	71015	09209	79157	24440	30244
28703	51709	94456	48396	73780	06436	86641	69239	57662	80181
68108	89266	94730	95761	75023	48464	65544	96583	18911	16391
99938	90704	93621	66330	33393	95261	95349	51769	91616	33238
91543	73196	34449	63513	83834	99411	58826	40456	69268	48562
42103	02781	73920	56297	72678	12249	25270	36678	21313	75767

TABLE 13-5 2500 Five Digit Random Numbers (Continued)

17138	27584	25296	28387	51350	61664	37893	05363	44143	42677
28297	14280	54524	21618	95320	38174	60579	08089	94999	78460
09331	56712	51333	06289	75345	08811	82711	57392	25252	30333
31295	04204	93712	51287	05754	79396	87399	51773	33075	97061
36146	15560	27592	42089	99281	59640	15221	96079	09961	05371
29553	18432	13630	05529	02791	81017	49027	79031	50912	09399
23501	22642	63081	08191	89420	67800	55137	54707	32945	64522
57888	85846	67967	07835	11314	01545	48535	17142	08552	67457
55336	71264	88472	04334	63919	36394	11196	92470	70543	29776
10087	10072	55980	64688	68239	20461	89381	93809	00796	95945
34101	81277	66090	88872	37818	72142	67140	50785	21380	16703
53362	44940	60430	22834	14130	96593	23298	56203	92671	15925
82975	66158	84731	19436	55790	69229	28661	13675	99318	76873
54827	84673	22898	08094	14326	87038	42892	21127	30712	48489
25464	59098	27436	89421	80754	89924	19097	67737	80368	08795
67609	60214	41475	84950	40133	02546	09570	45682	50165	15609
44921	70924	61295	51137	47596	86735	35561	76649	18217	63446
33170	30972	98130	95828	49786	13301	36081	80761	33985	68621
84687	85445	06208	17654	51333	02878	35010	67578	61574	20749
71886	56450	36567	09395	96951	35507	17555	35212	69106	01679
00475	02224	74722	14721	40215	21351	08596	45625	83981	63748
25993	38881	68361	59560	41274	69742	40703	37993	03435	18873

92882	53178	99195	93803	56985	53089	15305	50522	55900	43026
25138	26810	07093	15677	60688	04410	24505	37890	67186	62829
84631	71882	12991	83028	82484	90339	91950	74579	03539	90122
34003	92326	12793	61453	48121	74271	28363	66561	75220	35908
53775	45749	05734	86169	42762	70175	97310	73894	88606	19994
59316	97885	72807	54966	60859	11932	35265	71601	55577	67715
20479	66557	50705	26999	09854	52591	14063	30214	19890	19292
86180	84931	25455	26044	02227	52015	21820	50599	51671	65411
21451	68001	72710	40261	61281	13172	63819	48970	51732	54113
98062	68375	80089	24135	72355	95428	11808	29740	81644	86610
01788	64429	14430	94575	75153	94576	61393	96192	03227	32258
62465	04841	43272	68702	01274	05437	22953	18946	99053	41690
94324	31089	84159	92933	99989	89500	91586	02802	69471	68274
05797	43984	21575	09908	70221	19791	51578	36432	33494	79888
10395	14289	52185	09721	25789	38562	54794	04897	59012	89251
35177	56986	25549	59730	64718	52630	31100	62384	49483	11409
25633	89619	75882	98256	02126	72099	57183	55887	09320	73463
16464	48280	94254	45777	45150	68865	11382	11782	22695	41988

Source: Reprinted from *Handbook of Mathematical Functions*, Abramowitz and Stegun, National Bureau of Standards.

14

Symbols

14-1 GREEK ALPHABET

See Table 14-1.

TABLE 14-1 Greek Alphabet

Greek letter		Name	English equivalent
Α	α	Alpha	a
Β	β	Beta	b
Γ	γ	Gamma	g
Δ	δ	Delta	d
Ε	ϵ	Epsilon	ĕ
Ζ	ζ	Zeta	z
Η	η	Eta	ē
Θ	θ	Theta	th
Ι	ι	Iota	i
Κ	κ	Kappa	k
Λ	λ	Lambda	l

TABLE 14-1 Greek Alphabet (Continued)

Greek letter	Name	English equivalent
Μ μ	Mu	m
Ν ν	Nu	n
Ξ ξ	Xi	x
Ο ο	Omicron	ŏ
Π π	Pi	p
Ρ ρ	Rho	r
Σ σ	Sigma	s
Τ τ	Tau	t
Υ υ	Upsilon	u
Φ φ	Phi	ph
Χ χ	Chi	ch
Ψ ψ	Psi	ps
Ω ω	Omega	ō

14-2 ELECTRONICS SYMBOLS

See Table 14-2.

TABLE 14-2 Electronic Symbols

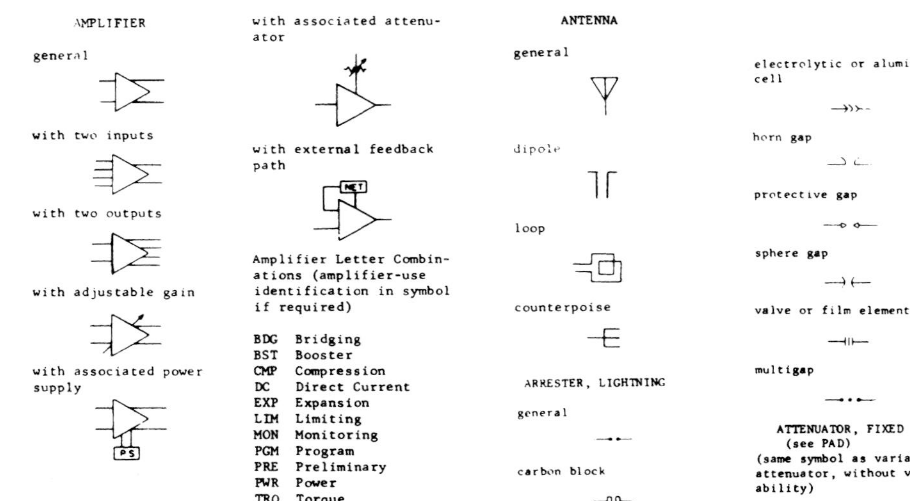

AMPLIFIER
general
with two inputs
with two outputs
with adjustable gain
with associated power supply
PS
with associated attenuator
with external feedback path
NET
Amplifier Letter Combinations (amplifier-use identification in symbol if required)
BDG Bridging
BST Booster
CMP Compression
DC Direct Current
EXP Expansion
LIM Limiting
MON Monitoring
PGM Program
PRE Preliminary
PWR Power
TRQ Torque
ANTENNA
general
dipole
loop
counterpoise
ARRESTER, LIGHTNING
general
carbon block
electrolytic or aluminum cell
horn gap
protective gap
sphere gap
valve or film element
multigap
ATTENUATOR, FIXED (see PAD)
(same symbol as variable attenuator, without variability)

TABLE 14-2 Electronic Symbols (Continued)

ATTENUATOR, VARIABLE

balanced

unbalanced

AUDIBLE SIGNALING DEVICE

bell, electrical; ringer, telephone

buzzer

horn, electrical; loudspeaker; siren; underwater sound hydrophone, projector or transducer

CAPACITOR

general

polarized

adjustable or variable

continuously adjustable or variable differential

phase-shifter

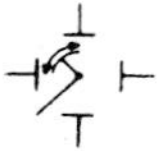

CIRCUIT BREAKER

general

with magnetic overload

OR

drawout type

CIRCUIT ELEMENT

general

Circuit Element Letter Combinations (replaces (*) asterisk)

EG	Equalizer
FAX	Facsimile set
FL	Filter
FL-BE	Filter, band elimination
FL-BP	Filter, band pass
FL-HP	Filter, high pass

CLUTCH; BRAKE

disengaged when operating means is de-energized

OR

engaged when operating means is de-energized

OR

COIL, RELAY and OPERATING

OR

semicircular dot indicates inner end of wiring

OR

CONNECTOR

assembly, movable or stationary portion; jack, plug, or receptacle

OR

jack or receptacle

OR

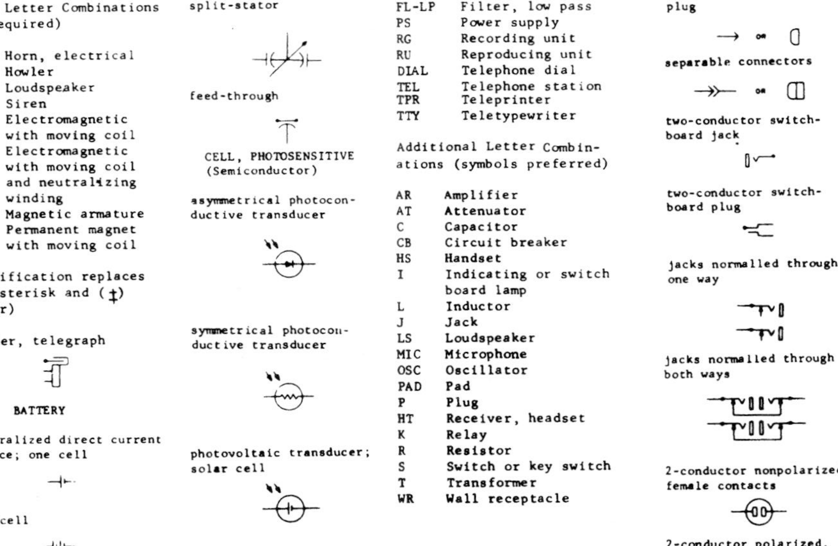

Horn, Letter Combinations (if required)
*HN Horn, electrical
*HW Howler
*LS Loudspeaker
*SN Siren
‡EM Electromagnetic with moving coil
‡EMN Electromagnetic with moving coil and neutralizing winding
‡MG Magnetic armature
‡PM Permanent magnet with moving coil
identification replaces (*) asterisk and (‡) dagger)
sounder, telegraph
BATTERY
generalized direct current source; one cell
multicell
split-stator
feed-through
CELL, PHOTOSENSITIVE (Semiconductor)
asymmetrical photoconductive transducer
symmetrical photoconductive transducer
photovoltaic transducer; solar cell
FL-LP Filter, low pass
PS Power supply
RG Recording unit
RU Reproducing unit
DIAL Telephone dial
TEL Telephone station
TPR Teleprinter
TTY Teletypewriter
Additional Letter Combinations (symbols preferred)
AR Amplifier
AT Attenuator
C Capacitor
CB Circuit breaker
HS Handset
I Indicating or switch board lamp
L Inductor
J Jack
LS Loudspeaker
MIC Microphone
OSC Oscillator
PAD Pad
P Plug
HT Receiver, headset
K Relay
R Resistor
S Switch or key switch
T Transformer
WR Wall receptacle
plug
OR
separable connectors
OR
two-conductor switchboard jack
two-conductor switchboard plug
jacks normalled through one way
jacks normalled through both ways
2-conductor nonpolarized, female contacts
2-conductor polarized, male contacts

TABLE 14-2 Electronic Symbols (Continued)

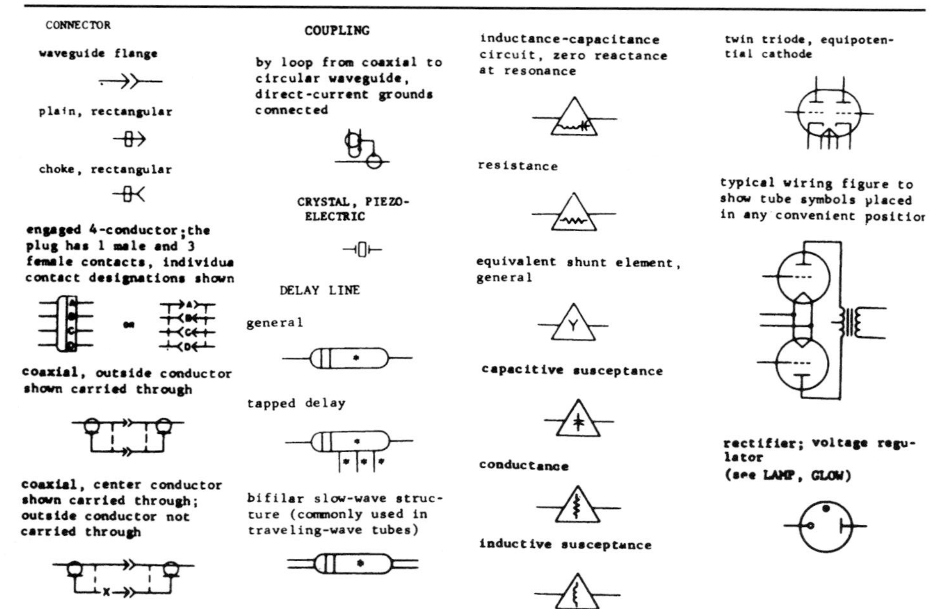

CONNECTOR
waveguide flange
plain, rectangular
choke, rectangular
engaged 4-conductor;the plug has 1 male and 3 female contacts, individua contact designations shown
A
B
C
D
or
coaxial, outside conductor shown carried through
coaxial, center conductor shown carried through; outside conductor not carried through
X
COUPLING
by loop from coaxial to circular waveguide, direct-current grounds connected
CRYSTAL, PIEZO-ELECTRIC
DELAY LINE
general
tapped delay
bifilar slow-wave structure (commonly used in traveling-wave tubes)
inductance-capacitance circuit, zero reactance at resonance
resistance
equivalent shunt element, general
Y
capacitive susceptance
conductance
inductive susceptance
twin triode, equipotential cathode
typical wiring figure to show tube symbols placed in any convenient positior
rectifier; voltage regulator
(see LAMP, GLOW)

mated choke flanges in rectangular waveguide

COUNTER, ELECTROMAGNETIC; MESSAGE REGISTER

general

with a make contact

COUPLER, DIRECTIONAL (common coaxial/waveguide usage)

(common coaxial/waveguide usage)

E-plane aperture-coupling, 30-decibel transmission loss

(length of delay indication replaces (*) asterisk)

DETECTOR, PRIMARY; MEASURING TRANSDUCER (see HALL GENERATOR and THERMAL CONVERTER)

DISCONTINUITY (common coaxial/waveguide usage)

equivalent series element, general

capacitive reactance

inductive reactance

inductance-capacitance circuit, infinite reactance at resonance

inductance-capacitance circuit, infinite susceptance at resonance

inductance-capacitance circuit, zero susceptance at resonance

ELECTRON TUBE

triode

pentode, envelope connected to base terminal

phototube, single and multiplier

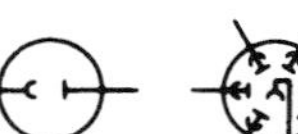

cathode-ray tube, electrostatic and magnetic deflection

mercury-pool tube, ignitor and control grid (see RECTIFIER)

resonant magnetron, coaxial output and permanent magnet

TABLE 14-2 Electronic Symbols (Continued)

ELECTRON TUBE

reflex klystron, integral cavity, aperture coupled

transmit-receive (TR) tube gas filled, tunable integral cavity, aperture coupled, with starter

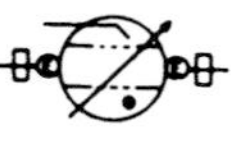

traveling-wave tube (typical)

forward-wave traveling-wave-tube amplifier shown with four grids, having slow-wave structure with attenuation, magnetic focusing by external permanent magnet, rf input and rf output coupling each E-plane aperture to external rectangular waveguide

GOVERNOR (Contact-making)

contacts shown here as closed

HALL GENERATOR

HANDSET

general

operator's set with pushto talk switch

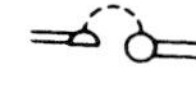

HYBRID

general

junction (common coaxial/waveguide usage)

KEY, TELEGRAPH

LAMP (44)

ballast lamp; ballast tube

lamp, fluorescent, 2 and 4 terminal

lamp, glow; neon lamp a-c

d-c

lamp, incandescent

indicating lamp; switchboard lamp (see VISUAL SIGNALING DEVICE)

electric inverter

(elec. invtr. output becomes 1-state if and only if the input is 1-state) (elec. invtr. output is more pos. if and only if input is less pos.)

level (relative)

1-state is less +   1-state is more +

(symbol is a rt. triangle pointing in direction of flow)

an AND func. with input 1-states at more pos. level and output 1-state at less pos. level

single shot (one output)

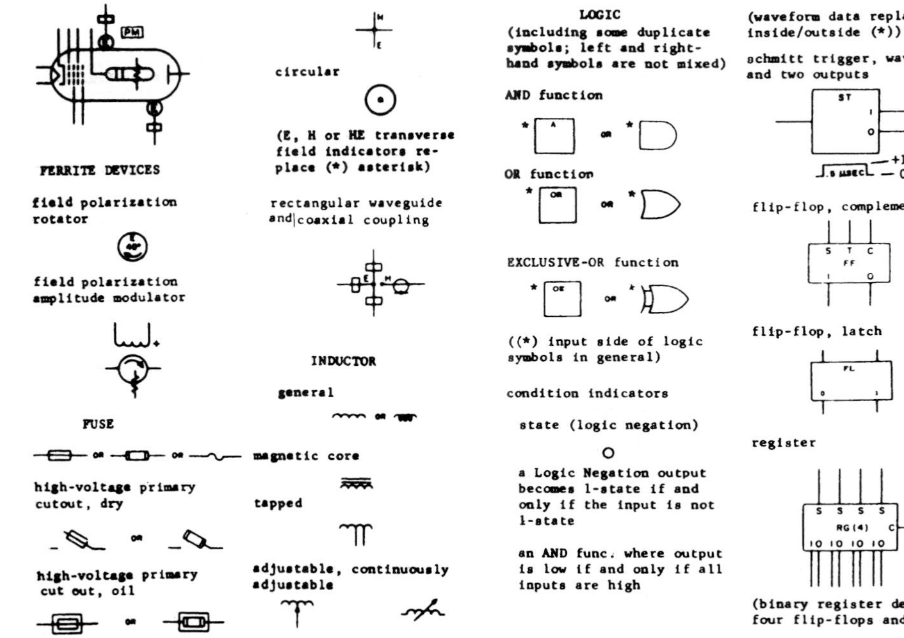

PM
FERRITE DEVICES
field polarization rotator
field polarization amplitude modulator
FUSE
OR
high-voltage primary cutout, dry
high-voltage primary cut out, oil
H
E
circular
(E, H or HE transverse field indicators replace (*) asterisk)
rectangular waveguide and coaxial coupling
INDUCTOR
general
magnetic core
tapped
adjustable, continuously adjustable
LOGIC
(including some duplicate symbols; left and right-hand symbols are not mixed)
AND function
A
OR function
EXCLUSIVE-OR function
((*) input side of logic symbols in general)
condition indicators
state (logic negation)
a Logic Negation output becomes 1-state if and only if the input is not 1-state
an AND func. where output is low if and only if all inputs are high
(waveform data replaces inside/outside (*))
schmitt trigger, waveform and two outputs
ST
+13v
0v
flip-flop, complementary
S T C
FF
flip-flop, latch
FL
register
S S S S
RG(4)
C
4
10 10 10 10
(binary register denoting four flip-flops and bits)

TABLE 14-2 Electronic Symbols (Continued)

LOGIC

amplifier (see AMPLIFIER)

AR OR

channel path(s) (see PATH, TRANSMISSION)

magnetic heads (see PICKUP HEAD)

oscillator (see OSCILLATOR)

OSC OR

relay, contacts (see CONTACT, ELECTRICAL)
relay, electromagnetic (see RELAY COIL RECOGNITION)

signal flow (see DIRECTION OF FLOW)

time delay (see DELAY LINE)

TD OR

time delay with typical delay taps:

METER, INSTRUMENT

identification replaces (*) asterisk)

Meter Letter Combinations

A	Ammeter
AH	Ampere-hour
CMA	Contact-making (or breaking) ammeter
CMC	Contact-making (or breaking) clock
CMV	Contact-making (or breaking) voltmeter
CRO	Oscilloscope or cathode-ray oscillograph
DB	DB (decibel) meter
DBM	DBM (decibels referred to 1 milliwatt) meter
DM	Demand meter
DTR	Demand-totalizing relay
F	Frequency meter
G	Galvanometer
GD	Ground detector
I	Indicating
INT	Integrating
µA or UA	Microammeter
MA	Milliammeter
NM	Noise meter

TUBE (50)
(see RADIATION-SENSITIVITY INDICATOR)

PATH, TRANSMISSION

cable; 2-conductor, shield grounded and 5-conductor shielded

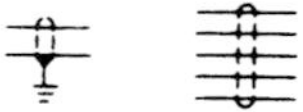

PICKUP HEAD

general

writing; recording

reading; playback

erasing

tapped

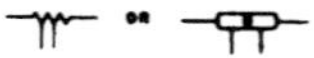

OR

heating

OR

symmetrical varistor resistor, voltage sensitive (silicon carbide, etc.)

OR OR

(identification marks replace (*) asterisk)

with adjustable contact

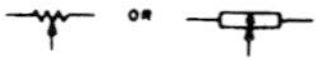

OR

adjustable or continuously adjustable (variable)

OR

(identification replaces (*) asterisk)

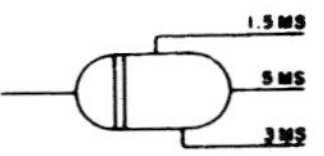

functions not otherwise symbolized

*

(identification replaces (*))

Logic Letter Combinations

S	set
C	clear (reset)
T	toggle (trigger)
(N)	number of bits
BO	blocking oscillator
CF	cathode follower
EF	emitter follower
FF	flip-flop
SS	single shot
ST	schmitt trigger
RG(N)	register (N stages)
SR	shift register

MACHINE, ROTATING

generator

GEN

motor

MOT

OHM	Ohmmeter
OP	Oil pressure

MODE TRANSDUCER

(common coaxial/waveguide usage)

transducer from rectangular waveguide to coaxial with mode suppression, direct-current grounds connected

MOTION, MECHANICAL

rotation applied to a resistor

CW or CW

(identification replaces (*) asterisk)

NUCLEAR-RADIATION DETECTOR, gas filled; IONIZATION CHAMBER; PROPORTIONAL COUNTER TUBE; GEIGER-MULLER COUNTER

writing, reading, and erasing

stereo

RECTIFIER

semiconductor diode; metallic rectifier; electrolytic rectifier; asymmetrical varistor

mercury-pool tube power rectifier

fullwave bridge-type

RESISTOR

general

or

RESONATOR, TUNED CAVITY

(common coaxial/waveguide usage)

resonator with mode suppression coupled by an E-plane aperture to a guided transmission path and by a loop to a coaxial path

tunable resonator with direct-current ground connected to an electron device and adjustably coupled by an E-plane aperture to a rectangular waveguide

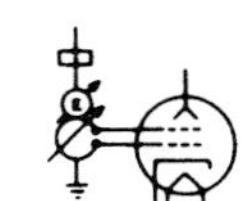

TABLE 14-2 Electronic Symbols (Continued)

ROTARY JOINT, RF (COUPLER)

general; with rectangular waveguide

(transmission path recognition symbol replaces (*) asterisk)

coaxial type in rectangular waveguide

circular waveguide type in rectangular waveguide

SEMICONDUCTOR DEVICE (Two Terminal, diode)

semiconductor diode; rectifier

capacitive diode (also Varicap, Varactor, reactance diode, parametric diode)

semiconductor diode, **PNPN** switch (also Shockley diode, four-layer diode and **SCR**).

(Multi-Terminal, transistor, etc.)

PNP transistor

NPN transistor

unijunction transistor, N-type base

NPN transistor, transverse-biased base

PNIP transistor, ohmic connection to the intrinsic region

NPIN transistor, ohmic connection to the intrinsic region

PNIN transistor, ohmic connection to the intrinsic region

NPIP transistor, ohmic connection to the intrinsic region

transfer

locking, circuit closing (make)

locking, circuit opening (break)

transfer, 3-position

wafer

(example shown: 3-pole 3-circuit with 2 nonshorting and 1 shorting moving contacts)

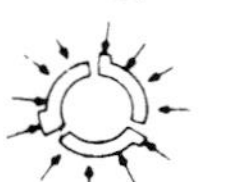

safety interlock, circuit opening and closing

breakdown diode, unidirectional (also backward diode, avalanche diode, voltage regulator diode, Zener diode, voltage reference diode)

breakdown diode, bidirectional and backward diode (also bipolar voltage limiter)

tunnel diode (also Esaki diode)

temperature-dependent diode

photodiode (also solar cell)

unijunction transistor, P-type base

field-effect transistor, N-type base

field-effect transistor, P-type base

semiconductor triode, PNPN-type switch

semiconductor triode, NPNP-type switch

SQUIB

explosive

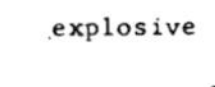

igniter

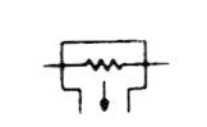

sensing link; fusible link operated

SWITCH

push button, circuit closing (make)

push button, circuit opening (break)

nonlocking; momentary circuit closing (make)

nonlocking; momentary circuit opening (break)

2-pole field-discharge knife, with terminals and discharge resistor

(identification replaces (*) asterisk)

SYNCHRO

Synchro Letter Combinations

CDX	Control-differential transmitter
CT	Control transformer
CX	Control transmitter
TDR	Torque-differential receiver
TDX	Torque-differential transmitter
TR	Torque receiver
TX	Torque transmitter
RS	Resolver
B	Outer winding rotatable in bearings

THERMAL ELEMENT

actuating device

TABLE 14-2 Electronic Symbols (Continued)

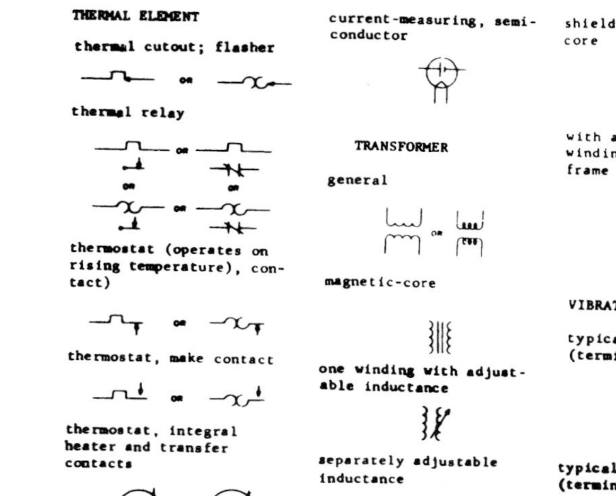

THERMAL ELEMENT

thermal cutout; flasher

thermal relay

thermostat (operates on rising temperature), contact)

thermostat, make contact

thermostat, integral heater and transfer contacts

current-measuring, semiconductor

TRANSFORMER

general

magnetic-core

one winding with adjustable inductance

separately adjustable inductance

adjustable mutual inductor, constant-current

shielded, with magnetic core

with a shield between windings, connected to the frame

VIBRATOR; INTERRUPTER

typical shunt drive (terminals shown)

typical seperate drive (terminals shown)

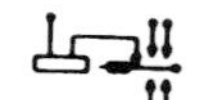

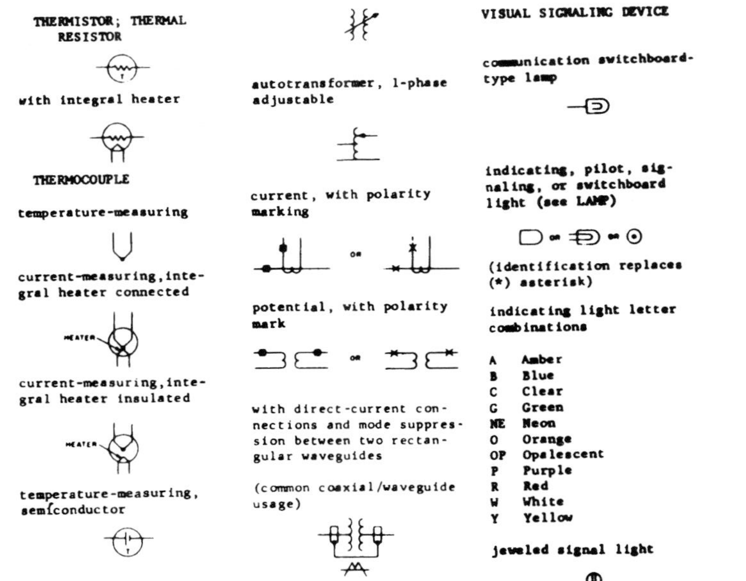

Source: Reprinted from *Basic Electronics*, Bureau of Naval Personnel, Dover Publications, Inc.

15

Conversion Formulas and Tables

15-1 SI UNITS

SI units are described in Tables 15-1 through 15-3.

TABLE 15-1 SI Base Units

Quantity	*Unit*
length	meter (m)
mass	kilogram (kg)
time	second (s)
electric current	ampere (A)
temperature (thermodynamic)	kelvin (K)
amount of substance	mole (mol)
luminous intensity	candela (cd)

In Tables 15-4 and 15-5, the first two digits of each numerical entry represent a power of 10. An asterisk follows each number, which expresses an exact definition. For example, the entry "−02 2.54*" expresses that 1 inch = 2.54×10^{-2} meter, exactly, by

TABLE 15-2 Prefixes for SI Units

Multiple and submultiple	*Prefix*	*Symbol*
1,000,000,000,000 = 10^{12}	tera	T
1,000,000,000 = 10^{9}	giga	G
1,000,000 = 10^{6}	mega	M
1,000 = 10^{3}	kilo	k
100 = 10^{2}	hecto	h
10 = 10	deka	da
0.1 = 10^{-1}	deci	d
0.01 = 10^{-2}	centi	c
0.001 = 10^{-3}	milli	m
0.000 001 = 10^{-6}	micro	μ
0.000 000 001 = 10^{-9}	nano	n
0.000 000 000 001 = 10^{-12}	pico	p
0.000 000 000 000 001 = 10^{-15}	femto	f
0.000 000 000 000 000 001 = 10^{-18}	atto	a

definition. Most of the definitions are extracted from National Bureau of Standards documents. Numbers not followed by an asterisk are only approximate representations of definitions or are the results of physical measurements. The conversion factors are listed alphabetically (Table 15-4) and by physical quantity (Table 15-5).

TABLE 15-3 Derived Units of the International System

Quantity	Name of unit	Unit symbol or abbreviation, where differing from basic form	Unit expressed in terms of base or supplementary units*
area	square meter		m^2
volume	cubic meter		m^3
frequency	hertz, cycle per second†	Hz	s^{-1}
density	kilogram per cubic meter		kg/m^3
velocity	meter per second		m/s
angular velocity	radian per second		rad/s
acceleration	meter per second squared		m/s^2
angular acceleration	radian per second squared		rad/s^2
volumetric flow rate	cubic meter per second		m^3/s
force	newton	N	$kg \cdot m/s^2$
surface tension	newton per meter, joule per square meter	N/m, J/m^2	kg/s^2
pressure	newton per square meter, pascal†	N/m^2, Pa†	$kg/m \cdot s^2$
viscosity, dynamic	newton-second per square meter, poiseuille†	$N\,s/m^2$, Pl†	$kg/m \cdot s$
viscosity, kinematic	meter squared per second		m^2/s
work, torque, energy, quantity of heat	joule, newton-meter, watt-second	J, $N \cdot m$, $W \cdot s$	$kg \cdot m^2/s^2$
power, heat flux	watt, joule per second	W, J/s	$kg \cdot m^2/s^3$
heat flux density	watt per square meter	W/m^2	kg/s^3
volumetric heat release rate	watt per cubic meter	W/m^3	$kg/m \cdot s^3$
heat transfer coefficient	watt per square meter degree	$W/m^2 \cdot deg$	$kg/s^3 \cdot deg$

TABLE 15-3 Derived Units of the International System (Continued)

Quantity	Name of unit	Unit symbol or abbreviation, where differing from basic form	Unit expressed in terms of base or supplementary units*
heat capacity (specific)	joule per kilogram degree	J/kg·deg	$m^2/s^2 \cdot deg$
capacity rate	watt per degree	W/deg	$kg \cdot m^2/s^3 \cdot deg$
thermal conductivity	watt per meter degree	W/m·deg, $\frac{Jm}{s \cdot m^2 \cdot deg}$	$kg \cdot m/s^3 \cdot deg$
quantity of electricity	coulomb	C	A·s
electromotive force	volt	V, W/A	$kg \cdot m^2/A \cdot s^3$
electric field strength	volt per meter		V/m
electric resistance	ohm	Ω, V/A	$kg \cdot m^2/A^2 \cdot s^3$
electric conductivity	ampere per volt meter	A/V·m	$A^2 s^3/kg \cdot m^3$
electric capacitance	farad	F, A·s/V	$A^3 s^1/kg \cdot m^2$
magnetic flux	weber	Wb, V·s	$kg \cdot m^2/A \cdot s^2$
inductance	henry	H, V·s/A	$kg \cdot m^2/A^2 s^2$
magnetic permeability	henry per meter	H/m	$kg \cdot m/A^2 s^2$
magnetic flux density	tesla, weber per square meter	T, Wb/m^2	$kg/A \cdot s^2$
magnetic field strength	ampere per meter		A/m
magnetomotive force	ampere		A
luminous flux	lumen	lm	cd sr
luminance	candela per square meter		cd/m^2
illumination	lux, lumen per square meter	lx, lm/m^2	$cd \cdot sr/m^2$

*Supplementary units are plane angle, radian (rad); solid angle, steradian (sr).

†Not used in all countries.

TABLE 15-4 Conversion Factors in Alphabetical Order

To convert from	*to*	*multiply by*
abampere	ampere	+01 1.00*
abcoulomb	coulomb	+01 1.00*
abfarad	farad	+09 1.00*
abhenry	henry	−09 1.00*
abmho	mho	+09 1.00*
abohm	ohm	−09 1.00*
abvolt	volt	−08 1.00*
acre	meter^2	+03 4.046 856 422 4*
ampere (international of 1948)	ampere	−01 9.998 35
angstrom	meter	−10 1.00*
are	meter^2	+02 1.00*
astronomical unit	meter	+11 1.495 978 9
atmosphere	newton/meter^2	+05 1.013 25*
bar	newton/meter^2	+05 1.00*
barn	meter^2	−28 1.00*
barrel (petroleum, 42 gallons)	meter^3	−01 1.589 873
barye	newton/meter^2	−01 1.00*
British thermal unit (ISO/TC 12)	joule	+03 1.055 06
British thermal unit (International Steam Table)	joule	+03 1.055 04
British thermal unit (mean)	joule	+03 1.055 87
British thermal unit (thermochemical)	joule	+03 1.054 350 264 488
British thermal unit (39° F)	joule	+03 1.059 67
British thermal unit (60° F)	joule	+03 1.054 68
bushel (U.S.)	meter^3	−02 3.523 907 016 688*

TABLE 15-4 Derived Units of the International System (Continued)

To convert from	*to*	*multiply by*
cable	meter	+02 2.194 56*
caliber	meter	−04 2.54*
calorie (International Steam Table)	joule	+00 4.1868
calorie (mean)	joule	+00 4.190 02
calorie (thermochemical)	joule	+00 4.184*
calorie (15° C)	joule	+00 4.185 80
calorie (20° C)	joule	+00 4.181 90
calorie (kilogram, International Steam Table)	joule	+03 4.1868
calorie (kilogram, mean)	joule	+03 4.190 02
calorie (kilogram, thermochemical)	joule	+03 4.184*
carat (metric)	kilogram	−04 2.00*
Celsius (temperature)	kelvin	$t_K = t_C + 273.15$
centimeter of mercury (0° C)	newton/meter^2	+03 1.333 22
centimeter of water (4° C)	newton/meter^2	+01 9.806 38
chain (engineer or ramden)	meter	+01 3.048*
chain (surveyor or gunter)	meter	+01 2.011 68*
circular mil	meter^2	−10 5.067 074 8
cord	meter^3	+00 3.624 556 3
coulomb (international of 1948)	coulomb	−01 9.998 35
cubit	meter	−01 4.572*
cup	meter^3	−04 2.365 882 365*
curie	disintegration/second	+10 3.70*
day (mean solar)	second (mean solar)	+04 8.64*
day (sidereal)	second (mean solar)	+04 8.616 409 0

To convert from	to	multiply by
degree (angle)	radian	−02 1.745 329 251 994 3
denier (international)	kilogram/meter	−07 1.00*
dram (avoirdupois)	kilogram	−03 1.771 845 195 312 5*
dram (troy or apothecary)	kilogram	−03 3.887 934 6*
dram (U.S. fluid)	meter^3	−06 3.696 691 195 312 5*
dyne	newton	−05 1.00*
electron volt	joule	−19 1.602 10
erg	joule	−07 1.00*
Fahrenheit (temperature)	kelvin	$t_K = (5/9)\ (t_F + 459.67)$
Fahrenheit (temperature)	Celsius	$t_C = (5/9)\ (t_F - 32)$
farad (international of 1948)	farad	−01 9.995 05
faraday (based on carbon 12)	coulomb	+04 9.648 70
faraday (chemical)	coulomb	+04 9.649 57
faraday (physical)	coulomb	+04 9.652 19
fathom	meter	+00 1.828 8*
fermi (femtometer)	meter	−15 1.00*
fluid ounce (U.S.)	meter^3	−05 2.957 352 956 25*
foot	meter	−01 3.048*
foot (U.S. survey)	meter	+00 1200/3937*
foot (U.S. survey)	meter	−01 3.048 006 096
foot of water (39.2° F)	newton/meter^2	+03 2.988 98
foot-candle	lumen/meter^2	+01 1.076 391 0
foot-lambert	candela/meter^2	+00 3.426 259
furlong	meter	+02 2.011 68*
gal (galileo)	meter/second^2	−02 1.00*
gallon (U.K. liquid)	meter^3	−03 4.546 087

TABLE 15-4 Derived Units of the International System (Continued)

To convert from	*to*	*multiply by*
gallon (U.S. dry)	$meter^3$	−03 4.404 883 770 86*
gallon (U.S. liquid)	$meter^3$	−03 3.785 411 784*
gamma	tesla	−09 1.00*
gauss	tesla	−04 1.00*
gilbert	ampere turn	−01 7.957 747 2
gill (U.K.)	$meter^3$	−04 1.420 652
gill (U.S.)	$meter^3$	−04 1.182 941 2
grad	degree (angular)	−01 9.00*
grad	radian	−02 1.570 796 3
grain	kilogram	−05 6.479 891*
gram	kilogram	−03 1.00*
hand	meter	−01 1.016*
hectare	$meter^2$	+04 1.00*
henry (international of 1948)	henry	+00 1.000 495
hogshead (U.S.)	$meter^3$	−01 2.384 809 423 92*
horsepower (550 foot lbf/second)	watt	+02 7.456 998 7
horsepower (boiler)	watt	+03 9.809 50
horsepower (electric)	watt	+02 7.46*
horsepower (metric)	watt	+02 7.354 99
horsepower (U.K.)	watt	+02 7.457
horsepower (water)	watt	+02 7.460 43
hour (mean solar)	second (mean solar)	+03 3.60*
hour (sidereal)	second (mean solar)	+03 3.590 170 4
hundredweight (long)	kilogram	+01 5.080 234 544*
hundredweight (short)	kilogram	+01 4.535 923 7*

inch	meter	−02 2.54*
inch of mercury (32° F)	newton/meter^2	+03 3.386 389
inch of mercury (60° F)	newton/meter^2	+03 3.376 85
inch of water (39.2° F)	newton/meter^2	+02 2.490 82
inch of water (60° F)	newton/meter^2	+02 2.4884
joule (international of 1948)	joule	+00 1.000 165
kayser	1/meter	+02 1.00*
kilocalorie (International Steam Table)	joule	+03 4.186 74
kilocalorie (mean)	joule	+03 4.190 02
kilocalorie (thermochemical)	joule	+03 4.184*
kilogram mass	kilogram	+00 1.00*
kilogram force (kgf)	newton	+00 9.806 65*
kilopond force	newton	+00 9.806 65*
kip	newton	+03 4.448 221 615 260 5*
knot (international)	meter/second	−01 5.144 444 444
lambert	candela/meter^2	+04 $1/\pi$*
lambert	candela/meter^2	+03 3.183 098 8
langley	joule/meter^2	+04 4.184*
lbf (pound force, avoirdupois)	newton	+00 4.448 221 615 260 5*
lbm (pound mass, avoirdupois)	kilogram	−01 4.535 923 7*
league (British nautical)	meter	+03 5.559 552*
league (international nautical)	meter	+03 5.556*
league (statute)	meter	+03 4.828 032*
light year	meter	+15 9.460 55
link (engineer or ramden)	meter	−01 3.048*
link (surveyor or gunter)	meter	−01 2.011 68*

TABLE 15-4 Derived Units of the International System (Continued)

To convert from	*to*	*multiply by*
liter	$meter^3$	−03 1.00*
lux	$lumen/meter^2$	+00 1.00*
maxwell	weber	−08 1.00*
meter	wavelengths Kr 86	+06 1.650 763 73*
micron	meter	−06 1.00*
mil	meter	−05 2.54*
mile (U.S. statute)	meter	+03 1.609 344*
mile (U.K. nautical)	meter	+03 1.853 184*
mile (international nautical)	meter	+03 1.852*
mile (U.S. nautical)	meter	+03 1.852*
millibar	$newton/meter^2$	+02 1.00*
millimeter of mercury (0° C)	$newton/meter^2$	+02 1.333 224
minute (angle)	radian	−04 2.908 882 086 66
minute (mean solar)	second (mean solar)	+01 6.00*
minute (sidereal)	second (mean solar)	+01 5.983 617 4
month (mean calendar)	second (mean solar)	+06 2.628*
nautical mile (international)	meter	+03 1.852*
nautical mile (U.S.)	meter	+03 1.852*
nautical mile (U.K.)	meter	+03 1.853 184*
oersted	ampere/meter	+01 7.957 747 2
ohm (international of 1948)	ohm	+00 1.000 495
ounce force (avoirdupois)	newton	−01 2.780 138 5
ounce mass (avoirdupois)	kilogram	−02 2.834 952 312 5*
ounce mass (troy or apothecary)	kilogram	−02 3.110 347 68*

ounce (U.S. fluid)	$meter^3$	−05 2.957 352 956 25*
pace	meter	−01 7.62*
parsec	meter	+16 3.083 74
pascal	$newton/meter^2$	+00 1.00*
peck (U.S.)	$meter^3$	−03 8.809 767 541 72*
pennyweight	kilogram	−03 1.555 173 84*
perch	meter	+00 5.0292*
phot	$lumen/meter^2$	+04 1.00
pica (printers)	meter	−03 4.217 517 6*
pint (U.S. dry)	$meter^3$	−04 5.506 104 713 575*
pint (U.S. liquid)	$meter^3$	−04 4.731 764 73*
point (printers)	meter	−04 3.514 598*
poise	$newton\ second/meter^2$	−01 1.00*
pole	meter	+00 5.0292*
pound force (lbf avoirdupois)	newton	+00 4.448 221 615 260 5*
pound mass (lbm avoirdupois)	kilogram	−01 4.535 923 7*
pound mass (troy or apothecary)	kilogram	−01 3.732 417 216*
poundal	newton	−01 1.382 549 543 76*
quart (U.S. dry)	$meter^3$	−03 1.101 220 942 715*
quart (U.S. liquid)	$meter^3$	−04 9.463 529 5
rad (radiation dose absorbed)	joule/kilogram	−02 1.00*
Rankine (temperature)	kelvin	$t_K=(5/9)t_R$
rayleigh (rate of photon emission)	$1/second\ meter^2$	+10 1.00*
rhe	$meter^2/newton\ second$	+01 1.00*
rod	meter	+00 5.0292*
roentgen	coulomb/kilogram	−04 2.579 76*
rutherford	disintegration/second	+06 1.00*

TABLE 15-4 Derived Units of the International System (Continued)

To convert from	*to*	*multiply by*
second (angle)	radian	−06 4.848 136 811
second (ephemeris)	second	+00 1.000 000 000
second (mean solar)	second (ephemeris)	Consult American Ephemeris and Nautical Almanac
second (sidereal)	second (mean solar)	−01 9.972 695 7
section	$meter^2$	+06 2.589 988 110 336*
scruple (apothecary)	kilogram	−03 1.295 978 2*
shake	second	−08 1.00
skein	meter	+02 1.097 28*
slug	kilogram	+01 1.459 390 29
span	meter	−01 2.286*
statampere	ampere	−10 3.335 640
statcoulomb	coulomb	−10 3.335 640
statfarad	farad	−12 1.112 650
stathenry	henry	+11 8.987 554
statmho	mho	−12 1.112 650
statohm	ohm	+11 8.987 554
statute mile (U.S.)	meter	+03 1.609 344*
statvolt	volt	+02 2.997 925
stere	$meter^3$	+00 1.00*
stilb	$candela/meter^2$	+04 1.00
stoke	$meter^2/second$	−04 1.00*
tablespoon	$meter^3$	−05 1.478 676 478 125*
teaspoon	$meter^3$	−06 4.928 921 593 75*
ton (assay)	kilogram	−02 2.916 666 6
ton (long)	kilogram	+03 1.016 046 908 8*

ton (metric)	kilogram	+03	1.00*
ton (nuclear equivalent of TNT)	joule	+09	4.20
ton (register)	$meter^3$	+00	2.831 684 659 2*
ton (short, 2000 pound)	kilogram	+02	9.071 847 4*
tonne	kilogram	+03	1.00*
torr (0° C)	$newton/meter^2$	+02	1.333 22
township	$meter^2$	+07	9.323 957 2
unit pole	weber	−07	1.256 637
volt (international of 1948)	volt	+00	1.000 330
watt (international of 1948)	watt	+00	1.000 165
yard	meter	−01	9.144*
year (calendar)	second (mean solar)	+07	3.1536*
year (sidereal)	second (mean solar)	+07	3.155 815 0
year (tropical)	second (mean solar)	+07	3.155 692 6
year 1900, tropical, Jan., day 0, hour 12	second (ephemeris)	+07	3.155 692 597 47*
year 1900, tropical, Jan., day 0, hour 12	second	+07	3.155 692 597 47

TABLE 15-5 Conversion Factors Listed by Physical Quantity

To convert from	to	multiply by
ACCELERATION		
$foot/second^2$	$meter/second^2$	−01 3.048*
free fall, standard	$meter/second^2$	+00 9.806 65*
gal (galileo)	$meter/second^2$	−02 1.00*
$inch/second^2$	$meter/second^2$	−02 2.54*
AREA		
acre	$meter^2$	+03 4.046 856 422 4*
are	$meter^2$	+02 1.00*
barn	$meter^2$	−28 1.00*
circular mil	$meter^2$	−10 5.067 074 8
$foot^2$	$meter^2$	−02 9.290 304*
hectare	$meter^2$	+04 1.00*
$inch^2$	$meter^2$	−04 6.4516*
$mile^2$ (U.S. statute)	$meter^2$	+06 2.589 988 110 336*
section	$meter^2$	+06 2.589 988 110 336*
township	$meter^2$	+07 9.323 957 2
$yard^2$	$meter^7$	−01 8.361 273 6*
DENSITY		
$gram/centimeter^3$	$kilogram/meter^3$	+03 1.00*
$lbm/inch^3$	$kilogram/meter^3$	+04 2.767 990 5
$lbm/foot^3$	$kilogram/meter^3$	+01 1.601 846 3
$slug/foot^3$	$kilogram/meter^3$	+02 5.153 79

ENERGY

To convert from	to	multiply by
British thermal unit (ISO/TC 12)	joule	+03 1.055 06
British thermal unit (International Steam Table)	joule	+03 1.055 04
British thermal unit (mean)	joule	+03 1.055 87
British thermal unit (thermochemical)	joule	+03 1.054 350 264 488
British thermal unit (39° F)	joule	+03 1.059 67
British thermal unit (60° F)	joule	+03 1.054 68
calorie (International Steam Table)	joule	+00 4.1868
calorie (mean)	joule	+00 4.190 02
calorie (thermochemical)	joule	+00 4.184*
calorie (15° C)	joule	+00 4.185 80
calorie (20° C)	joule	+00 4.181 90
calorie (kilogram, International Steam Table)	joule	+03 4.1868
calorie (kilogram, mean)	joule	+03 4.190 02
calorie (kilogram, thermochemical)	joule	+03 4.184*
electron volt	joule	−19 1.602 10
erg	joule	−07 1.00*
foot lbf	joule	+00 1.355 817 9
foot poundal	joule	−02 4.214 011 0
joule (international of 1948)	joule	+00 1.000 165
kilocalorie (International Steam Table)	joule	+03 4.1868
kilocalorie (mean)	joule	+03 4.190 02
kilocalorie (thermochemical)	joule	+03 4.184*
kilowatt hour	joule	+06 3.60*
kilowatt hour (international of 1948)	joule	+06 3.600 59
ton (nuclear equivalent of TNT)	joule	+09 4.20
watt hour	joule	+03 3.60*

TABLE 15-5 Conversion Factors Listed by Physical Quantity (Continued)

To convert from	*to*	*multiply by*
ENERGY/AREA TIME		
Btu (thermochemical)/foot^2 second	watt/meter^2	+04 1.134 893 1
Btu (thermochemical)/foot^2 minute	watt/meter^2	+02 1.891 488 5
Btu (thermochemical)/foot^2 hour	watt/meter^2	+00 3.152 480 8
Btu (thermochemical)/inch^2 second	watt/meter^2	+06 1.634 246 2
calorie (thermochemical)/cm^2 minute	watt/meter^2	+02 6.973 333 3
erg/centimeter^2 second	watt/meter^2	−03 1.00*
watt/centimeter^2	watt/meter^2	+04 1.00*
FORCE		
dyne	newton	−05 1.00*
kilogram force (kgf)	newton	+00 9.806 65*
kilopond force	newton	+00 9.806 65*
kip	newton	+03 4.448 221 615 260 5*
lbf (pound force, avoirdupois)	newton	+00 4.448 221 615 260 5*
ounce force (avoirdupois)	newton	−01 2.780 138 5
pound force, lbf (avoirdupois)	newton	+00 4.448 221 615 260 5*
poundal	newton	−01 1.382 549 543 76*
LENGTH		
angstrom	meter	−10 1.00*
astronomical unit	meter	+11 1.495 978 9
cable	meter	+02 2.194 56*
caliber	meter	−04 2.54*

chain (surveyor or gunter)	meter	+01	2.011 68*
chain (engineer or ramden)	meter	+01	3.048*
cubit	meter	−01	4.572*
fathom	meter	+00	1.8288*
fermi (femtometer)	meter	−15	1.00*
foot	meter	−01	3.048*
foot (U.S. survey)	meter	+00	1200/3937*
foot (U.S. survey)	meter	−01	3.048 006 096
furlong	meter	+02	2.011 68*
hand	meter	−01	1.016*
inch	meter	−02	2.54*
league (U.K. nautical)	meter	+03	5.559 552*
league (international nautical)	meter	+03	5.556*
league (statute)	meter	+03	4.828 032*
light year	meter	+15	9.460 55
link (engineer or ramden)	meter	−01	3.048*
link (surveyor or gunter)	meter	−01	2.011 68*
meter	wavelengths Kr 86	+06	1.650 763 73*
micron	meter	−06	1.00*
mil	meter	−05	2.54*
mile (U.S. statute)	meter	+03	1.609 344*
mile (U.K. nautical)	meter	+03	1.853 184*
mile (international nautical)	meter	+03	1.852*
mile (U.S. nautical)	meter	+03	1.852*
nautical mile (U.K.)	meter	+03	1.853 184*
nautical mile (international)	meter	+03	1.852*
nautical mile (U.S.)	meter	+03	1.852*

TABLE 15-5 Conversion Factors Listed by Physical Quantity (Continued)

To convert from	*to*	*multiply by*
pace	meter	−01 7.62*
parsec	meter	+16 3.083 74
perch	meter	+00 5.0292*
pica (printers)	meter	−03 4.217 517 6*
point (printers)	meter	−04 3.514 598*
pole	meter	+00 5.0292*
rod	meter	+00 5.0292*
skein	meter	+02 1.097 28*
span	meter	−01 2.286*
statute mile (U.S.)	meter	+03 1.609 344*
yard	meter	−01 9.144*
	MASS	
carat (metric)	kilogram	−04 2.00*
dram (avoirdupois)	kilogram	−03 1.771 845 195 312 5*
dram (troy or apothecary)	kilogram	−03 3.887 934 6*
grain	kilogram	−05 6.479 891*
gram	kilogram	−03 1.00*
hundredweight (long)	kilogram	+01 5.080 234 544*
hundredweight (short)	kilogram	+01 4.535 923 7*
kgf second^2 meter (mass)	kilogram	+00 9.806 65*
kilogram mass	kilogram	+00 1.00*
lbm (pound mass, avoirdupois)	kilogram	−01 4.535 923 7*
ounce mass (avoirdupois)	kilogram	−02 2.834 952 312 5*
ounce mass (troy or apothecary)	kilogram	−02 3.110 347 68*
pennyweight	kilogram	−03 1.555 173 84*

pound mass, lbm (avoirdupois)	kilogram	−01 4.535 923 7*
pound mass (troy or apothecary)	kilogram	−01 3.732 417 216*
scruple (apothecary)	kilogram	−03 1.295 978 2*
slug	kilogram	+01 1.459 390 29
ton (assay)	kilogram	−02 2.916 666 6
ton (long)	kilogram	+03 1.016 046 908 8*
ton (metric)	kilogram	+03 1.00*
ton (short, 2000 pound)	kilogram	+02 9.071 847 4*
tonne	kilogram	+03 1.00*

POWER

Btu (thermochemical)/second	watt	+03 1.054 350 264 488
Btu (thermochemical)/minute	watt	+01 1.757 250 4
calorie (thermochemical)/second	watt	+00 4.184*
calorie (thermochemical)/minute	watt	−02 6.973 333 3
foot lbf/hour	watt	−04 3.766 161 0
foot lbf/minute	watt	−02 2.259 696 6
foot lbf/second	watt	+00 1.355 817 9
horsepower (550 foot lbf/second)	watt	+02 7.456 998 7
horsepower (boiler)	watt	+03 9.809 50
horsepower (electric)	watt	+02 7.46*
horsepower (metric)	watt	+02 7.354 99
horsepower (U.K.)	watt	+02 7.457
horsepower (water)	watt	+02 7.460 43
kilocalorie (thermochemical)/minute	watt	+01 6.973 333 3
kilocalorie (thermochemical)/second	watt	+03 4.184*
watt (international of 1948)	watt	+00 1.000 165

TABLE 15-5 Conversion Factors Listed by Physical Quantity (Continued)

To convert from	*to*	*multiply by*
	PRESSURE	
atmosphere	newton/meter^2	+05 1.013 25*
bar	newton/meter^2	+05 1.00*
barye	newton/meter^2	−01 1.00*
centimeter of mercury (0° C)	newton/meter^2	+03 1.333 22
centimeter of water (4° C)	newton/meter^2	+01 9.806 38
dyne/centimeter^2	newton/meter^2	−01 1.00*
foot of water (39.2° F)	newton/meter^2	+03 2.988 98
inch of mercury (32° F)	newton/meter^2	+03 3.386 389
inch of mercury (60° F)	newton/meter^2	+03 3.376 85
inch of water (39.2° F)	newton/meter^2	+02 2.490 82
inch of water (60° F)	newton/meter^2	+02 2.4884
kgf/centimeter^2	newton/meter^2	+04 9.806 65*
kgf/meter^2	newton/meter^2	+00 9.806 65*
lbf/foot^2	newton/meter^2	+01 4.788 025 8
lbf/inch^2 (psi)	newton/meter^2	+03 6.894 757 2
millibar	newton/meter^2	+02 1.00*
millimeter of mercury (0° C)	newton/meter^2	+02 1.333 224
pascal	newton/meter^2	+00 1.00*
psi (lbf/inch^2)	newton/meter^2	+03 6.894 757 2
torr (0° C)	newton/meter^2	+02 1.333 22
	SPEED	
foot/hour	meter/second	−05 8.466 666 6

foot/minute	meter/second	−03 5.08*
foot/second	meter/second	−01 3.048*
inch/second	meter/second	−02 2.54*
kilometer/hour	meter/second	−01 2.777 777 8
knot (international)	meter/second	−01 5.144 444 444
mile/hour (U.S. statute)	meter/second	−01 4.4704*
mile/minute (U.S. statute)	meter/second	+01 2.682 24*
mile/second (U.S. statute)	meter/second	+03 1.609 344*

TEMPERATURE

Celsius	kelvin	$t_K = t_C + 273.15$
Fahrenheit	kelvin	$t_K = (5/9)(t_F + 459.67)$
Fahrenheit	Celsius	$t_C = (5/9)(t_F - 32)$
Rankine	kelvin	$t_K = (5/9)t_R$

TIME

day (mean solar)	second (mean solar)	+04 8.64*
day (sidereal)	second (mean solar)	+04 8.616 409 0
hour (mean solar)	second (mean solar)	+03 3.60*
hour (sidereal)	second (mean solar)	+03 3.590 170 4
minute (mean solar)	second (mean solar)	+01 6.00*
minute (sidereal)	second (mean solar)	+01 5.983 617 4
month (mean calendar)	second (mean solar)	+06 2.628*
second (ephemeris)	second	+00 1.000 000 000
second (mean solar)	second (ephemeris)	Consult American Ephemeris and Nautical Almanac

TABLE 15-5 Conversion Factors Listed by Physical Quantity (Continued)

To convert from	*to*	*multiply by*
second (sidereal)	second (mean solar)	−01 9.972 695 7
year (calendar)	second (mean solar)	+07 3.1536*
year (sidereal)	second (mean solar)	+07 3.155 815 0
year (tropical)	second (mean solar)	+07 3.155 692 6
year 1900, tropical, Jan., day 0, hour 12	second (ephemeris)	+07 3.155 692 597 47*
year 1900, tropical, Jan., day 0, hour 12	second	+07 3.155 692 597 47
VISCOSITY		
centistoke	$meter^2$/second	−06 1.00*
stoke	$meter^2$/second	−04 1.00*
$foot^2$/second	$meter^2$/second	−02 9.290 304*
centipoise	newton second/$meter^2$	−03 1.00*
lbm/foot second	newton second/$meter^2$	+00 1.488 163 9
lbf second/$foot^2$	newton second/$meter^2$	+01 4.788 025 8
poise	newton second/$meter^2$	−01 1.00*
poundal second/$foot^2$	newton second/$meter^2$	+00 1.488 163 9
slug/foot second	newton second/$meter^2$	+01 4.788 025 8
rhe	$meter^2$/newton second	+01 1.00*
VOLUME		
acre foot	$meter^3$	+03 1.233 481 9
barrel (petroleum, 42 gallons)	$meter^3$	−01 1.589 873
board foot	$meter^3$	−03 2.359 737 216*
bushel (U.S.)	$meter^3$	−02 3.523 907 016 688*
cord	$meter^3$	+00 3.624 556 3

cup	$meter^3$	−04	2.365 882 365*
dram (U.S. fluid)	$meter^3$	−06	3.696 691 195 312 5*
fluid ounce (U.S.)	$meter^3$	−05	2.957 352 956 25*
$foot^3$	$meter^3$	−02	2.831 684 659 2*
gallon (U.K. liquid)	$meter^3$	−03	4.546 087
gallon (U.S. dry)	$meter^3$	−03	4.404 883 770 86*
gallon (U.S. liquid)	$meter^3$	−03	3.785 411 784*
gill (U.K.)	$meter^3$	−04	1.420 652
gill (U.S.)	$meter^3$	−04	1.182 941 2
hogshead (U.S.)	$meter^3$	−01	2.384 809 423 92*
$inch^3$	$meter^3$	−05	1.638 706 4*
liter	$meter^3$	−03	1.00*
ounce (U.S. fluid)	$meter^3$	−05	2.957 352 956 25*
peck (U.S.)	$meter^3$	−03	8.809 767 541 72*
pint (U.S. dry)	$meter^3$	−04	5.506 104 713 575*
pint (U.S. liquid)	$meter^3$	−04	4.731 764 73*
quart (U.S. dry)	$meter^3$	−03	1.101 220 942 715*
quart (U.S. liquid)	$meter^3$	−04	9.463 529 5
stere	$meter^3$	+00	1.00*
tablespoon	$meter^3$	−05	1.478 676 478 125*
teaspoon	$meter^3$	−06	4.928 921 593 75*
ton (register)	$meter^3$	+00	2.831 684 659 2*
$yard^3$	$meter^3$	−01	7.645 548 579 84*

TABLE 15-6 Conversion Factors for Converting Non-SI Metric Units to SI Units

Multiply	*by*	*to obtain*
LENGTH		
angstroms	10^{-10}	meters
ACCELERATION		
galileos	1.0	centimeters per second squared
milligalileos	0.001	centimeters per second squared
MASS		
metric tons	1,000	kilograms
quintals	100	kilograms
metric carats	0.2	grams
FORCE		
sthenes	1,000	newtons
kiloponds	9.8066	newtons
dynes	10^{-5}	newtons

PRESSURE, STRESS

atmospheres	101.325	kilonewtons per square meter
bars	10^5	newtons per square meter
piezes	10^3	newtons per square meter
millimeters Hg	133.32	newtons per square meter
torr	133.32	newtons per square meter
kilograms per square centimeter	98.066	kilonewtons per square meter
millimeters H_2O	9.8066	newtons per square meter
millitorr	0.1333	newtons per square meter

DYNAMIC VISCOSITY

poises	0.1	newton seconds per square meter
centipoises	0.001	newton seconds per square meter

KINEMATIC VISCOSITY

stokes	10^{-4}	meter squared per second
centistokes	10^{-6}	meter squared per second

TABLE 15-6 Conversion Factors for Converting Non-SI Metric Units to SI Units (Continued)

Multiply	*by*	*to obtain*
ENERGY, WORK, HEAT		
thermies	4.1855	megajoules
liter atmospheres	101.328	joules
kilogram meters	9.8066	joules
calories, I.T.	4.1868	joules
calories, 15°	4.1855	joules
thermochemical calories	4.184	joules
ergs	10^{-7}	joules
POWER		
metric horsepower	735.499	watts
ergs per second	10^{-7}	watts

15-2 TEMPERATURE CONVERSIONS

Table 15-7 provides ready conversion between Celsius (Centigrade) and Fahrenheit temperatures in common ranges. For other situations or temperature scales, use the following formulas:

$$C = \frac{5}{9}(F - 32) = K - 273.16 = \frac{5Re}{4} = \frac{Ra}{1.8} - 273.16$$

$$F = \frac{9}{5}C + 32 = \frac{9}{5}(K - 273.16) + 32 = \frac{9Re}{4} + 32 = Ra - 459.7$$

$$K = C + 273.16 = \frac{5}{9}(F - 32) + 273.16 = \frac{5Re}{4} + 273.16 = \frac{Ra}{1.8}$$

$$Re = \frac{4C}{5} = \frac{4}{9}(F - 32) = \frac{4}{5}(K - 273.16) = \frac{4}{9}(Ra - 491.7)$$

$$Ra = 1.8(C + 273.16) = F + 459.7 = 1.8K = \frac{9Re}{4} + 491.7$$

where C = Celsius temperature, degrees
F = Fahrenheit temperature, degrees
K = Kelvin (absolute) temperature, kelvins
Re = Réaumur temperature, degrees
Ra = Rankine temperature, degrees

TABLE 15-7 Temperature Conversion Table

Read known temperature in bold face type. Corresponding temperature in degrees Fahrenheit will be found in column to the right. Corresponding temperature in degrees Celsius will be found in column to the left.

−5 to 100			0 to 100			0 to 100			100 to 500		
°C		°F	°C		°F	°C		°F	°C		°F
−73.3	**−100**	−148	−17.8	**0**	32.0	10.0	**50**	122.0	38	**100**	212
−70.5	**− 95**	−139	−17.2	**1**	33.8	10.6	**51**	123.8	43	**110**	230
−67.8	**− 90**	−130	−16.7	**2**	35.6	11.1	**52**	125.6	49	**120**	248
−65.0	**− 85**	−121	−16.1	**3**	37.4	11.7	**53**	127.4	54	**130**	266
−62.2	**− 80**	−112	−15.6	**4**	39.2	12.2	**54**	129.2	60	**140**	284
−59.5	**− 75**	−103	−15.0	**5**	41.0	12.8	**55**	131.0	66	**150**	302
−56.7	**− 70**	− 94	−14.4	**6**	42.8	13.3	**56**	132.8	71	**160**	320
−53.9	**− 65**	− 85	−13.9	**7**	44.6	13.9	**57**	134.6	77	**170**	338
−51.1	**− 60**	− 76	−13.3	**8**	46.4	14.4	**58**	136.4	82	**180**	356
−48.3	**− 55**	− 67	−12.8	**9**	48.2	15.0	**59**	138.2	88	**190**	374
−45.6	**− 50**	− 58	−12.2	**10**	50.0	15.6	**60**	140.0	93	**200**	392
−42.8	**− 45**	− 49	−11.7	**11**	51.8	16.1	**61**	141.8	99	**210**	410
−40.0	**− 40**	− 40	−11.1	**12**	53.6	16.7	**62**	143.6	100	**212**	413
−37.2	**− 35**	− 31	−10.6	**13**	55.4	17.2	**63**	145.4	104	**220**	428
−34.4	**− 30**	− 22	−10.0	**14**	57.2	17.8	**64**	147.2	110	**230**	446
−31.6	**− 25**	− 13	− 9.44	**15**	59.0	18.3	**65**	149.0	116	**240**	464
−28.9	**− 20**	− 4	− 8.89	**16**	60.8	18.9	**66**	150.8	121	**250**	482
−26.1	**− 15**	5	− 8.33	**17**	62.6	19.4	**67**	152.6	127	**260**	500
−23.3	**− 10**	14	− 7.78	**18**	64.4	20.0	**68**	154.4	132	**270**	518
−20.5	**− 5**	23	− 7.22	**19**	66.2	20.6	**69**	156.2	138	**280**	536
			− 6.67	**20**	68.0	21.1	**70**	158.0	143	**290**	554
			− 6.11	**21**	69.8	21.7	**71**	159.8	149	**300**	572
			− 5.56	**22**	71.6	22.2	**72**	161.6	154	**310**	590
			− 5.00	**23**	73.4	22.8	**73**	163.4	160	**320**	608
			− 4.44	**24**	75.2	23.3	**74**	165.2	166	**330**	626

			−3.89	**25**	77.0	23.9	**75**	167.0	171	**340**	644
			−3.33	**26**	78.8	24.4	**76**	168.8	177	**350**	662
			−2.78	**27**	80.6	25.0	**77**	170.6	182	**360**	680
			−2.22	**28**	82.4	25.6	**78**	172.4	188	**370**	698
			−1.67	**29**	84.2	26.1	**79**	174.2	193	**380**	716
			−1.11	**30**	86.0	26.7	**80**	176.0	199	**390**	734
			−0.56	**31**	87.8	27.2	**81**	177.8	204	**400**	752
			0.	**32**	89.6	27.8	**82**	179.6	210	**410**	770
			0.56	**33**	91.4	28.3	**83**	181.4	216	**420**	788
			1.11	**34**	93.2	28.9	**84**	183.2	221	**430**	806
			1.67	**35**	95.0	29.4	**85**	185.0	227	**440**	824
			2.22	**36**	96.8	30.0	**86**	186.8	232	**450**	842
			2.78	**37**	98.6	30.6	**87**	188.6	238	**460**	860
			3.33	**38**	100.4	31.1	**88**	190.4	243	**470**	878
			3.89	**39**	102.2	31.7	**89**	192.2	249	**480**	896
			4.44	**40**	104.0	32.2	**90**	194.0	254	**490**	914
			5.00	**41**	105.8	32.8	**91**	195.8	260	**500**	932
			5.56	**42**	107.6	33.3	**92**	197.6			
			6.11	**43**	109.4	33.9	**93**	199.4			
			6.67	**44**	111.2	34.4	**94**	201.2			
			7.22	**45**	113.0	35.0	**95**	203.0			
			7.78	**46**	114.8	35.6	**96**	204.8			
			8.33	**47**	116.6	36.1	**97**	206.6			
			8.89	**48**	118.4	36.7	**98**	208.4			
			9.44	**49**	120.2	37.2	**99**	210.2			
						37.8	**100**	212.0			

Interpolation Factors

°C		°F	°C		°F	°C		°F
0.56	1	1.8	2.22	4	7.2	3.89	7	12.6
1.11	2	3.6	2.78	5	9.0	4.44	8	14.4
1.67	3	5.4	3.33	6	10.8	5.00	9	16.2

Source: Reprinted by permission of Brand-Rex Electronics and Industrial Cable Division, Willimantic, Conn.

16

Properties of Materials

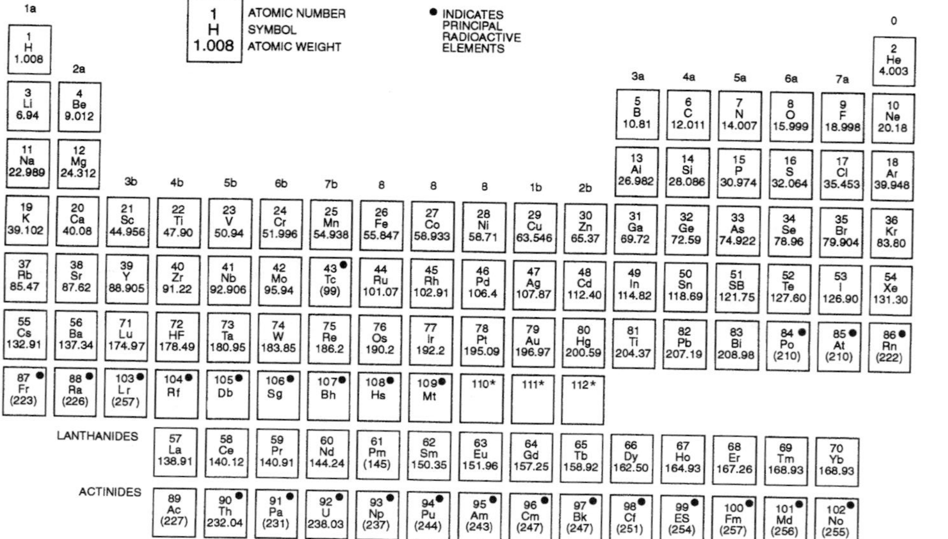

FIG. 16-1 Periodic table of the elements. See Table 16-1 for interpretation of symbols.

TABLE 16-1 Periodic Table of the Elements*

Symbol	Name	Atomic number	Atomic weight
Ac	Actinium	89	(227)
Ag	Silver	47	107.868
Al	Aluminum	13	26.982
Am	Americium	95	(243)
Ar	Argon	18	39.948
As	Arsenic	33	74.922
At	Astatine	85	(210)
Au	Gold	79	196.967
B	Boron	5	10.81
Ba	Barium	56	137.34
Be	Beryllium	4	9.012
Bh	Bohrium	107	—
Bi	Bismuth	83	208.980
Bk	Berkelium	97	(247)
Br	Bromine	35	79.904
C	Carbon	6	12.011
Ca	Calcium	20	40.08
Cd	Cadmium	48	112.40
Ce	Cerium	58	140.12
Cf	Californium	98	(251)
Cl	Chlorine	17	35.453
Cm	Curium	96	(247)
Co	Cobalt	27	58.933
Cr	Chromium	24	51.996
Cs	Cesium	55	132.905
Cu	Copper	29	63.546
Db	Dubnium	105	—
Dy	Dysprosium	66	162.50
Es	Einsteinium	99	(254)
Er	Erbium	68	167.26
Eu	Europium	63	151.96
F	Fluorine	9	18.998
Fe	Iron	26	55.847
Fm	Fermium	100	(257)
Fr	Francium	87	(223)
Ga	Gallium	31	69.72

TABLE 16-1 Periodic Table of the Elements* (Continued)

Symbol	Name	Atomic number	Atomic weight
Gd	Gadolinium	64	157.25
Ge	Germanium	32	72.59
H	Hydrogen	1	1.008
He	Helium	2	4.003
Hf	Hafnium	72	178.49
Hg	Mercury	80	200.59
Ho	Holmium	67	164.930
Hs	Hassium	108	—
I	Iodine	53	126.904
In	Indium	49	114.82
Ir	Iridium	77	192.2
K	Potassium	19	39.102
Kr	Krypton	36	83.80
La	Lanthanum	57	138.91
Li	Lithium	3	6.94
Lr	Lawrencium	103	(256)
Lu	Lutetium	71	174.97
Md	Mendelevium	101	(256)
Mg	Magnesium	12	24.312
Mn	Manganese	25	54.938
Mo	Molybdenum	42	95.94
Mt	Meitnerium	109	—
N	Nitrogen	7	14.007
Na	Sodium	11	22.989
Nb	Niobium	41	92.906
Nd	Neodymium	60	144.24
Ne	Neon	10	20.18
Ni	Nickel	28	58.71
No	Nobelium	102	(255)
Np	Neptunium	93	(237)
O	Oxygen	8	15.999
Os	Osmium	76	190.2
P	Phosphorus	15	30.974
Pa	Protactinium	91	(231)
Pb	Lead	82	207.19
Pd	Palladium	46	106.4

TABLE 16-1 Periodic Table of the Elements* (Continued)

Symbol	Name	Atomic number	Atomic weight
Pm	Promethium	61	(145)
Po	Polonium	84	(210)
Pr	Praseodymium	59	140.907
Pt	Platinum	78	195.09
Pu	Plutonium	94	(244)
Ra	Radium	88	(226)
Rb	Rubidium	37	85.47
Re	Rhenium	75	186.2
Rf	Rutherfordium	104	—
Rh	Rhodium	45	102.905
Rn	Radon	86	(222)
Ru	Ruthenium	44	101.07
S	Sulfur	16	32.06
Sb	Antimony	51	121.75
Sc	Scandium	21	44.956
Se	Selenium	34	78.96
Sg	Seaborgium	106	—
Si	Silicon	14	28.086
Sm	Samarium	62	150.35
Sn	Tin	50	118.69
Sr	Strontium	38	87.62
Ta	Tantalum	73	180.948
Tb	Terbium	65	158.924
Tc	Technetium	43	(99)
Te	Tellurium	52	127.60
Th	Thorium	90	232.038
Ti	Titanium	22	47.90
Tl	Thallium	81	204.37
Tm	Thulium	69	158.934
U	Uranium	92	238.03
V	Vanadium	23	50.942
W	Tungsten	74	183.85
Xe	Xenon	54	131.30
Y	Yttrium	39	88.905
Yb	Ytterbium	70	173.04
Zn	Zinc	30	65.37
Zr	Zirconium	40	91.22

* Elements 110, 111, and 112 have been discovered but not officially named.

TABLE 16-2 Physical Constants and Conversion Factors

Constant	Symbol	Value	Uncertainty ‡	Unit: Systeme International (SI)		Unit: Centimeter-gram-second (CGS)	
Speed of light in vacuum	c	2.997 925 0	±10	$\times 10^{8}$	m/s	$\times 10^{10}$	cm/s
Elementary charge	e	1.602 191 7	70	10^{-19}	C	10^{-20}	$cm^{1/2}g^{1/2}$ *
		4.803 250	21			10^{-10}	$cm^{3/2}g^{1/2}s^{-1}$ †
Avogadro constant	N_A	6.022 169	40	10^{23}	mol^{-1}	10^{23}	mol^{-1}
Atomic mass unit	u	1.660 531	11	10^{-27}	kg	10^{-24}	g
Electron rest mass	m_e	9.109 558	54	10^{-31}	kg	10^{-28}	g
		5.485 930	34	10^{-4}	u	10^{-4}	u
Proton rest mass	m_p	1.672 614	11	10^{-27}	kg	10^{-24}	g
		1.007 276 61	8	10^{0}	u	10^{0}	u
Neutron rest mass	m_n	1.674 920	11	10^{-27}	kg	10^{-24}	g
		1.008 665 20	10	10^{0}	u	10^{0}	u
Faraday constant	F	9.648 670	54	10^{4}	C/mol	10^{3}	$cm^{1/2}g^{1/2}mol^{-1}$ *
		2.892 599	16			10^{14}	$cm^{3/2}g^{1/2}s^{-1}mol^{-1}$ †
Planck constant	h	6.626 196	50	10^{-34}	J · s	10^{-27}	erg · s
	$\hbar$	1.054 591 9	80	10^{-34}	J · s	10^{-27}	erg · s
Fine structure constant	α	7.297 351	11	10^{-3}		10^{-3}	
	$1/\alpha$	1.370 360 2	21	10^{2}		10^{2}	
Charge to mass ratio for electron	e/m_e	1.758 802 8	54	10^{11}	C/kg	10^{7}	$cm^{1/2}/g^{1/2}$ *
		5.272 759	16			10^{17}	$cm^{3/2}g^{-1/2}s^{-1}$ †
Quantum-charge ratio	h/e	4.135 708	14	10^{-15}	J · s/C	10^{-7}	$cm^{3/2}g^{1/2}s^{-1}$ *
		1.379 523 4	46			10^{-17}	$cm^{1/2}g^{1/2}$ †
Compton wavelength of electron	λ_C	2.426 309 6	74	10^{-12}	m	10^{-10}	cm
	$\lambda_C/2\pi$	3.861 592	12	10^{-13}	m	10^{-11}	cm
Compton wavelength of proton	$\lambda_{C,p}$	1.321 440 9	90	10^{-15}	m	10^{-13}	cm
	$\lambda_{C,p}/2\pi$	2.103 139	14	10^{-16}	m	10^{-14}	cm
Rydberg constant	R_∞	1.097 373 12	11	10^{7}	m^{-1}	10^{5}	cm^{-1}
Bohr radius	a_0	5.291 771 5	81	10^{-11}	m	10^{-9}	cm

Electron radius	r_e	2.817 939	13	10^{-15}	m	10^{-13}	cm
Gyromagnetic ratio of proton	γ	2.675 196 5	82	10^{8}	$rad \cdot s^{-1}T^{-1}$	10^{4}	$rad \cdot s^{-1}G^{-1}$ *
	$\gamma/2\pi$	4.257 707	13	10^{7}	Hz/T	10^{3}	$s^{-1}G^{-1}$ *
(uncorrected for diamagnetism, H_2O)	γ'	2.675 127 0	82	10^{8}	$rad \cdot s^{-1}T^{-1}$	10^{4}	$rad \cdot s^{-1}G^{-1}$ *
	$\gamma'/2\pi$	4.257 597	13	10^{7}	Hz/T	10^{3}	$s^{-1}G^{-1}$ *
Bohr magneton	μ_B	9.274 096	65	10^{-24}	J/T	10^{-21}	erg/G *
Nuclear magneton	μ_N	5.050 951	50	10^{-27}	J/T	10^{-24}	erg/G *
Proton moment	μ_p	1.410 620 3	99	10^{-26}	J/T	10^{-23}	erg/G *
	μ_p/μ_N	2.792 782	17	10^{0}	....	10^{0}	....
(uncorrected for diamagnetism, H_2O)	μ'_p/μ_N	2.792 709	17	10^{0}	....	10^{0}	....
Gas constant	R	8.314 34	35	10^{0}	$J \cdot K^{-1} mol^{-1}$	10^{7}	$erg \cdot K^{-1} mol^{-1}$
Normal volume perfect gas	V_0	2.241 36	39	10^{-2}	m^3/mol	10^{4}	cm^3/mol
Boltzmann constant	k	1.380 622	59	10^{-23}	J/K	10^{-16}	erg/K
First radiation constant ($8\pi hc$)	c_1	4.992 579	38	10^{-24}	$J \cdot m$	10^{-15}	$erg \cdot cm$
Second radiation constant	c_2	1.438 833	61	10^{-2}	$m \cdot K$	10^{0}	$cm \cdot K$
Stefan-Boltzmann constant	σ	5.669 61	96	10^{-8}	$W \cdot m^{-2}K^{-4}$	10^{-5}	$erg \cdot cm^{-2}s^{-1}K^{-4}$
Gravitational constant	G	6.673 2	31	10^{-11}	$N \cdot m^2/kg^2$	10^{-8}	$dyn \cdot cm^2/g^2$

‡Based on 1 std. dev; applies to last digits in preceding column. *Electromagnetic system. †Electrostatic system.

Source: *Handbook of Mathematical Functions*, M. Abramowitz and I. A. Stegun, National Bureau of Standards.

TABLE 16-3 Properties of Selected Extrudable Primary Insulations

Property	Test Method	POLYOLEFINS Polyethylene	
		Low Density	High Density
Physical			
Specific Gravity	ASTM D-792	0.92	0.95
Durometer Hardness	ASTM D-2240	D45	D58
Tensile Strength, psi (min.)	ASTM D-412	2200	3000
Elongation, % (min.)	ASTM D-412	600	250
Environmental Stress Cracking	–	E	E
Abrasion Resistance†	–	G	G
Cut-Thru Resistance†	–	G	G
Water Resistance†	–	E	E
Thermal			
Max. Operating Temperature, °C		80	80
Brittle Temp., 50% non-failure, °C	ASTM D-476	−65	−65
Flame Resistance†	–	P	P
Solder Iron Resistance†		P	P
Electrical			
Dielectric Constant @ 1 MHz	ASTM D-150	2.27	2.27
Dissipation Factor @ 1 MHz	ASTM D-150	0.0002	0.0003
Volume Resistivity, ohm-cm	ASTM D-257	$>10^{16}$	$>10^{16}$
Dry Dielectric v/mil	ASTM D-149	1200	1200
Wet Dielectric v/mil	ASTM D-149	900-1000	1050

Property	Test Method	POLYVINYL CHLORIDE		
		Irradiated	Kynar	Tefzel ETFE
Physical				
Specific Gravity	ASTM D-792	1.34	1.76	1.70
Durometer Hardness	ASTM D-2240	A94-96	D70-80	D75
Tensile Strength, psi (min.)	ASTM D-412	4500	5000	6000
Elongation, % (min.)	ASTM D-412	150-200	250	150
Environmental Stress Cracking	–	–	E	–
Abrasion Resistance†	–	E	E	G
Cut-Thru Resistance†	–	E	E	E
Water Resistance†	–	G	G	E
Thermal				
Max. Operating Temperature, °C	–	105	135	150
Brittle Temp., 50% non-failure, °C	ASTM D-476	55	−65	−80
Flame Resistance†	–	G	G	G
Solder Iron Resistance†	–	E	P	P
Electrical				
Dielectric Constant @ 1 MHz	ASTM D-150	2.7§	6.4	2.6
Dissipation Factor @ 1 MHz	ASTM D-150	–	0.159	0.005
Volume Resistivity, ohm-cm	ASTM D-257	2×10^{12}	2×10^{14}	$>10^{16}$
Dry Dielectric v/mil	ASTM D-149	800-900	1000	1100-1300
Wet Dielectric v/mil	ASTM D-149	700-850	950	1000

*Varies with degree of foaming
†E – Excellent, G – Good, F – Fair, P – Poor
‡Flame retardant formulations are available
§Average value of several formulations with braided copper shield

TABLE 16-3 Properties of Selected Extrudable Primary Insulations (Continued)

POLYOLEFINS				POLYVINYL CHLORIDE	
Polyethylene			Ethylene Propylene Copolymer	Standard	Semi-Rigid
Flame Retard.	Cross-Linked	Cellular			
1.30	1.17-1.26	0.50*	0.90	1.25-1.38	1.38
D55	D51	—	—	A85-90	A90-96
1800	2200	600	3600	2100-2700	3200-4000
250	500	300	> 600	250-350	150-250
E	E	G	—	—	—
G	F	P	G	F-G	G
G	F	P	G	G	G
E	E	P	E	G	G
80	150	80	80	60-105	80
−65	−65	−65	−15	−55	55
G	P	P	F	G	G
P	G	P	P	P	P
2.50	2.45	1.50*	2.236	2.7	4-6
0.0015	0.0003	0.0002	0.0003	0.06-0.10	0.08-0.085
> 10^{16}	10^{13}	—	10^{17}	5×10^{13}	2×10^{14}
1000	1000	500	850	800-900	800-900
850	800	500	800	700-800	700-800

FLUOROCARBONS				MISCELLANEOUS		
Teflon FEP	Teflon TFE	Teflon PFA	Halar ECTFE	Polysulfone	Surlyn A	Silicone Rubber
2.14-2.17	2.13-2.20	2.12-2.17	1.68	1.24	0.96	1.2-1.5
D59	52	D60	D75	D60	D60	A50-60
2700-3100	4500	3000-3500	7000	7500	3800	1000-1400
250-300	300	250-300	200	100	450	175-500
—	—	—	G	—	E	E
F	F	F-G	F	G	G	G
F	F	F	G	G	E	E-G
E	E	E	G	E	E	G
200	260	250	150	130	75	200
−80	−80	−80	< −76	−65	−50	−100
E	E	E	E	F	P	E
P	E	P	P	F	P	G
2.1	2.1	2.06	2.5	3.1	2.36	3.1
0.0007	< 0.0002	0.0002	0.013	0.0056	0.0019	< 0.005
> 2×10^{18}	> 10^{18}	10^{18}	> 10^{15}	5×10^{16}	10^{18}	10^{15}-10^{16}
1200	1200	1200	500	1200-1400	1100	400-600
1000	1000	1000	—	1000	1000	300-500

Source: Reprinted by permission of Brand-Rex Electronic and Industrial Cable Division, Willimantic, Conn.

TABLE 16-4 Physical Properties of Conductor Materials

Conductor Material	Min. Conduct. (%)	Tensile Strength		Elong.* (%)	Oper. Temp. (°C)	Resistance to		Solder-ability	Relative Weight
		(p.s.i.)	(kg/cm²)			Oxidation	Galvanic Corrosion		
Copper									
Annealed, bare	100.	35,000	2461	15-30	150	Poor	Good	Fair	1.000
Annealed, tinned	97.16	35,000	2461	10-25	150	Good	Good	Good	1.000
Annealed, silver coated	100.	35,000	2461	15-30	200	Good	Poor	Good	1.000
Annealed, nickel coated	96.	35,000	2461	15-25	260	Good	Good	Poor	1.000
Medium Hard Drawn, bare	96.66	55,000	3867	0.88-1.08	150	Poor	Good	Fair	1.000
Hard Drawn, bare	96.16	65,000	4570	0.85-1.06	150	Poor	Good	Fair	1.000
Copper-Covered Steel									
Annealed, bare	40.	47,000	3304	10-15	200	Good	Good	Fair	.925
Annealed, silver coated	40.	47,000	3304	10-15	200	Good	Fair	Good	.925
Hard Drawn, bare	40.	115,000	8085	1-1.5	200	Poor	Good	Fair	.925
Hard Drawn, bare	30.	135,000	9491	1-1.5	200	Poor	Good	Fair	.925
High Strength Alloy, silver-coated									
Cadmium-Chromium Copper†	85.	55,000	3867	6	200	Good	Poor	Good	.980
Cadmium Copper	80.	85,000	5976	0.82-1.06	200	Good	Poor	Good	.980
Aluminum, EC Grade									
Bare	62.	12,000	844	23	150	Poor	Good	Poor	.304

*Values are for solid wires 8 AWG and smaller, and for individual strands before stranding. Varies with diameter.

†Alloy 135

Source: Reprinted by permission of Brand-Rex Electronic and Industrial Cable Division, Willimantic, Conn.

TABLE 16-5 Preferred AWG-Size Copper Conductors

Conductor		Diameter[2]		Area		Approx. Weight		Nom. D.C. Resistance @ 20°C[3]			
								Tin Coated		Bare or Silver Coated	
AWG[1]	Stranding	in.	mm.	circ. mils	sq. mm.	lbs./M′	kg./km.	ohms/M′	ohms/km.	ohms/M′	ohms/kg.
40	Solid	.0031	.079	9.61	.0049	.029	.043	1158	3799	1080	3540
38	Solid	.0040	.102	16.0	.0081	.048	.072	696	2283	648	2130
36	Solid	.0050	.127	25.0	.0127	.076	.113	445	1460	415	1360
34	Solid	.0063	.160	39.7	.0201	.120	.179	280	918	261	857
33	Solid	.0071	.180	50.4	.0255	.153	.228	221	725	206	675
32	Solid	.0080	.203	64.0	.0324	.194	.289	174	571	162	532
30	Solid	.010	.254	100	.0507	.30	.45	113	365	104	340
	7/38	.012	.305	112	.0568	.35	.52	102	336	95.4	313
28	Solid	.0126	.320	159	.0804	.48	.72	70.8	232	65.3	214
	7/36	.0150	.381	175	.0887	.55	.82	65.5	215	61.0	200
27	7/35	.018	.457	220	.1112	.69	1.03	52.2	171	48.7	160
26	Solid	.0159	.404	253	.1281	.77	1.14	43.6	143	41.0	135
	7/34	.0190	.483	278	.1408	.87	1.29	41.3	135	38.5	126
	19/38	.0209	.533	304	.1540	.97	1.44	38.1	125	35.5	116
24	Solid	.0201	.511	404	.205	1.22	1.82	27.3	89.4	25.7	84.2
	7/32	.024	.610	448	.227	1.40	2.08	25.6	84.0	23.8	78.3
	19/36	.025	.635	475	.241	1.50	2.23	24.4	80.0	22.7	74.5
22	Solid	.0253	.643	640	.324	1.95	2.91	16.8	55.3	16.2	53.2
	7/30	.031	.787	700	.355	2.18	3.25	16.4	53.7	15.3	50.1
	19/34	.032	.813	754	.382	2.35	3.50	15.4	50.4	14.3	46.9

TABLE 16-5 Preferred AWG-Size Copper Conductors (Continued)

Conductor		Diameter[2]		Area		Approx. Weight		Nom. D.C. Resistance @ 20°C[3]			
								Tin Coated		Bare or Silver Coated	
AWG[1]	Stranding	in.	mm.	circ. mils	sq. mm.	lbs./M′	kg./km.	ohms/M′	ohms/km.	ohms/M′	ohms/kg.
20	Solid	.0320	.813	1,020	.519	3.10	4.61	10.5	34.6	10.1	33.2
	7/28	.038	.965	1,111	.563	3.47	5.16	10.2	33.5	9.6	31.5
	10/30	.036	.914	1,000	.507	3.14	4.67	11.5	37.8	10.7	35.2
	19/32	.040	1.02	1,216	.616	3.84	5.71	9.5	31.2	8.9	29.1
18	Solid	.0403	1.02	1,620	.823	4.92	7.32	6.64	21.8	6.39	21.0
	7/26	.048	1.22	1,770	.897	5.52	8.21	6.41	21.0	6.04	19.8
	16/30	.046	1.17	1,600	.811	5.04	7.51	7.24	23.7	6.74	22.1
	19/30	.050	1.27	1,900	.963	5.98	8.90	6.09	20.0	5.68	18.6
16	Solid	.0508	1.29	2,580	1.31	7.81	11.6	4.18	13.7	4.02	13.2
	19/29	.057	1.45	2,426	1.23	7.64	11.4	4.72	15.5	4.45	14.6
	26/30	.060	1.52	2,600	1.32	8.27	12.3	4.50	14.8	4.19	13.7
14	Solid	.0641	1.63	4,110	2.08	12.4	18.5	2.62	8.61	2.52	8.28
	19/27	.071	1.80	3,831	1.94	12.1	18.0	2.99	9.81	2.82	9.24
	19/.0147	.074	1.88	4,106	2.08	12.9	18.9	—	—	2.63	8.62
	41/30	.069	1.75	4,100	2.08	13.1	19.4	2.85	9.35	2.66	8.71
12	Solid	.0808	2.05	6,530	3.31	19.8	29.5	1.65	5.42	1.59	5.21
	19/25	.090	2.29	6,088	3.08	19.4	28.9	1.88	6.17	1.77	5.81
	19/.0185	.093	2.36	6,503	3.30	20.5	30.1	—	—	1.66	5.44
	65/30	.091	2.31	6,500	3.29	20.8	31.1	1.81	5.93	1.68	5.52

10	Solid	.1019	2.59	10,380	5.26	31.4	46.8	1.04	3.41	.999	3.28
	19/.0234	.117	2.97	10,404	5.27	32.8	48.8	—	—	1.04	3.40
	37/26	.112	2.84	9,354	4.74	28.3	42.2	1.23	4.04	1.16	3.82
	105/30	.130	3.30	10,500	5.32	33.1	49.2	1.12	3.69	1.05	3.43
8	19/.0295	.144	3.36	16,535	8.38	50.0	74.4	—	—	.652	2.14
	133/29	.167	4.24	16,983	8.61	54.5	80.4	.688	2.26	—	—
	168/30	.174	4.42	16,800	8.51	54.0	80.3	.709	2.23	—	—
6	19/.0372	.186	4.72	26,293	13.32	82.8	123	—	—	.410	1.35
	133/27	.210	5.33	26,818	13.60	86.1	128	.435	1.43	—	—
	266/30	.204	5.18	26,600	13.49	85.4	127	.448	1.47	—	—
4	133/25	.257	6.53	42,615	21.61	137	204	.274	.899	—	—
	420/30	.257	6.53	42,000	21.29	136	202	.287	.941	—	—
2	19/.0591	.292	7.42	66,363	33.67	209	311	—	—	.163	.533
	665/30	.338	8.59	66,500	33.72	215	321	.183	.600	—	—
1/0	259/24	.424	10.77	104,639	53.05	339	504	—	—	.108	.355
	1045/30	.425	10.80	104,500	52.95	316	479	.114	.373	.106	.348
2/0	37/.0600	.420	10.66	133,200	67.49	423	630	—	—	.0808	.268
	1330/30	.475	12.07	133,000	67.39	403	599	.089	.292	.0833	.274
3/0	37/.0673	.470	11.93	167,584	84.92	533	793	—	—	.0650	.213
4/0	37/.0756	.530	13.46	211,468	107	673	1000	—	—	.0515	.169
	2107/30	.608	15.44	210,700	107	682	1016	.0589	.193	—	—

[1] In stranded conductors, nearest AWG size.

[2] Actual nominal diameters for solid wires; theoretical average diameters for stranded wires.

[3] Typical D.C. Resistance values for uninsulated wire. Multiply by 1.04 for typical values after insulation.

Source: Reprinted by permission of Brand-Rex Electronic and Industrial Cable Division, Willimantic, Conn.

TABLE 16-6 MIL-C-17 Coaxial Cables◊Physical Data*

MIL Designation	Nom. Imped.† (ohms)	Inner Cond.‡ (AWG)	Core		Shield O.D. (max.)	Jacket	
			Mat.§	O.D. (max.)		Material	O.D. (max.)
RG-6A/U	75	.0285 BCW	P.E.	.189	.264[1]	PVC IIa	.336
RG-8A/U[2]	52	7/.0285 BC	P.E.	.295	.340	PVC IIa	.415
RG-11A/U	75	7/.0159 TC	P.E.	.292	.340	PVC IIa	.412
RG-13A/U[3]	74	7/.0159 TC	P.E.	.290	.355[1]	PVC IIa	.430
RG-58C/U	50	19/.0071 TC	P.E.	.120	.150	PVC IIa	.199
RG-59B/U	75	.023 BCW	P.E.	.150	.191	PVC IIa	.246
RG-62A/U	93	.0253 BCW	P.E.[4]	.151	.191	PVC IIa	.249
RG-63B/U	125	.0253 BCW	P.E.[4]	.295	.340	PVC IIa	.415
RG-71B/U	93	.0253 BCW	P.E.[5]	.151	.208[5]	PE IIIa	.250
RG-122/U	50	27/36 TC	P.E.	.099	.126	PVC IIa	.165
RG-174A/U	50	7/34 BCW	P.E.	.063	.088	PVC IIa	.105
RG-212/U	50	.0556 SCC	P.E.	.189	.265[1]	PVC IIa	.336
RG-213/U	50	7/.0296 BC	P.E.	.292	.340	PVC IIa	.412
RG-216/U	75	7/.0159 TC	P.E.	.292	.360[1]	PVC IIa	.432
RG-222/U	50	.0556 HR	P.E.	.189	.264[1]	PVC IIa	.336
RG-223/U	50	.035 SCC	P.E.	.120	.176[1]	PVC IIa	.216

◊Can be supplied with UL labels where required
*Except as otherwise indicated, cable construction is as in Figure 1
†For reference only. To facilitate location of electrical data
‡BC—Bare copper
BCW—Bare copper-covered steel
HR—High resistance wire
SCC—Silver coated copper
TC—Tinned copper
§P.E.—Polyethylene
[1]Double braided (Figure 2)
[2]Replaced by RG-213/U. Not listed in MIL-C-17
[3]Replaced by RG-216/U. Not listed in MIL-C-17
[4]Air-spaced core (Figure 3)
[5]Air-spaced core, double braided (Figure 4)

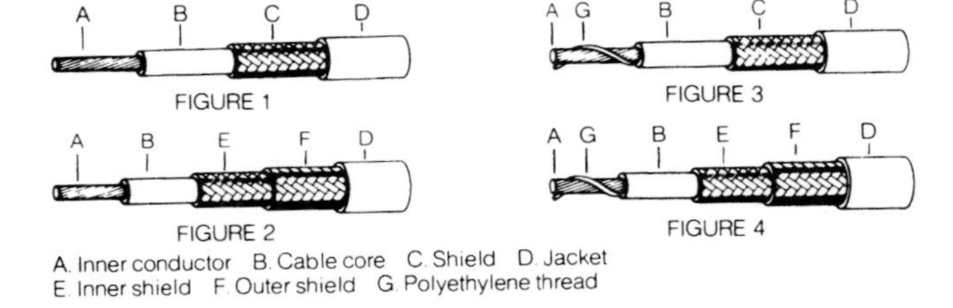

Source: Reprinted by permission of Brand-Rex Electronic and Industrial Cable Division, Willimantic, Conn.

TABLE 16-7 Electrical Data for MIL-C-17 Coaxial Cables

Nom. Imped.	MIL Designation	Nom. Capacitance		V.P. %
50	RG-58C/U	pf/ft pf/m	30.8 101.0	65.5
50	RG-122/U	pf/ft pf/m	29.4 96.5	65.5
50	RG-174A/U	pf/ft pf/m	30.8 101.0	65.5
50	RG-212/U	pf/ft pf/m	29.4 96.5	65.5
50	RG-213/U	pf/ft pf/m	30.8 101.0	65.5
50	RG-222/U	pf/ft pf/m	30.8 101.0	65.5
50	RG-223/U	pf/ft pf/m	30.8 101.0	65.5
52	RG-8A/U	pf/ft pf/m	29.5 96.8	65.5
74	RG-13A/U	pf/ft pf/m	20.8 68.2	65.5
75	RG-6A/U	pf/ft pf/m	20.6 67.6	65.5
75	RG-11A/U	pf/ft pf/m	20.6 67.6	65.5
75	RG-59B/U	pf/ft pf/m	20.6 67.6	65.5
75	RG-216/U	pf/ft pf/m	20.6 67.6	65.5
93	RG-62A/U	pf/ft pf/m	13.5 44.3	84
93	RG-71B/U	pf/ft pf/m	13.5 44.3	84
125	RG-63B/U	pf/ft pf/m	10.0 32.8	84

Source: Reprinted by permission of Brand-Rex Electronic and Industrial Cable Division, Willimantic, Conn.

TABLE 16-7 Electrical Data for MIL-C-17 Coaxial Cables (Continued)

	Nom. Attenuation @ Frequencies (MHz)					
	10	50	100	400	1000	3000
dB/100 ft dB/100 m	1.4 4.6	3.3 10.8	4.9 16.1	11.0 36.1	20.0 65.6	41.0 134.5
dB/100 ft dB/100 m	1.6 5.2	4.4 14.4	6.9 22.6	16.6 54.5	29.2 95.8	57.2 187.7
dB/100 ft dB/100 m	3.8 12.5	6.5 21.3	8.9 29.2	17.5 57.4	31.0 101.7	64.3 211.0
dB/100 ft dB/100 m	.80 2.6	1.4 4.6	2.9 9.5	6.4 21.0	11.0 36.1	22.0 72.2
dB/100 ft dB/100 m	.66 2.2	1.5 4.9	2.2 7.2	4.6 15.1	9.0 29.5	19.0 62.3
dB/100 ft dB/100 m	4.4 14.4	9.4 30.8	12.9 42.3	26.5 86.9	44.0 144.4	87.0 285.4
dB/100 ft dB/100 m	1.4 4.4	3.0 9.8	4.3 14.1	8.8 28.9	16.5 54.1	36.0 118.1
dB/100 ft dB/100 m	.66 2.2	1.5 4.9	2.2 7.2	4.6 15.1	9.0 29.5	19.0 62.3
dB/100 ft dB/100 m	.66 2.2	1.5 4.9	2.2 7.2	4.6 15.1	9.0 29.5	19.0 62.3
dB/100 ft dB/100 m	.80 2.6	1.4 4.6	2.9 9.5	6.4 21.0	11.0 36.1	22.0 72.2
dB/100 ft dB/100 m	.66 2.2	1.5 4.9	2.2 7.2	4.6 15.1	9.0 29.5	19.0 62.3
dB/100 ft dB/100 m	1.1 3.6	2.3 7.5	3.3 10.8	6.7 22.0	11.5 37.7	25.5 83.7
dB/100 ft dB/100 m	.66 2.2	1.5 4.9	2.2 7.2	4.6 15.1	9.0 29.5	19.0 62.3
dB/100 ft dB/100 m	.90 3.0	1.9 6.2	2.8 9.2	5.2 17.1	8.5 27.9	18.4 60.4
dB/100 ft dB/100 m	.90 3.0	1.9 6.2	2.8 9.2	5.2 17.1	8.5 27.9	18.4 60.4
dB/100 ft dB/100 m	.50 1.6	1.1 3.6	1.5 4.9	3.4 11.2	5.7 18.7	12.2 40.0

Current Carrying Capacity *

Current carrying capacity (ampacity) is the maximum amount of current a conductor can carry without heating beyond a safe limit. Among other things it is influenced by:

1. Conductor Material. Ampacity is affected by conductivity. Thus the ampacity of aluminum is approximately .80 of the same size copper conductor.
2. Ambient Temperature. The higher the surrounding temperature, the less heat required to reach the maximum allowable temperature.
3. Insulation Type. The degree to which heat is conducted through the insulation.
4. Installation Method. In air, conduit, duct, tray or direct burial. Bundling, stacking and spacing all affect heat dissipation.
5. Installation Environment. Heat dissipation by conduction, convection, forced air flow, air conditioning, etc.
6. Number of Conductors. Single conductors have a higher ampacity rating than equivalent size conductors in a cable.
7. Amperage. Heat rise varies as the square of the applied current.

TABLE 16-8 Current Carrying Capacity of Copper Conductors

Single Conductor in Free Air – Ambient Temperature 30°C

AWG Size	Amperes per Conductor — Copper Temperature				
	80°C	90°C	105°C	125°C	200°C
30	2	3	3	3	4
28	3	4	4	5	6
26	4	5	5	6	7
24	6	7	7	8	10
22	8	9	10	11	13
20	10	12	13	14	17
18	15	17	18	20	24
16	19	22	24	26	32
14	27	30	33	40	45
12	36	40	45	50	55
10	47	55	58	70	75
8	65	70	75	90	100
6	95	100	105	125	135
4	125	135	145	170	180
2	170	180	200	225	240

* Reprinted by permission of Brand-Rex Electronic and Industrial Cable Division, Willimantic, Conn.

TABLE 16-9 Correction Factors—Ambient Temperatures Over 30°C

°C	Conductor Temperature				
	80°C	90°C	105°C	125°C	200°C
40	.88	.90	.92	.95	–
45	.82	.85	.87	.92	–
50	.75	.80	.82	.89	–
55	.67	.74	.78	.86	–
60	.58	.67	.73	.83	.91
70	.35	.52	.61	.76	.87
80	–	.30	.46	.69	.84
90	–	–	.30	.61	.80
100	–	–	–	.51	.77
120	–	–	–	–	.69
140	–	–	–	–	.59

TABLE 16-10 Correction Factors for Current Carrying Capacity

No. of Conductors in Bundles	Multiplying Factor
1	1.6
2-3	1.0
4-5	0.8
6-15	0.7
16-30	0.5

16-1 CABLE DESIGN FORMULAS*

Conductors

Weight

$$W = wKN$$

Resistance

$$R = \frac{10.371K}{10NC\,(d)^2}$$

where C = minimum conductivity in percent permitted by the applicable A.S.T.M. specification for the diameter and material involved.

d = minimum single strand diameter in inches permitted by the applicable A.S.T.M. specification.

K = stranding factor as follows:

No. of Strands	7	19	37	133	over 133
Factor	1.03	1.04	1.05	1.06	1.07

N = number of strands in the conductor.

R = maximum conductor resistance of stranded conductor in ohms/1000 feet.

w = nominal weight of a single strand in pounds/1000 feet.

W = nominal weight of finished conductor in pounds/1000 feet.

* Reprinted by permission of Brand-Rex Electronic and Industrial Cable Division, Willimantic, Conn.

Insulation (Primary or Jacket)

Weight

$$W = 340.5\,(D^2 - d^2)G$$

where d = diameter over conductor, inches (primary insulation)
= diameter over cable core (jacket)
D = diameter over insulation or jacket, inches
G = specific gravity of insulation
W = weight, pounds/1000 feet

Shield

Weight

$$W = \frac{1.03\,NwC}{\cos a}$$

Resistance

$$DCR\text{ (single shield)} = \frac{R_C}{\cos a\,(NC)}$$

$$DCR\text{ (double shield)} = \frac{S_1 \times S_2}{S_1 + S_2}$$

where a = braid angle

$$\tan a = \frac{2\pi\,(D + 2d)P}{C}$$

C = number of carriers
d = diameter of individual braid wire, inches
D = diameter of cable under the shield, inches
N = number of ends per carrier
P = picks per inch
R_C = dc resistance of individual braid wire (See Table 16-5)
S_1 = dc resistance of inner shield
S_2 = dc resistance of outer shield
w = weight of a single end, pounds/1000 feet

Shield Electrical Characteristics Impedance:

$$Z_0 = \frac{138.2}{\sqrt{K}} \log_{10} \frac{D + 1.5w}{ad}$$

Capacitance

$$C = \frac{7.354\ K}{\log_{10} D + \frac{1.5w}{ad}}$$

Attenuation:

$$A = \frac{0.435}{Z_0}\left(\frac{R_s}{d} + \frac{R_b}{D}\right)\sqrt{F} + 2.87P \quad \sqrt{KF}$$

Velocity of Propagation:

$$VP = \frac{100}{\sqrt{K}}$$

Time Delay

$$TD = \frac{1}{VP} = 1.018 \quad \sqrt{K}$$

Inductance

$$L = Z_0^2C = 140 \log_{10} \frac{D + 1.5w}{ad}$$

where a = conductor stranding factor

No. of Strands	7	19	37	61	91
Factor	939	970	980	985	988

A = attenuation, db/100 ft.
C = capacitance, pF/feet.
d = diameter of inner conductor, inches
D = diameter of cable under the shield, inches
F = frequency, MHz
K = effective dielectric constant
L = inductance, μH
P = power factor
R_b = braiding factor †
R_s = conductor stranding factor †
TD = time delay, ns/foot.
VP = velocity of propagation, percent
w = diameter of individual braid wire, inches
Z_0 = characteristic impedance, Ω

† Since R_b and Rs are approximately 1 in most cases, usable attenuation values can be obtained by substituting 1 for these items in the equation. More exact values of R_b and R_s are obtained empirically and are influenced by construction, wire and shield strand sizes, etc.

Cable

Weight

$$W = 1.02\ Nw$$

where N = number of wires

w = weight of individual wires, pounds/1000 feet

W = weight of cable, pounds/1000 feet

Cable Diameter

$$\text{Diameter} = \frac{5}{92^{(G-36/39)}}\ \text{mils}$$

where G = AWG gauge (0000 to 36)

Cable Space Filling

$$\text{Turns/In} = \frac{1000\ (1 - G/180)}{1.025D}$$

where G = AWG gauge

D = diameter, mils

17

Safety and First Aid

When installing, maintaining, or repairing electric and electronic equipment you may be near dangerous high voltages. This work is often done in confined spaces as well. Among the hazards of this work are electric shock, electrical fires, harmful gases, which are sometimes generated by faulty electric and electronic equipment, and injuries that may be caused by improper use of tools.

Because of these dangers, you should form safe and intelligent work habits, which are as important as your knowledge of electrical and electronics principles. Your primary objectives should be to recognize and correct dangerous conditions and to avoid unsafe acts. You should know the proper methods for dealing with electrical fires, providing first aid for burns, and giving rescue breathing to someone suffering from electric shock.

17-1 ELECTRIC SHOCK

Electric shock may cause burns of varying seriousness, stoppage of breathing, unconsciousness, cardiac arrest, or death. If a 60-Hz alternating current passes through a person from hand to hand or hand to foot, the results depend on the amount of current:

1 mA: The shock will be felt.

10 mA: The shock is severe enough to paralyze muscles and a person may be unable to release the conductor.

100 mA: The shock is usually fatal if it lasts more than 1 second.

The resistance of your body will vary. If the skin is dry and unbroken, body resistance will be as high as 500 kΩ. If the skin is wet, however, body resistance can drop to as low as 300 Ω. A voltage of only 30 V can cause a fatal current to flow, so any circuit with more than 30 V must be considered dangerous.

Electric shock is a jarring, shaking sensation resulting from contact with electric circuits or a lightning strike. The victim usually experiences the sensation of a sudden blow and, if the voltage is high enough, unconsciousness. Severe burns may appear at the point of contact with the skin. Muscle spasms may cause the victim to clasp the wire that caused the shock, and he or she may be unable to let go. Electric shock can kill its victim by stopping the heart or breathing, or both. The following procedures are recommended by the American Red Cross for initial rescue and care of shock victims. If you are not trained in these methods, try to locate someone who has received instruction in cardiopulmonary resuscitation (CPR).

1. Remove the victim from electric contact at once, but *do not endanger yourself*. This can be done by throwing the switch or using a *dry* stick, rope, leather belt, coat, blanket, or other insulated item. In any case *do not directly touch the victim, or you will also be subject to shock.*
2. Shout for help and ask that the emergency medical service be called.
3. Determine whether the victim is breathing. If not, and you are familiar with CPR techniques, position the victim on his or her back and open the airway with the head-tilt/chin-lift method. Check for breathing, and if not present, begin rescue breathing. Then check for a pulse (Fig. 17-1).

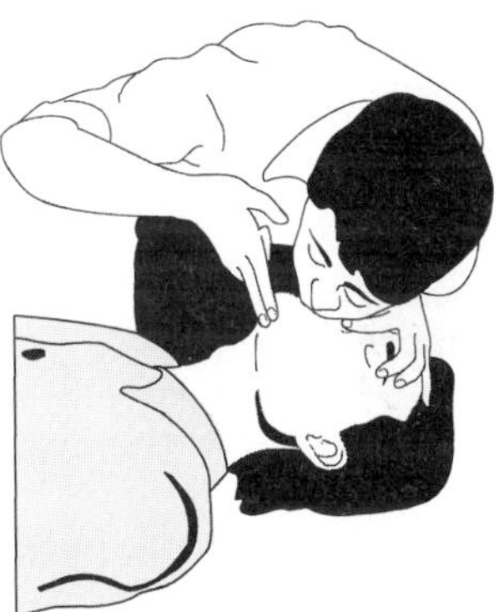

FIG. 17-1 Emergency breathing is given to a victim of electric shock who cannot breathe alone.

4. If there is no pulse, give chest compressions. Then resume rescue breathing and recheck the pulse. Continue the compression and breathing cycle until a pulse is detected (Fig. 17-2).

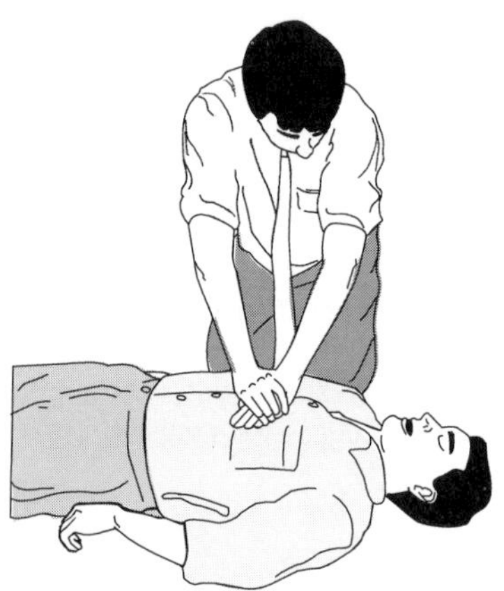

FIG. 17-2 Chest compression first aid for victims without a pulse.

5. Continue to give rescue breathing until the victim begins breathing by himself or herself.
6. Once the victim is breathing, monitor the victim until medical help arrives. If there is severe bleeding, treat it.

17-2 SAFETY SHORT-CIRCUITING PROBE

Always assume that there is a voltage present when working on circuits that have high capacitance, even when the circuit has been disconnected from its power source. Capacitors in such circuits should be discharged individually. High-value capacitors can retain a charge for a considerable time after the power is turned off. Use an approved type of short-circuiting probe. Some equipment has built-in short-circuiting devices. When this is the case, use these devices instead of the probe (Fig. 17-3).

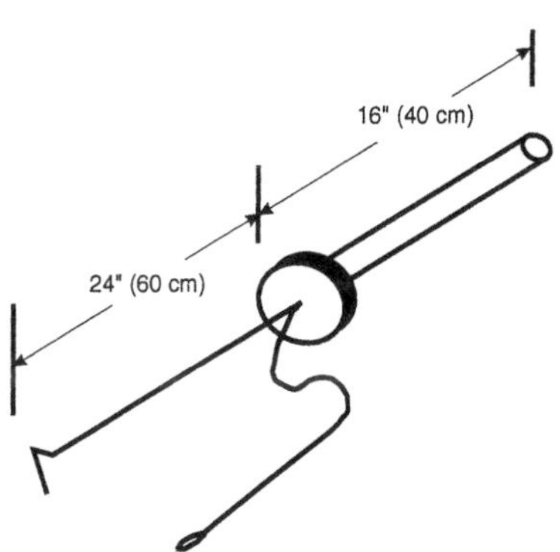

FIG. 17-3 Safety short-circuiting probe.

When using the short-circuiting probe, first connect the clip to a good ground. (If necessary, scrape the paint of the grounding metal to make good contact.) Then hold the probe by the insulated handle and touch the probe end to the point to be short-circuited. The probe end is fashioned so that it can be hooked over the terminal to provide a constant contact by the weight of the device alone. Always take care not to touch any of the metal parts of the probe while it is touching a "hot" terminal.

17-3 WORKING ON DEENERGIZED CIRCUITS

When repairing electronic equipment, safety precautions should be observed. These practices include the following.

1. Remember that electric and electronic circuits can have more than one source of power. Take time to study schematics or wiring diagrams of the entire system to ensure that all power sources have been disconnected.
2. If pertinent, inform the remote station that you are working on the circuit.
3. All circuit breakers and switches from the power source should be secured in an open position and tagged. After the work is completed, tags should be removed only by the person who originally installed them.
4. Use one hand when turning switches on or off.
5. Safety devices such as interlocks, overload relays, and fuses should never be altered or disconnected except for replacement. In addition, they should never be changed or modified without specific authorization.
6. Fuses should be removed and replaced only after the circuit has been deenergized. When a fuse blows, the replacement should be of the same type and have identical voltage and current ratings. Use a fuse puller to remove and replace cartridge fuses.
7. Keep clothing, hands, and feet dry. When you work in a wet or damp location, use a dry platform to sit or stand on and place rubber matting on top of the platform. Use insulated tools and flashlights.

17-4 WORKING ON ENERGIZED CIRCUITS

As much as practical, you should not attempt to repair equipment and circuits while they are energized. If this work is necessary, carefully observe safety precautions such as the following.

1. Ensure that there is adequate lighting. You must be able to see clearly to perform the job properly and safely.
2. Be sure that you are insulated from ground by use of an approved type of rubber matting or other material. Check that the matting is dry and without holes.
3. When practical, use only one hand, keeping the other hand either behind you or in your pocket.
4. If the voltage exceeds 150 V, wear rubber gloves.
5. Have an assistant near the main switch or circuit breaker so that in case of an emergency the equipment can be deenergized immediately.
6. Someone qualified in first aid for electric shock should be nearby during the entire operation.
7. Do no work alone.
8. Do not work on any electric equipment when wearing wet clothing or when your hands are wet. Do not wear loose or flapping clothing. Flammable articles should not be worn.
9. Remove all rings, wristwatches, bracelets, and similar metal items. Ensure that clothing does not contain exposed metal fasteners such as zippers, snaps, buttons, or pins.
10. Do not tamper with interlock switches by attempting to defeat their purpose by short-circuiting them or blocking them open.

11. Be certain that the equipment is properly grounded before energizing.
12. Deenergize equipment and short-circuit or ground the terminals of all components capable of retaining a charge before attaching clips or probes to any circuit.
13. Use only approved meters and other indicating devices to check for the presence of voltage.

17-5 ELECTRICAL FIRES

Electric or electronic equipment fires can result from overheating, short circuits, part failures, or radio-frequency arcs. Fires involving insulation and other combustible materials in electric or electronic equipment are designated class C. Class C fires have the added hazard of electric shock. (Class A fires involve wood, paper, fabrics, and rubbish; class B fires consume oil, grease, gasoline, fuels, and paints.) Electric equipment can be ignited by exposure to class A or B fires. When possible, electric equipment exposed to class A or B fires should be deenergized immediately. If the equipment cannot be deenergized, protective measures should be taken to avoid shock.

Extinguishing agents other than gases will contaminate delicate instruments, contacts, and similar electronic devices. Carbon dioxide or other gas extinguishers which evaporate rapidly and leave little or no residue are, therefore, the preferred extinguishing agents for class C fires. A dry chemical agent composed chiefly of potassium bicarbonate (such as Purple K) is suitable for electrical fires because it is a nonconductor and protects against electric shock. Damage to equipment or components may result, however, from the use of this agent.

A stream of water should never be used to extinguish electrical fires in energized equipment. The impurities in the water can make it a conductor and result in electric shock. Foam is not recommended for electrical fires because of equipment damage and the possible shock hazard. In an extreme emergency foam may be used on deenergized circuits. When a blanket of foam is applied to a fire, the flames are smothered by having the air supply cut off.

The general procedures for handling an electrical fire begin with cutting off the power to the circuit or equipment involved. Then an alarm should be sounded. Ventilation to the area should be secured by closing doors, windows, and vents. If you can do so safely, extinguish the fire using a carbon dioxide extinguisher. Avoid prolonged exposure to high concentrations of carbon dioxide in confined spaces as there is danger of suffocation unless oxygen breathing apparatus is used.

Index

About the Author

Ed Pasahow lives in Southern California where he consults, writes, and teaches. Formerly, he was a technical director at PRC, Inc. and a group vice president at Systems Exploration, Inc. He is currently pursuing efforts involving advanced communications networks and environmental protection measures. He holds degrees in electronic engineering and management science.